PROGRAMMING ROBOT CONTROLLERS

■ Other Books in the Robot DNA Series

Constructing Robot Bases by Gordon McComb (ISBN0-07-140852-5)

Building Robot Drive Trains by Dennis Clark and Michael Owings (ISBN 0-07-140850-9)

Programming Robot Controllers

Myke Predko

McGraw-Hill

New York Chicago San Francisco Lisbon
London Madrid Mexico City Milan
New Delhi San Juan Seoul Singapore
Sydney Toronto

The McGraw·Hill Companies

Cataloging-in-Publication Data is on file with the Library of Congress.

1 2 3 4 5 6 7 8 9 0 DOC/DOC 0 9 8 7 6 5 4 3 2

P/N 141098-8
PART OF
ISBN 0-07-140851-7

The sponsoring editor for this book was Scott Grillo and the production supervisor was Sherri Souffrance. It was set in Century Schoolbook by MacAllister Publishing Services, LLC.

Printed and bound by RR Donnelley.

McGraw-Hill books are available at special quantity discounts to use as premiums
and sales promotions, or for use in corporate training programs. For more information, please write to the Director of Special Sales, Professional Publishing, McGraw-
Hill, Two Penn Plaza, New York, NY 10121-2298. Or contact your local bookstore.

 This book is printed on recycled, acid-free paper containing a minimum of 50
percent recycled de-inked fiber.

■ CONTENTS ■

Contents

INTRODUCTION

Every year at Halloween, there is a special "Simpsons" episode in which spoofs are made of different horror and science fiction movies and stories. One of my favorite one involves Mr. Burns, the evil nuclear plant operator, building a robot to replace his employees. The only thing missing from his design is a controller. The quick solution to this problem is to use Homer Simpson's brain, which has its own quirks and priorities. Despairing, Mr. Burns abandons the project and pulls Homer's brain from the robot.

Although this is a seemingly trivial introduction to the topic of programming robot controllers, this spoof actually illustrates what I see happening all the time with robot developers. I have found that most people trying to create their own robots do not put as much thought, consideration, and planning into their robot's controller as they do for its mechanical structure, sensors, and drive train. As part of my research for this book, I discovered many beautifully designed and built robots that languished on the shelf because their developers had not planned properly for the controller or did not design the application in such a way that the controller could be modified and enhanced.

The goal of this book (along with the other books in the Robot DNA series) is to provide a robot designer with the knowledge and tools that will help guarantee that his or her robot will perform to expectation and specification and can be easily modified. I consider the controller to be at least as important as the other major subsystems that make up the robot, and they should be specified and designed for the robot as it begins to take shape in the designer's mind, and not left as something that will be "gotten around to."

In this book, I will introduce you to how robot controllers are integrated into mobile robots. An important part of this is showing how different sensors, outputs, controllers, and peripherals can be wired to a microcontroller as well as interfaced to. Along with introducing you to the different subsystem functions that the controller will have to perform, I will go into quite a bit of detail about the programming techniques for integrating the different functions together as well as creating a "high-level" control program for the robot.

Although a number of robots are controlled by PC motherboards, PC laptops, specially designed controller boards, neural networks, and simple light sensors, I will focus on using microcontrollers for robot control. This does not mean that my examples cannot be scaled up or down to different controller methodologies and hardware; it's just that I would like to take advantage of the amazing power and ease of use that are available from these devices. Along with these capabilities, modern *complementary metal oxide semiconductor* (CMOS) devices are remarkably robust and able to survive an astonishing amount of abuse. You will find that virtually all of the example circuits presented in this book can be built in five minutes or less.

Along with introducing you to high-level control strategies for robots, I will also be working through quite a few common and different input/output (I/O) methodologies that you can use as part of your robot. I will be describing the basic theory behind the different devices as well as sample projects to help you understand how the different interfaces are actually implemented in a microcontroller. I have broken up the different I/O devices into three categories (output devices, input devices/sensors, and motor control) and I have designed the controlling code in such a way that different devices can be integrated together with a minimal opportunity for problems.

I will be focusing on the Microchip PICmicro PIC16F627, an 18-pin device, providing 1K instruction space and 68 bytes of variable memory. The chip has up to 16 I/O pins available (some of which can act as voltage comparators) along with two 8-bit timers, a 16-bit timer, serial communications, and a simple single-vector interrupt capability. The PIC16F627 is Flash based, which means that it can be erased and reprogrammed without an ultraviolet light source and can be programmed using very simple hardware. Included on the CD-ROM is a simple programmer circuit that I have developed that is quite fast and takes the output of the software development tools directly. Despite the apparently modest capabilities of the microcontroller, it is extremely popular, has an amazing amount of resources available for it free of charge on the Internet, and has quite a few books written about it. In an informal survey of user-developed robots, the PIC16F84 and the obsolete PIC16C84, which are precursors to the PIC16F627, were clearly the most popular microcontrollers used in home-built robots.

The sample interfaces and robot control application code presented in this book are designed for easy "porting" to other microcontrollers with a minimum of changes. There are literally hundreds (and maybe *thousands*) of different microcontrollers that can be used to control your robot. The C code used in this book is designed to avoid using assembly language or hardware features that are specific to one microcontroller and instead use capabilities that are common to many different chips. Although the concept of real-time operating systems (RTOS) is introduced in this book, the example code presented is designed for a simple, single-interrupt-driven environment. The PIC16F627 cannot implement a "true" RTOS (which is the case for many other microcontrollers), but other programming techniques provide similar ease of application development. My goal for this book is to give you the skills/knowledge/sample software to allow you to "mix 'n match" interfaces in your robot and allow the applications to work without significant problems.

Making the PIC16F627 even more attractive as a demonstration/experimental tool is the availability of Microchip's outstanding MPLAB integrated development environment (IDE) as well as HI-TECH Software's PICC Lite C programming language compiler, which integrates with the MPLAB IDE. These two tools, which are provided on the CD-ROM that comes with this book, gives you development capabilities that would cost a thousand dollars or more if a different microcontroller was used. I'm sure you will find that the two tools will make robot software development a lot easier than what is available with other tools that are not as well integrated.

In describing how software is written for robots, I work at compartmentalizing the work into one of three different spectrums that I call biologic, mechalogic, and elelogic, with each term representing what the software is doing. The different spectrums, along with the functions they provide, are shown in Figure I-1.

Biologic programming is the high-level decision making being done by the robot's controller. Commands from this code spectrum are not required to respond exceptionally quickly—responses to multiple inputs that take 50 msecs (1/20th of a second) or more are not an issue. This code provides the high-level "artificial intelligence" demonstrated by the robot. It is often the most challenging code and requires the most computing resources.

Biologic programming can be compared to the functions provided by the central nervous system in the human body and assumes that the actual hardware interfaces are provided by other functions. In the human body, the peripheral nervous system that controls muscles, process inputs, and reflexes can be compared to the mechalogic and elelogic code spectrums.

The mechalogic programming spectrum executes in the range of 100 usecs to 100 msecs and is primarily responsible for controlling the mechanical devices built into the robot. The primary mechalogic devices controlled by mechalogic software include motors (both DC and servo) along with their speed controls and collision sensors ("whiskers" connected to micro switches). Mechalogic programming is generally the simplest of the three spectrums and, depending on the controller selected, can use built-in interfaces to simplify the interface design.

In this book, I will tend to represent mechalogic code as being interrupt driven. The reason for this is that while the PIC16F627 has some advanced hardware features that simply the task of controlling motors or providing other hardware functions, they may not all be available to a robot developer for a specific application.

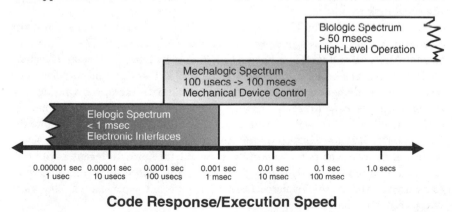

Code Response/Execution Speed

Figure I-1   Robot Software timin spectrums

The last programming spectrum, elelogic provides intercomputer communications as well as some interface and output functions. Elelogic code runs at the full speed of the processor to minimize data transfer times and care must be taken to ensure that the elelogic code does not "starve" the other spectrum's code of processor cycles. Along with this, elelogic code should be designed to be tolerant of interrupts required for the mechalogic functions. Elelogic code is generally not very difficult to develop, but care must be taken to make sure it does not interfere with the other processes executing in the robot's controller.

The three spectrums (biologic, mechalogic and elelogic) are terms that *I* have come up with and I have never seen them in any robot literature. Breaking up the software into these three groups is my way of rationalizing how robot controller applications are designed and helps me visualize different software requirements within the robot and how they interact. Note that I consider that the biologic code is the high-level operation of the robot, and I treat the mechalogic and elelogic code as functions and subroutines that can be called without regard to how the tasks are actually carried out. Breaking the software into the three parts allows for code to be reused simply between robots and allows for quite simple experimentation with the biologic code.

As part of this book, I will go through much of the process of developing a complete robot application. This information will be from the perspective of the robot's controller, but as I indicated, I believe that all the major parts of the robot should be specified and designed together. This means that the techniques I discuss in this book can be applied to the information provided in other books in the Robot DNA series.

I also discuss how to find and resolve the problems that will inevitably come up as you work with your robot. I believe very strongly in testing your application code in a simulator before trying it out in actual hardware. The MPLAB IDE simulator will allow you to observe exactly how the application is executing and if there are any errors in programming or the strategy that was used to implement the robot functions.

For information on building robot structures and learning more about the electrical characteristics of motors, I recommend *Constructing Robot Bases* by Gordon McComb and *Building Robot Drive Trains* by Dennis Clark and Michael Owings, two other books in the Robot DNA series. The three books have been written to complement each other and by working with them, you will have all the material you need to create a mobile robot with a minimum of problems.

I wouldn't be surprised if you were wary that a simple microcontroller like the PIC16F627, which has about the same amount of memory and processing power as the first Apple computer, can effectively control many different peripheral functions in a robot as well as its high-level control. Before going on, I just want to assure you that a simple microcontroller like this one can control a robot more effectively than Homer Simpson's brain.

■ Prerequisites for This Book

Like the other books in this series, *Programming Robot Controllers* was written for robot developers with some experience in developing robots. I will not be going into detail explaining basic programming, electronics, or PC operation, but you will have to be familiar with these areas of study. If you can write and execute a simple program on your PC (as simple as "Hello World!") and wire a circuit consisting of some digital electronic chips, you should not have any difficulty with the concepts and information presented in this book.

All the software presented in this book will be written in the C high-level language. I have worked at making the code as device independent as possible—you should be able to take the code presented in this book and port easily to another compiler (or language) and microcontroller. The basic requirement for a microcontroller to be able to execute the applications presented here is just a simple interrupt (based on timers as well as changes in I/O pin input) capability. The timer interrupt function of the microcontroller is used to sequence the different elelogic and mechalogic interfaces and allow the biologic code to execute as if it were the only application running in the microcontroller (and the interfaces are all hardware, rather than software based).

If you aren't familiar with C, you should get a copy of *The C Programming Language* by Brian Kernighan and Dennis Ritchie—this book is considered to be the best introduction to the C programming language and explains the operation of the language, which is implemented faithfully in the PICC Lite compiler. There are a variety of C compilers available for other microcontrollers and there is a version of the GNU open source/open license C compiler (GCC) available for the PC that you can download for free to better understand how C programs are written.

The C programming language has a reputation for being difficult to learn and work with. I have seen it derisively called the "universal assembly language" because of its strength in manipulating low-level data. Although C is not an ideal language for writing database queries or Internet web page applications, it is well suited for providing control of hardware interfaces. C is a relatively simple language that is very strongly structured and typed, which makes it very attractive for electronics programming in general and robot programming in particular.

There is one aspect of C that can be abused, making it very difficult for people to read and understand other people's code. C has the capability to combine statements, in order to lessen the amount of data entry required for an application. For example, the three statements

```
A = A + 1;
B = A * 4;
if (B == 16) {
```

could be written as

```
if ((B = (++A * 4)) == 16) {
```

The second block of code, one statement long, is identical to the first which consists of three statements (A is incremented, B is assigned the value of A [after A was incremented] times four and conditional code is executed if B is equal to 16). In this book, I will refrain from aggressively combining statements as this hurts the readability of the code and makes it difficult to port the code to another language like BASIC.

Another concern with C is its use of pointers. Right now, there is a trend away from programming languages that implement pointers—this is one of the selling points of Java. Although I believe that it is important to remember that C strings are implemented as a pointer to an array of characters, there are few requirements for pointers when creating robot applications.

To build the sample applications presented in this book, you must have some basic understanding of electronics. The reason for saying this is because some of the parts I have used in developing the applications came from junk boxes, while some others came from surplus stores, and others came from electronics distributors only available in North America. I doubt you will be able to get exactly all of the same parts that I have used. This should not be an issue because I have explained the theory and use of these devices as well as any issues that you should be aware of when using them in robot applications.

If you have an understanding of the basic DC voltage/current laws (Ohm's law, Kirchoff's law, Thevinin's equivalence), along with some transistor theory and an understanding of how to work with digital electronic devices, then you will not have any problems with the circuits presented in this book.

Of greater importance than understanding basic electronic theory is the ability to build prototype electronic circuits. For many of the example circuits presented in this book, I have shown how they can be wired on a simple "breadboard." For an actual robot, I recommend that soldered or wire-wrapped circuitry be used on your robot for reliability.

You will have to be comfortable with loading applications and developing code on your PC. I highly recommend that the files for different interfaces and applications be kept in separate folders (or subdirectories) on your PC's hard disk. Before starting to work through the applications presented in this book, you should be comfortable with navigating and using the different resources available within the PC, including the ability to work with a typical Windows-based editor like WordPad.

Lastly, you should have access to the Internet. Not only will it be useful for accessing my web page (for updates and errata), but also for looking up information on different components and circuits. There are quite a few different web sites, list servers, bulletin boards, and FAQs devoted to robots and this additional information will help you in deciding how you should design your robot.

Conventions Used in This Book

Ω	Ohms
k	Thousands of ohms
MΩ	Millions of ohms
µF	microfarads (1/1,000,000 farads)
pF	picofarads (1/1,000,000,000,000 farads)
secs	Seconds
msecs	Milliseconds (1/1,000 of a second)
µsecs	Microseconds (1/1,000,000 of a second)
nsecs	Nanoseconds (1/1,000,000,000 of a second)
Hz	Hertz (Number of cycles per second)
kHz	Kilohertz (1,000 cycles per second)
MHz	Megahertz (1,000,000 cycles per second)
GHz	Gigahertz (1,000,000,000 cycles per second)
####	Decimal number
-####	Negative decimal number
0x0####	Hexadecimal number
0b0####	Binary number
n.nn × 10^e	Real number. The number n.nn times 10 to the power of "e"
[ ]	Optional parameters within bold square brackets
|	Either/or parameters
_	Underscore indicating that the statement is continued to next line
_Label	Negatively active signal/bit
Register.Bit	Specific bit in a register
Monospace Font	Example code

//	Text/code/information that follows is commented out
or . . .	"And so on." This is put in to avoid having to put in meaningless (and confusing to the discussion) text
&	Two input bitwise AND

Truth table:

```
Inputs | Output
A    B  |
-------+-------
0    0  |    0
0    1  |    0
1    0  |    0
1    1  |    1
```

| AND, && | Logical AND |
| \| | Two input bitwise OR |

Truth table:

```
Inputs | Output
A    B  |
-------+-------
0    0  |    0
0    1  |    1
1    0  |    1
1    1  |    1
```

| OR, \|\| | Logical OR |
| ^ | Two input bitwise XOR |

Truth table:

```
Inputs | Output
A    B  |
-------+-------
0    0  |    0
0    1  |    1
1    0  |    1
1    1  |    0
```

| XOR | Logical XOR |
| ! | Single input bitwise inversion |

Truth table:

```
Input | Output
A     |
------+-------
```

0	1
1	0

NOT	Logical inversion
+	Addition
−	Subtraction or negation of a decimal
*, ×	Multiplication
/	Division
%	Modulus of two numbers. The modulus is the "remainder" of integer division
<< #	Shift value to the left "#" times
>> #	Shift value to the right "#" times
->	Click the specified pull-down/button from the previous pull-down/control

The CD-ROM That Comes with This Book

Once you are ready to start going through this book and build the projects I have presented, I highly recommend that you put the CD-ROM that comes with this book into your PC and take a look at what's available to you. The CD-ROM contains a lot of information that will help you to better understand the data as well as quickly build the sample circuits presented in this book.

The CD-ROM contains the following information and materials:

- All software source files
- Development tools, including copies of HI-TECH Software's PICC Lite C compiler for the Microchip PIC16F627 microcontroller, Microchip's MPLAB, IDE, and my El Cheapo PICmicro MCU programmer circuitry and software
- PDF datasheets for the PICmicro MCU used in the book as well as the manuals for the PICC Lite compiler and the MPLAB IDE
- Web resources, including page references to different information

The CD-ROM also includes the necessary instructions and tools for loading the files from the CD-ROM, as well as instructions for installing the tools included on the CD-ROM.

Registered Marks

Microchip is the owner of the following trademarks: PIC, PICmicro, ICSP, KEELOQ, MPLAB, PICSTART, PRO MATE, and PICMASTER. microEngineering Labs, Inc. is the owner of PicBasic. PICC and PICC-Lite are trademarks owned by Microchip but licensed exclusively to HI-TECH Software. HI-TECH C is a trademark of HI-TECH Software. Microsoft is the owner of Windows 95, Windows 98, Windows NT, Windows 2000 and Visual Basic. All other copyrights and trademarks not listed are the property of their respective manufacturers and owners.

ACKNOWLEDGMENTS

This book (and the series that it is part of) wouldn't have been possible except for the enthusiasm of my publisher at McGraw-Hill, Scott Grillo. As always, what I appreciated most of all was the time Scott has devoted to my projects (supporting *Programming and Customizing PICmicro® Micrcontrollers* and the *TAB Electronics Build Your Own Robot Kit* as well as this book) this past year. I also have to thank him for the time he spent with Gordon McComb and me, moderating the discussions and helping us through the process of defining the book series and helping us to conceptualize what this book and the others in the series would become.

Gordon McComb was the originator of the idea for this series and I appreciate the opportunity to take part in it. The two of us have spent many hours over the phone trying to define the book series and compartmentalize the topics that each title would encompass. Gordon is an amazing resource on robots and somebody that I learn something new from each time I talk to him.

As always, Ben Wirz, the codesigner of the *TAB Electronics Build Your Own Robot Kit*, has long been an invaluable resource to me for writing books. Ben was instrumental in helping me articulate the terms "biologic," "mechalogic," and "elelogic" and what they meant to the robot developer. I always enjoy our long conversations trying to define what would be the "perfect" robot and maybe one day we'll get the chance to build it.

The example interfaces and applications could have been created for almost any microcontroller on the market today, but I choose the Microchip PICmicro MCU because of the support and tools available from Microchip and HI-Tech Software. Microchip's datasheets, the MPLAB *integrated development environment* (IDE), their supported distributors, and their wide range of products are superior to those of any other manufacturers and I feel that Microchip should be the first supplier you consider when you are choosing a microcontroller for any project. This book would not have been possible without the help of Greg Anderson, Al Lovrich, and Kim Van Herk of Microchip. Their help and support, although typical of Microchip employees, is a standard by which the rest of the industry should be judged.

I must also thank everyone I talked to at HI-Tech software, especially Clyde Stubbs Smythe and Mark Luckman. The quality of the PICC Lite compiler and their generosity in terms of support meets the standard achieved by Microchip. I have been very impressed with the PICC PICmicro C and PICC Lite compilers and I can honestly recommend them without reservation to people interested in developing applications for the PICmicro microcontroller without having to learn PICmicro assembly language programming.

As always, I must recognize everyone on the PICList. The 2,000 or so individuals subscribed to this list server have aided Microchip in making the PICmicro probably the best supported and most interesting chip available on the market today. I could probably fill several pages with regular contributors that have answered questions or made suggestions. Everyone on the list is to be commended for their generosity in helping others and providing support when the chips are (literally) down.

Over the past year, I have been introduced to the joys of having a dog. I must thank Susi (our Siberian Husky) for all the walks she has taken me on to help distract me from the problems of writing the book and getting the interface code to work. I must confess that on many of these excursions about Don Mills I have been studying her walk and trying to figure out how to design a robot that moves the same way she does.

Now, we have a new husky puppy, Lobo, to help my cardiovascular system as well as give me another example of canine locomotion to study.

For my daughter Marya, thank you for your questions, suggestions, enthusiasm, and (destructively) testing the example interfaces and sample robots. I don't know of too many seven-year-olds that want to be geneticists and can still find time for their dad's robot projects.

As always, the biggest "thank you" goes to my aptly named wife, Patience. Thank you for letting me spend all those hours in front of my PC and coming down at three in the morning to tell me to come to bed. I really appreciate your recognition that robots are a "guy thing" and while you don't understand it, you still support and love me.

I don't think anybody looks at life exactly the same way since September 11th, 2001. I would like to recognize and thank all the men and women who have traveled throughout the world trying to make it a safer and more just place. Your efforts are appreciated and I sincerely hope you all return home safely.

MYKE PREDKO
TORONTO, CANADA

Micro-controllers in Robots

When I was planning this book, I did an informal survey of over 360 autonomously controlled mobile robots that were published on the Internet to see how they were built, how they were controlled, and what peripherals were used. An overwhelming majority of the robots that I sampled were controlled by microcontrollers (MCUs). As I looked at the different schematic diagrams and source code, I discovered that there were many microcontroller implementations that could have been done more efficiently and, in many cases, the robot designers had not yet even approached the task of designing their controllers or writing software.

When I looked at many of the robots where the controller circuits were not completed or even started, it seemed that the designers either ran out of steam and couldn't complete the robot or didn't know how to proceed in designing a controller. This survey confirmed a feeling that I had when I started planning this book—many robot designers are not familiar with how controllers are designed as part of a robot.

Before discussing how microcontrollers are designed for robots and how software is developed for them, I wanted to step back and look at microcontrollers to see what functions and features they provide to the robot designer. In this chapter, I will provide you with a basic understanding of what a microcontroller is and what features you should consider when selecting a microcontroller for your robot application.

Controlling a Robot

There is no hard definition of the term *controller* that can be agreed on by individuals. If you look in different manufacturer catalogs, you will see an amazing number of devices (not all are single-chip devices) with different characteristics that are all called *controllers* or *microcontrollers*. If you try to find common features, you will discover that the only common feature between the different devices is that they perform a preprogrammed electronic output based on an electronic input. You might think that this is an overly broad definition, like calling amoebas human because both creatures eat, excrete, reproduce, and are carbon-based life forms, but there is a subtlety to this definition that needs to be considered when discussing how a mobile robot is controlled. To understand this subtlety, I will use the example of a simple light-seeking, differentially driven robot that is implemented using the circuit shown in Figure 1-1.

The *whisker* is a simple microswitch that opens (turns off the power to the motors) when it collides with an object. The photodiodes, when exposed to light, generate a current that is injected into the base of the two NPN transistors, which control the flow of current to the two motors. Each of the photodiodes is connected to the motor on the opposite side to help control which way the robot turns; the motor, which is controlled by the photodiode receiving the most light, receives the most current and turns the robot so that it is facing the light. Inci-

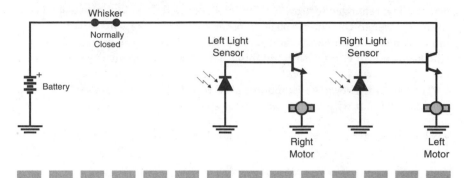

Figure 1-1   Basic robot controller circuit

dentally, this circuit is very similar to the ones that were originally used in the first experimental robots.

This robot, when started, turns toward the brightest source of light in the room. When the light falling on both photodiodes is equal, then both motors receive the same amount of current and the robot moves forward toward the source of light. The microswitch connected to the whisker opens when the robot collides with the light source. The only concern I have with this circuit is that it will start moving faster as it gets closer to the light. This is because as it gets closer the light will be brighter, more photons strike the photodiode, which causes the photodiode to produce more current. As the photodiode output current increases, the current available to the motors increases, speeding up the robot. There is the potential that the robot will hit the light like a freight train.

Under some definitions, this circuit is a controller because it performs a purposeful operation (it moves the robot toward the light). Going back to my original definition, the circuit meets the criteria of outputting an electronic response based on the electronic input. However, this circuit does not meet the definition entirely because its operation is not programmable—if you want to hold the robot to a constant speed (and not accelerate as the robot approaches the light), you have to redesign the circuit. There is no method of programming that can be changed that regulates the speed of the motors.

To be able to change the operation of the robot, the controller circuit must have some set of saved operational rules that can be changed. When the term *operating rules* is used today, most people think of a computer, which executes a set of rules called a *program*.

Using a computer to control the robot with a program is an ideal solution to the problem of regulating the operation of the motors. Adding a computer to the circuit involves converting the current output of the photodiodes to voltages by passing their produced current through a resistor and measuring the voltage across the resistor using an *analog-to-digital converter* (ADC). The whisker input also has to be changed to provide a digital signal to the computer instead of controlling power to the entire robot. In order to convert the whisker to a digital input, it must be connected to a pull-up that is tied to the ground when the whisker is not in contact with anything. The resulting circuit is shown in Figure 1-2.

The following shows the computer code that will implement the same function as Figure 1-1, but with a maximum motor speed:

```
void main(void)                 //  Application mainline
{

    while (Whisker != Collision) {
                            //  Loop until robot hits something

        if (LeftLightSensor >= MaximumSpeed)
            RightMotor = MaximumSpeed;
        else                //  Run Motor at less than full speed
            RightMotor = LeftLightSensor;

        if (RightLightSensor >= MaximumSpeed)
            LefttMotor = MaximumSpeed;
        else                //  Run Motor at less than full speed
            LefttMotor = RighttLightSensor;

    } //  endwhile

    RightMotor = 0;         //  Turn off the Motors
    LeftMotor = 0;

} //  End mainline
```

The previous code will not allow the motors to turn faster than a set MaximumSpeed. If additional functions are required for the robot, the previous software application code can be updated. This computer circuit meets the definition of a controller that I gave at the beginning of this section.

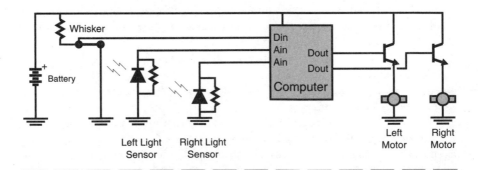

Figure 1-2 Robot computer controller circuit

Saying that a computer can control a robot is another very broad statement that could be misunderstood. I have a number of PCs in my house that I really like, but I wouldn't consider any of them appropriate for use in a robot. For the rest of this chapter, I will discuss the characteristics of computer controllers that are appropriate for use in a robot.

Support Components

When designing the controller circuit that you are going to use for your robot, you must consider what kind of support is required to run the controller. Fortunately, most modern microprocessors and microcontrollers require very little support circuitry and you can usually design the controller circuit without seriously impacting the size of the robot or decreasing its operating life due to the power required by the controller. Although the requirements for supporting the controller are quite modest, a few issues must be considered and planned for in the design of the robot.

I would like to emphasize the importance of ensuring that the requirements of the controller (for itself as well as its support) must be included in the basic design of the robot. If you do not consider the requirements of the controller, you may find that your robot has an unacceptably short operating life, the controller might execute code unreliably, or the controller might not be able to be physically built into the robot.

Power for robot controllers is almost always provided by on-board batteries. Some use photovoltaic cells, but in these cases, power is stored in capacitors before being passed to the control circuit. The current being passed to the controller only powers it for a few hundred milliseconds to a few seconds. You will find that the robot will have a very jerky operation that is often expected and part of the overall concept of the robot. Other robots, such as BattleBots, use a gasoline engine for motive power and use a small generator that is attached to the motor for electrical power for the controller. In this case, powering the controller by the motor minimizes the opportunity for batteries to be damaged during combat. These two cases where batteries are not used to power a robot controller are somewhat unusual and application specific.

Before selecting the batteries to be used to power your robot controller, you have to decide how to power the motors and any peripheral electronics. Many robots have two battery packs: one for motors and one for electronics. The reason for using two battery packs is to minimize the power fluctuations experienced when the motors are turned on and off. Personally, I prefer to power the robot motors and the controller using one set of batteries.

For a small robot, the single battery pack minimizes the cost and weight of the robot. You will find that as you minimize the cost and weight of your robot, something amazing happens. By going to a single battery pack, the overall weight of the robot is decreased, which means that smaller motors are required. Smaller motors are usually cheaper and require less current, which means they require a smaller battery. A smaller battery weighs and costs less than a larger one, enabling you to use a smaller motor, and so on.

This loop of decreasing the cost and weight is known as a *supereffect*, which can help you create the smallest, lightest, and cheapest possible robot for a given set of requirements. I do want to caution you about one thing—it is very easy to "design down" your robot to a point where it does not fulfill the original requirements because of the attractiveness of using smaller and cheaper motors and batteries.

You can choose to use alkaline radio batteries, *nickel metal hydride / nickel cadmium* (NiMH/NiCad) rechargeable batteries, and lead-acid (motorcycle or car) rechargeable batteries for your robot. The correct choice of battery is important because it affects the following things:

- Size and weight of the robot
- Cell voltage
- Operational life of robot
- Speed of movement
- Cost
- Recharge time

Different battery types output different voltage levels per cell and discharge at different rates. For carbon-based batteries, you can expect 1.5 volts per cell, whereas alkaline radio batteries produce 1.7 volts per cell or more. Rechargeable battery cells (such as NiCad and NiMH) output 1.2 volts per cell. As shown in Figure 1-3, rechargeable batteries tend to output a constant voltage, whereas the output of single-use cells (carbon and alkaline) decreases linearly as they are used.

When considering which batteries to use in a robot, it is important to remember that the critical parameter for choosing the battery type and manufacturer is the internal resistance of the batteries themselves. The higher the internal resistance, the less current is available for the motors and the higher the amount of power that is lost within the batteries. Often when batteries heat up (due to power being dissipated within them), you will find that their capability to source current diminishes. The higher the internal resistance of a battery, the more the output voltage and current are affected by voltage transients, resulting in increased demands for filtering the power being passed to the controller. Like the supereffects that appear when a single battery pack and the smallest possible motor is used, minimizing the internal resistance of the battery allows for a smaller (and ultimately cheaper) battery and minimizes the cost and weight of the total robot.

Often people (and experts) say that you should only buy cheap carbon batteries instead of expensive alkaline (or rechargeable) batteries because the Ampere-

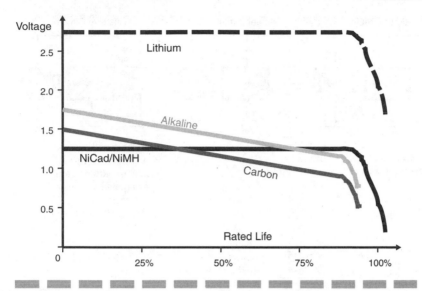

Figure 1-3 Different battery type operation

Hour (A-H) rating between the two is very similar. This statement is true, but I want to caution you about using these batteries in robots because they tend to have very high internal resistances. You can usually find out what a battery's internal resistance is by reading the datasheet for the battery, which is available on the manufacturer's web page or specified in their industrial customer catalogs.

The problem with very high internal battery resistances is illustrated in Figure 1-4. Most battery packs used in robots consist of multiple cells in series and you generally visualize the circuit as being the idealized circuit that is shown in Figure 1-4. The actual circuit in the diagram with each battery symbol and resistor representative of a single battery cell and its internal resistance is much closer to what is actually experienced. In the effective circuit of Figure 1-4, I have lumped the internal resistances together to show that as the current drawn from the battery increases, the voltage across the internal battery resistances also increases. According to Kirchoff's law, the voltage drops within the circuit are equal to the voltage applied to it. In this case, as the voltage across the internal battery resistance increases, the voltage (and current) available to the circuit decreases.

When you look around, you will find that the lowest internal resistance batteries available are either premium alkaline radio battery cells or rechargeable batteries designed for high-current applications. The 9.6-volt NiCad batteries used for remote-controlled electric racers have very low internal resistances and can be bought with chargers at a reasonable price.

Once the robot power source has been specified (one or two power sources for the robot and the type of batteries), you can design the microcontroller power subsystem. The most common method of powering robots is to use linear voltage

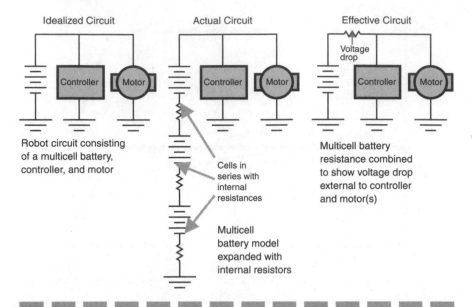

Figure 1-4 Battery operation in a robot

regulators such as the 78(L)05 to regulate an input voltage to a level that is useful by the controller. If you are using a single battery, you might want to use a DC-DC step-up converter to convert the voltage to a level that your controller can use. The reason why I went into such detail about describing how batteries can be used in a robot is because there are situations where you might not want to put in a voltage regulator at all. If you run your robot with individual AAA, AA, C, or D cells and place enough in a series to get a working voltage for the microcontroller (anywhere from 2.5 to 6.0 volts), you may be able to eliminate the need of a regulator altogether. There are many different options that you can consider for powering your controller circuit from the robot's batteries.

Although I would consider several different methods for powering the controller circuitry in the robot, I would not be so open minded with regards to the power-filtering circuitry built into the application. At a minimum, I would suggest that you place a high-value (10 μF or greater) and a medium-value (0.001 to 0 μF) capacitor on the controller power input. The large capacitor is responsible for filtering out low-frequency power upsets, whereas the small capacitor helps take care of high-frequency transients (such as those caused by motors turning on and off). Along with the capacitors, you might consider using a series inductor to filter out any current transients.

Most microcontrollers require very simple circuits for reset control; most chips just require a resistor that ties the device's reset pin to Vcc or Gnd. Some microcontrollers have the capability to reset themselves if the input voltage is too low (which is known as *brownout detection*). Other chips have an internal reset that enables the microcontroller to start executing when the power is at a preset level. If you want to use a controller that does not have these capabilities internally, a number of reset control chips are available on the market that will give you these capabilities in a small (transistor-like) package.

For many microcontroller chips, the reset pin is often used as a programming control pin and when a specific voltage (which is normally higher than Vcc/Vdd) is placed on this pin, the microcontroller goes into programming mode. For this reason, I recommend that you always explicitly tie this pin high or low to make sure the microcontroller does not accidentally go into programming mode during normal operation.

Along with the possibility for power upsets, the physical environment of a robot is also quite challenging. There is a very good chance that the microcontroller in the robot will be exposed to significant mechanical shocks and temperature extremes. In light of this, the clocking circuitry used for the robot's controller should be more physically robust than something you would consider using in a PC or other computer system which will not move.

The most basic clocking circuit is the *reflex oscillator* (see Figure 1-5), which uses a resistor/capacitor network to provide a repeatable delay through an inverter/control circuit. Many microcontrollers have reflex oscillator hardware built into them, which eliminates the need for the inverter circuit (or even the resistor/capacitor). The advantage of this type of oscillator is that it is very robust and cheap. The disadvantage is that it is also quite inaccurate and cannot be counted on to be accurate within ±10 percent of the desired frequency. For many

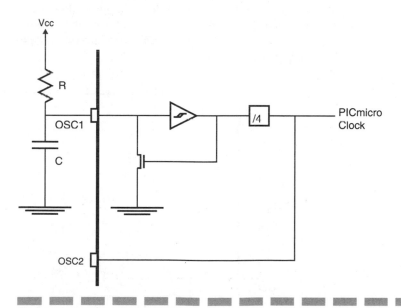

Figure 1-5 PICMicro MCU RC "Reflex" oscillator

applications, this is not a problem, but if (asynchronous) communications with another device are required, then there can be significant difficulties working through the timing issues.

Crystals can be used in a robot, but they are not as physically robust as a ceramic resonator. A ceramic resonator is wired to a controller in a similar manner to a crystal (see Figure 1-6), but it is more physically robust and often cheaper, and some parts are available with built-in capacitors, simplifying the task of wiring the controller's oscillator.

If an external ceramic resonator is used as a frequency reference for a microcontroller, make sure that the wiring between the microcontroller and ceramic resonator is as short as possible to avoid the signal being changed by inducted noise or radiating noise on its own which can affect other circuits.

The only downside to ceramic resonators is that they are generally less accurate than a crystal. Whereas an accuracy of 0.01 percent can be expected with a crystal, a ceramic resonator provides a signal accuracy of 0.5 to 2 percent. This decreased accuracy is generally not a problem with respect to intercomputer communications, but it could be a problem with some applications and circuits.

The last support component to consider is the robot's power switch. This feature is often forgotten. Either that or the designer just expects the user to pull the batteries out of the robot or thinks that the programming is so perfect that when it's not in use, it will stop. There should always be some method of interrupting the power to the robot's controller. I call this function the *big red switch* (BRS), which removes all power from the robot. Without the BRS, you might forget that the robot is on and it ends up running off the table it is stored on and its batteries wear out due to the unexpected current drain. It just takes a few moments to add the power switch; your customers will thank you and you will have eliminated the possibility that a stray control signal will cause the robot to run off your bench and fall on the floor.

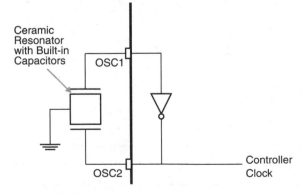

Figure 1-6
Ceramic resonator with internal capacitors used in an oscillator

■ Memory and Device Programming

When you are specifying the controller that will be used in your robot, you should make sure you understand the controller's requirements in terms of memory and how applications are loaded onto it. Most people think that a robot is just programmed once and finished, but this is never the case—the controller will be reprogrammed many times during the initial application development and will need to be updated when problems with execution are discovered and if there are updates to the application.

There are two types of memory to consider when you are specifying the robot controller. The first (and probably the most important) type of memory is the *nonvolatile program memory* that the application code executes from. You can use a number of different program memory technologies, which each have unique characteristics.

The term *nonvolatile* is used to describe memory devices that do not lose their contents when power is taken away—each time the robot is started up, the application code must be available to the controller. If you are only familiar with developing applications on a PC, you may be confused by the importance of having nonvolatile memory for program storage. You have to remember that most microcontroller chips and controller boards do not have a disk from which to retrieve code and data. When the controller boots up, it will start executing the robot application immediately.

Not having a disk provides you with the advantage of not having to understand an operating system, its interfaces, and *application programming interfaces* (APIs). The disadvantage is that there are many low-level functions that you will have to provide in the application yourself that would normally be provided by an operating system. For a traditional computer system, I would say the disadvantages outweigh the advantages, but for a robot application, with its nonstandard hardware interfaces, the advantage of not having the overhead of a full operating system far outweighs the lack of standard APIs (most of which will not be required).

Program memory can be built onto a microcontroller chip or implemented as separate chips that are connected to the controller. For the sample applications presented in this book, I will be using a microcontroller chip that has the program memory built in.

The following lists some different types of program memory that can be used:

■ Mask *read-only memory* (ROM) in which the application code has been built into the ROM. This type of memory generally requires close to 10,000 units in order to be cost effective and cannot be reprogrammed. This type of memory is used when your robot is going into mass production.

■ PROM is an acronym for *programmable read-only memory*, which is similar to the mask ROM because its contents cannot be changed once they have been stored in the chip. The concept of *burning* data into a chip for programming comes from PROM chips that have *fuse links* that can be melted, or burned, by applying a high current across them. PROMs are rarely used today because of the propensity of the broken fuse links to grow back over time, invalidating the stored data.

■ Standard *random access memory* (RAM) (which is similar to the variable memory described in the following list) can be used with a battery backup. Modern *complementary metal-oxide semiconductor* (CMOS) memories can require current in the μA range to retain information, providing months or years of operation with simple batteries providing the backup power.

■ Virtually all modern microcontroller program memory is based on *erasable programmable read-only memory* (EPROM) technology. This technology consists of individual cells that can be programmed using relatively modest amounts of current and erased by applying *ultraviolet* (UV) light to the chip. To enable UV light to reach the chip, many EPROM-based microcontrollers and EPROM chips have a large quartz window in the chip packaging (as shown in Figure 1-7). In the last 10 years, EPROM chips have become available in standard black plastic packages (with no quartz window) and are known as *one-time programmable* (OTP) parts. OTP parts offer the same advantages of PROMs without their unreliable nature and are the cheapest method of implementing program memory for lower quantities of products.

■ The last type of program memory and the one that I recommend that you use for your robot controllers is known as *Flash*, or *electrically erasable programmable read-only memory* (EEPROM, which is often pronounced "double-EPROM"). This type of memory has the programmability of EPROM, but does not have to be exposed to UV light to be erased; an electrical signal to the memory deletes it. Figure 1-8 shows a side view of how an EEPROM/Flash cell is constructed. The float(ing) gate can be charged or discharged from the control gate. When the floating gate is charged, there is a depletion zone in the silicon substrate, enabling current to pass between the cell's source and drain.

There can be some confusion over whether or not a device is Flash or EEPROM; EEPROM can have any address within the memory changed at any time, whereas

Figure 1-7
Windowed
ceramic
package

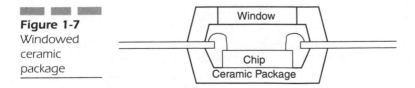

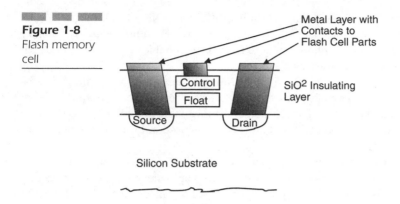

Figure 1-8
Flash memory
cell

Metal Layer with
Contacts to
Flash Cell Parts

Control
Float

SiO^2 Insulating
Layer

Source

Drain

Silicon Substrate

Flash is designed to have a block of memory erased before it can be reprogrammed. Flash tends to be somewhat cheaper than EEPROM because the erase circuitry does not have to be replicated for each cell in the memory and can be used for robot applications as individual memory cells do not have to be reprogrammed.

To program (or burn) EPROM/Flash/EEPROM chips, data has to be presented along with a series of control signals, which is known as a *programming algorithm*. The programming algorithm is normally unique to the part number of the chip being burned. Depending on the chip, data can be programmed serially (one bit at a time) or in parallel (a block of bits at a time), and the time required to program each memory location can range from hundreds of microseconds to tens of milliseconds. The total time for programming an individual chip ranging from a few seconds to a few minutes.

Commercial and home-built programmer designs are available for most EPROM/Flash/EEPROM chips and microcontrollers based on this technology. On the CD-ROM that comes with this book, I have provided you with a programmer design that you can build for Flash program memory Microchip PICmicro microcontrollers. This design is typical of what you will find on the Internet for home-built programmers; the electronic design is very simple and uses a PC to control the programming operation. Commercial programmers tend to be able to program a wide variety of different chips and can operate in a standalone mode (that is, not connected to another device). When choosing a programmer, it is important to remember to make sure that the program data file produced by your development tools can be read and burned into the chip using the programmer.

The second type of memory to consider when specifying is the RAM available to the controller for storing the robot application's variables. Whereas EPROM/Flash/EEPROM is nonvolatile, RAM is volatile because its contents are lost when you remove power from it.

Two types of variable memory are normally used: *dynamic random access memory* (DRAM) and *static random access memory* (SRAM). DRAM, which is used in most PCs, uses capacitors to store data as a varying-level charge. In your PC, you have probably bought *synchronous dynamic random access memory* (SDRAM), which is an enhancement to DRAM that enables your PC to work with faster processors and systems.

Over time, the charge in the (S)DRAM cell may leak away and must be refreshed periodically before any data is lost. This refresh requires quite a bit of current and makes the (S)DRAM memory unavailable periodically (up to 5 percent of the time for a PC). SDRAM and DRAM is used in PCs because of its very low cost per bit.

SRAM is the most common type of variable memory used in microcontrollers and robots. SRAM consists of a four- or six-transistor memory cell (as shown in Figure 1-9) that feeds back on itself to save a set value. The advantage of the SRAM cell is that it uses very little current to store data (no refreshing is required), although it uses up to six times of the silicon real estate of the (S)DRAM cell.

SRAM is the typical variable storage medium in microcontrollers, although there is usually very little of it in the chip. I would expect your PC to have memory measured in the tens of megabytes (or much more). It is not unusual to work with a microcontroller that has variable memory on the order of tens of bytes. This can be very confusing, and you must remember that the bulk of the application (the code) is located in the program memory, not the variable memory. In the PC, there is no physical differentiation between program memory and variable memory, and any SDRAM space within the PC can be used for either function. To further reduce the variable memory requirements for your robot application, I recommend that you minimize the number of text strings used in your application and keep them only in program memory. You will find that,

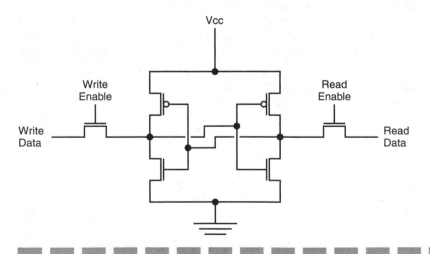

Figure 1-9 SRAM memory cell

despite the very limited variable memory in microcontrollers, you will be able to easily fit a comprehensive robot application into them.

Interrupts

I am always amazed at how many graduate engineers and computer scientists are afraid of adding interrupts to their applications. The root of this fear seems to go back to their basic programming courses when interrupts were presented as something the course instructor didn't know that well, but he or she had to teach and the students had to demonstrate at least one interrupt-based application that didn't crash. Personally, I feel that interrupts must be used in a robot application to allow for simple, focused biologic code.

Using interrupts in your application eliminates the need for complex polling loops and minimizes the chance that incoming data will be lost. It also enables the precise timing of events in high-level language applications. I feel so strongly about how interrupts can enhance a robot application that in every sample interface presented in this book, I have based the code's execution on the use of at least a timer interrupt.

Going right back to the first principles of interrupt operation, there are six steps to how an interrupt request (the processor can always refuse to acknowledge and respond to the interrupt) is serviced or handled. The execution flow of an interrupt is shown in Figure 1-10 and the following points explain what is happening in more detail:

1. The mainline code executing with hardware is set to acknowledge an interrupt request.

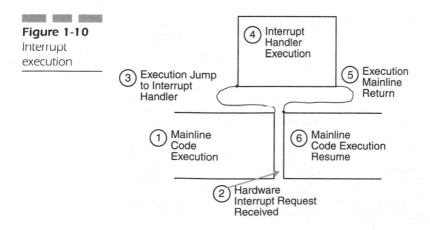

Figure 1-10
Interrupt execution

2. An interrupt request is received.

3. Execution jumps to the interrupt handler *vector* (the address where the code that services the interrupts is located). As part of this process, the *context registers* (the registers that are used during normal program execution to save temporary values) are saved.

4. The interrupt handler executes. The first operation carried out is to reset the interrupt requesting hardware.

5. Return to the mainline execution. Before this can take place, the context registers must be restored to the same values as they had when the interrupt request was received.

6. Resume the mainline execution.

Depending on the controller processor hardware you are working with, you should understand how interrupts are enabled and how they are reset from within the interrupt handler. Figure 1-11 shows how interrupts execute in the Microchip PICmicro microcontroller used in this book for demonstrations. Notice that this diagram shows how execution jumps to the interrupt vector (at address 0x00004 with the interrupt request flag active.

If the interrupt-requesting hardware is not reset in the interrupt handler, once the interrupt has completed and the return from interrupt instruction is executed, you will discover that execution returns right back to the handler without executing any of the mainline code. This can be seen in Figure 1-11—if the interrupt request flag is not reset in the interrupt handler, then upon returning

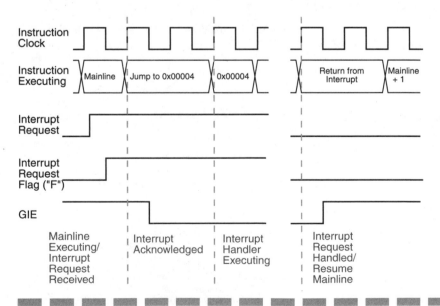

Figure 1-11   PICmicro microcontroller interrupt execution

from the interrupt handler, the interrupt controller will recognize that the interrupt request flag is still active and immediately jump back to address 0x00004.

Another important point you should understand is how multiple interrupt hardware requests from different sources are handled. In some processors (like the PICmicro MCU), there is only one interrupt vector and the code must first determine which interrupt request is active. In other processors (like the 8088 used in the IBM PC), many different interrupt vectors are built into the hardware and you may find it necessary to allow multiple interrupts to be handled simultaneously.

Depending on the processor you are working with, you may discover that it is impossible for an interrupt handler to be invoked while it or another interrupt handler is executing. Again, the PICmicro MCU does not allow this, but the IBM PC has hardware built in to enable interrupt code to execute reenterently (that is, start executing while a previous version of the handler is executing).

Interrupt handlers should be kept as simple and execute as quickly as possible. The reason for these requirements is due to the asynchronous (that is, you don't know when they are going to happen) nature of interrupt requests and it is possible to get two very close together. Rather than developing the code infrastructure to support multiple interrupts at the same time, it is much easier to make the handlers as short as possible so there is no opportunity for one of the requests to wait an unreasonably long period of time. Keeping interrupt handlers very short also eliminates the need to have interrupt handlers that can execute reenterently.

At the start of this section, I noted how difficult it was for most people to understand interrupts and then went on to say that I considered it necessary for robot designers to use them in their application code because it makes it much easier to develop the applications. I hope that what I've had to say about interrupts has not left you with a sinking feeling because using interrupts in a practical application with a high-level language is not that difficult. To give you an idea of how simple it is, consider the basic interrupt handler that is used for all the applications presented in this book:

```
void interrupt tmr0_int(void)       //  TMR0 Interrupt Handler
{

    if (TOIF) {                     //  Timer Overflow Interrupt?

        TOIF = 0;                   //  Reset Interrupt Flag

        RTC++;                      //  Increment the Clock

    } // endif

} // End Interrupt Handler
```

Going through this code, the first statement (void interrupt tmr0_int(void)) indicates that the function is located at the interrupt vector. Next, the code checks to see what type of interrupt request is being made. If the timer 0 (TMR0) has overflowed and, if it has, it resets the requesting hardware and increments a counter, indicating that the interrupt handler has executed. As I said previously, it's pretty simple and even when I add additional functions to the interrupt handler, this basic format doesn't change. I think you will be surprised to discover that adding interrupts using a compiler, like in this book, is very simiple.

In Chapter 4, "Microcontroller Connections," I will show how interrupts are used to sequence the elelogic and mechalogic applications, how to eliminate the need for assembly language programming, and, most importantly, how interrupt-based *input/output* (I/O) enables the high-level biologic code to be written as simply as possible. This enables the application code presented in this book to be ported to different controllers with very few modifications.

Built-In Peripherals

All microcontrollers are built with some kind of I/O capability along with some hardware features that are available for use by the application code to simplify the execution of an application. Basic microcontrollers just have digital I/O pins and ports along with a timer. Advanced MCUs have analog voltage I/O, advanced intercomputer communications capabilities, and external device interfaces built in them. Part of the robot definition exercise is to choose the controller for the robot with the appropriate peripherals.

The most basic peripheral is the digital I/O pin. This can be as simple as the circuit used by the Intel 8051 microcontroller, which is shown in Figure 1-12. This type of pin acts as an input pin when the open drain transistor output is off and

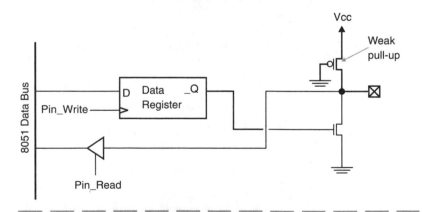

Figure 1-12 8051 parallel I/O pins

weak pull-up is active. The weak pull-up is easily overpowered when the pin is driven low by an external device. When the pin is used as a digital output, it can sink current when the open drain transistor is on—the weak pull-up can only source a few tens of microamps.

In many other types of microcontrollers (including the Microchip PICmicro MCU used in the sample applications presented in this book), each digital I/O pin has a tristate driver that converts the pin from a basic CMOS output to a CMOS input. The basic circuit drawing is shown in Figure 1-13, and although the output driver is loaded internally to the chip, the output value is always read from the external pin. Reading from the actual pin instead of from the output of the pin data latch enables your application to determine whether the I/O pin is being overpowered by an external circuit.

In addition to providing the interface for different I/O peripherals, some microcontrollers' I/O pins have the capability to measure analog voltages or output them. There are many different ways in which analog signals are measured and produced within a microcontroller. These functions can be extremely useful in a robot's environment when external light or sound levels are to be monitored by the robot.

Before you start thinking of a voice-controlled robot, let me note that most microcontrollers do not have the processing capabilities to decode speech (especially from across a noisy room). Voice-controlled robots come under the heading of projects that may seem easy when you first look at them, but there are teams of engineers investing millions of dollars trying to develop them. You should instead be looking at a robot that responds to simple external sounds like clapping.

I/O pins may be given the capability to output microprocessor parallel bus read/write waveforms. This is sometimes referred to as *microcontroller mode* because the bus can be used to implement expanded program and variable memory as well as standard microprocessor peripherals. I consider this type of feature

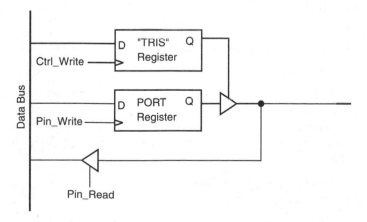

Figure 1-13
Standard
PICmicro MCU
I/O pin block
diagram

to be very useful as it gives you the best of both worlds. As a partial microcontroller, you have access to the chip's on-board memory (RAM and ROM), which simplifies the final application circuit, and you have access to the built-in peripherals of the microcontroller without having to add these functions as external chips. Accessing these devices is generally accomplished by a simple read and write, so by having the resources built into the microcontroller, you can add any needed standard microprocessor peripherals without a significant software penalty.

An interesting capability of a microcontroller is that it can be used as an external peripheral to another microcontroller. In this case, the external peripheral microcontroller is providing the elelogic or mechalogic functions, while the other microcontroller can execute the biologic code. Using multiple microcontrollers in this manner is not that difficult because many devices have built-in peripheral functions (such as synchronous serial I/O or *Inter-Intercomputer Communications* [I2C]) that enable the devices to communicate without a significant time or software penalty.

Internal timers are an important resource for microcontroller and robot applications. Timers (with enhancing functions) can provide both advanced I/O features as well as software-sequencing capabilities that may not be obvious when you first look at a chip's datasheet. Timers generally run continuously and when they roll over from the highest possible value to zero, an interrupt request or external output can be generated. To further enhance the capabilities of the timers, many implementations include *prescalers* and *postscalers*, which count input and rollover events externally to the timer—this enables timer intervals to be increased.

Many, if not most, microcontroller part numbers have some capability for serial communications with a host computer. The most basic serial communications consist of *nonreturn to zero* (NRZ) for RS-232- or RS-485-based communications. Along with this capability, many microcontrollers are built with other serial communications interfaces, including the following:

- Serial Peripheral Interface (SPI)
- Microwire
- I2C
- Common Automotive Network (CAN)
- *Universal serial bus* (USB)

Finally, chances are if you look hard enough, you will find a microcontroller part number that is designed with built-in peripheral features that are perfect for your application. After a quick search on the Internet, I was able to find microcontrollers with the following features that could be useful in a robot application:

- Stepper motor controller
- DC motor controllers

- Single-phase AC motor controllers
- Three-phase AC motor controllers
- *National Television Standards Committee* (NTSC) composite video output
- NTSC composite video onscreen display output
- Audio I/O
- *Liquid crystal display* (LCD) segment controllers
- PC mouse/keyboard interfaces

Several of these peripherals may not seem well suited for a robot application, but there will be situations where you will want to take advantage of these features or see how they can be used to provide functions that will be useful for your robot application.

Interfacing the Controller to the Robot

Once you decide what controller you are going to use on your robot and what functions it is going to perform, it is time to wire the controller with its neat and orderly operations into the very "dirty" and disorganized realm of the robot PC. Computer code, by its very nature, is very "clean" with basic functions built in and already programmed. This cleanliness is lost when the controller is connected to the robot with noisy power supplies varying according to load and battery charge, high vibration, and intermittent signals from inputs. You will have to make allowances in the wiring of the interfaces as well as the controller's code for this type of environment. As time goes on, you will discover that your vision of an efficient interface between the controller and your robot has become a seemingly disorganized mess of response to hardware problems when you have it working.

Fortunately, modern microcontrollers are very reliable and robust, which simplifies your task of integrating them into the robot. Commercial robot controller boards generally have features designed to minimize the potential for problems as well. You will find that the most significant problem you will have is developing the software that integrates the different robot peripherals into your application.

When you are designing the electronic interface between your controller, I have a few suggestions that will help you ensure that the controller runs reliably. This means that when you have a problem that requires debugging, you can almost always cross off the microcontroller's basic operations and interfaces as a source of the problem.

Before even attempting to get the interfaces built into your robot and the full code to work properly, make sure that the controller's support (that is, the power, clock, and reset) is working reliably. If you have an oscilloscope at your disposal, make sure that all signals appear with minimum noise or transients when different functions are turned on and off in the robot. The following tasks will assist you in this process:

- It is a good idea to keep the clock lines as short as possible. This minimizes the chance that magnetic fields within the robot will cause an induced signal that can affect the clock circuitry. Short clock wires also minimize the noise that could be picked up by other circuits in the robot.

- Make sure that you have a good ground built into your controller circuitry and there are not ground loops or Vcc loops, which can pick up induced noise. A ground loop consists of multiple ground connections between two points. These multiple connections are actually inductive loops that can pick up stray electromagnetic signals.

- Ideally, the controller should have an isolated power supply from the robot's motors. This can be as simple as providing a high-voltage source to the robot's motors and regulating this voltage for the controller and circuitry it is connected to.

- If you have multiple power supplies for different circuits in the robot, make sure they are connected by a single ground that links the components together. You may find it advantageous to distribute power from a central board in the robot (such as the central controller board) rather than via a power panel. By routing power through the central controller, you eliminate any chance of having ground loops and could conceivably simplify the wiring of the robot. For example, if a circuit is on a moving part, by routing power with the connector to the main controller, you eliminate the need for two wiring harnesses, along with making sure that neither harness will cause the moving part to bind.

- You may want to consider having the robot's controller execute at the slowest possible speed to minimize the power draw and transients coming from the controller. In most robots, neither point is an issue, but there will be cases where the controller will draw a significant fraction of the power required by the robot.

- I recommend that all electronic devices built into the robot be powered by 5-volt power supplies. Virtually all electronic devices (chips and so on) are available as 5 volt power versions, which makes part selection easier 5 volts also has a larger noise tolerance than 2.5- or 3.3-volt parts.

- Design the microcontroller to be reprogrammed. This can be as simple as making sure that the controller can be pulled in and out easily of its socket (and without disconnecting any wires) or it can be as sophisticated as building a microcontroller programmer into the robot circuit. In any case, you should recognize that you will be changing the robot application code during the development and life of the robot and you should make sure that changes are made with a minimum impact to the robot's structure and circuitry.

Software Development

In the introduction to this book, I presented the concept of the three spectrums of programming for robots. They were the biologic, mechalogic, and elelogic and were designed to take advantage of the robot's microcontroller in different situations. The purpose of these different spectrums of code is twofold. The first reason for breaking these functions into the different spectrums is to compartmentalize the different functions of code and simplify the task of writing them. The second reason is to allow much more substantial mixing and matching than would be possible with code that has all the functions built in line. When I present some sample mechalogic and elelogic functions later in this book, you will see that they have been written in such a way that they can be integrated together easily. Later, I will present samples of code written for the different spectrums and discuss how they are integrated together to create a robot application.

In this chapter, I will present the different tools for software development that are available to you. I will start with the most basic tools and work my way up to the more sophisticated ones. The software development tools that are presented here are general-purpose software development tools and not specific to robot applications. Later in the book, I will share with you a source code template that can be used to start off your robot applications. Note that although there are some dedicated robot software development tools, most robots use general tools because they are usually cheaper and enable different robot programming techniques as well as much more varied robots and hardware.

I will introduce the different software development tools from a very high level. The reason for this is to give you a basic understanding of what these tools are capable of as well as their advantages and disadvantages. Once I start sharing the sample software, I will be assuming that you will be able to install the PICC lite compiler and MPLAB *integrated development environment* (IDE) tools provided on the CD-ROM without a significant amount of support. If this is not the case, then you should look for introductory resource information regarding operating PCs and regard as well as a reference explaining application programming using the C programming language.

For the example applications presented in this book, the HI-TECH Software PICC Lite compiler with the Microchip MPLAB IDE will be used. The MPLAB IDE (explained in more detail in Chapter 3, "The Microchip PICmicro® Microcontroller") provides a basic framework for developing applications as part of projects in which all the source code for an application can be developed, assembled or compiled, simulated, and then programmed into a Microchip PICmicro microcontroller.

The PICC Lite compiler is a free, full-function C language compiler for the Microchip PIC16F84 microcontroller, which will be used in the example interfaces and robot applications shown later in the book. PICC Lite coupled with the MPLAB IDE is an impressive and powerful package that enables you to easily develop your own robot application code, and it gives you capabilities that would cost a thousand dollars or more if you were to buy tools with similar capabilities.

As is often lamented by workstation, Linux, and Apple users, most popular microcontroller software development tools are written for the IBM *personal computers* (PCs) running Windows (often referred to as Wintel PCs). Microcontroller software development tools written for workstations are usually expensive and part of other hardware or software development packages. At the other end of the scale, there are still quite a few tools capable of running under Windows 3.x and MS-DOS, and they are often available as free downloads. Unfortunately, these tools don't offer a lot of advantages in terms of usability or updates for the latest target processors and microcontrollers.

Linux tools are becoming more and more available with compiler and linker front ends that can be customized to specific microcontrollers. A number of microcontrollers have versions of GNUC compiler (GCC) written for them, which will give you many of the PICC Lite capabilities used in this book. Some UNIX workstation tools exist, as well as Apple iMac- or Macintosh-based tools, but there is not as complete a set as is available for the Windows-based PCs.

To avoid problems with trying to find tools for a specific microcontroller, I would recommend that you only use Wintel tools to minimize costs and maximize your ability to share code between yourself and others. Used PCs capable of running Windows 95 can be purchased for as little as $50 or $60. The cost of the PC will be more than made up by the selection of development tools available, along with an increased selection of microcontrollers that can be used in your robot.

Source Files, Object Files, Libraries, Linkers, and Hex Files

Before explaining the different development tools that can be used to process application code for your robot, I'll explain the various files that will be used with the different tools. These different files will come from a variety of sources and their presence or spontaneous appearance can be confusing.

When you create your robot application source code, you will store the program in what is known as a *source file*. This file will be in a basic *American Standard Code for Information Interchange* (ASCII) format and will contain the program statements and comments (which are used by humans to better understand how the program is to execute). The source file can be created in a standard editor (like NotePad or WordPad in the PC) or it can be built in an IDE's editor to avoid the need to load and execute multiple applications.

This source file will be used as input for an assembler, interpreter, or compiler to produce the instructions necessary to control the robot. In the case of an assembler or compiler, the source code will be converted into an object file or hex file, which will ultimately be loaded into the microcontroller that is built into the robot. The interpreter is built into the robot's microcontroller and will execute the source file as a series of instructions and command the robot directly without any intermediate processes.

The source file could include other files that provide additional source code, or prototypes for library functions or data required by the application. The include files for assembly language programs generally end in .inc as a file system extension. The include files for C language programs have a .h extension. The assembler or compiler developer will provide these include files to make the task of developing applications easier and you can develop your own, based on the requirements of your application.

At the end of the assembly or compilation operation, either an object file (which usually ends in .obj) or a hex file (which usually ends in .hex) is produced. The hex file is the completed application, ready to be programmed into the microcontroller. The object file is linked with other object files to produce the hex file.

The most popular hex file format is the Intel 8-bit format known as INHXM8. This data format is required by the El Cheapo programmer circuit and software included in this book, as well as output by most development tools. Motorola microcontrollers and programmers use a different format known as S9.

A sample INHXM8 hex file (for Microchip PICmicro MCUs) looks like the following:

```
:10000000FF308600831686018312A001A101A00B98
:0A0010000728A10B07288603072824
:02400E00F13F80
:00000001FF
```

Each line consists of a starting address and data to be placed starting at this address. The different positions of each line are defined as shown in the following table:

Byte	Function
First (1)	Always : to indicate the start of the line.
2–3	Two times the number of instructions on the line.
4–7	Two times the starting address for the instructions on the line. This is in Big Endian format.
8–9	The line type (00 – Data, 01 End).
10–13	The first instruction to be loaded into the PICmicro MCU at the specified address. This data (along with all the instructions) is in Little Endian format.
End–4	The checksum of the line.
End–2	ASCII carriage return (0x00D) and line feed (0x00A) characters.

The checksum is calculated by summing each byte in a line and subtracting the least significant bits from 0x0100. For the second line in the previous example hex file, the sum of the line is as follows:

```
    0A
    00
    10
    00
    07
    28
    A1
    0B
    07
    28
    86
    03
    07
 +  28
 ------
    1DC
```

To find the checksum, the last two digits of the sum are subtracted from 0x0100:

```
   100
 - DC
 ------
    24
```

The value 0x024 is the same as the last two bytes in the second line.

When the hex file is produced, either as a single step by the compiler or the linker, library files may be added to the application to provide functions required by the application. For a high-level application, the library files will include multiplication, division, and other complex mathematical operations: string manipulation function and device *input/output* (I/O) functions. The library is generally produced by the compiler provider, but there will be cases when you want to use one developed by somebody else or one that you have created on your own.

The linker usually provides many different options for creating hex files. Generally, a simple set of defaults is built into the linker to provide a basic hex file, but you will find that most linkers have myriad different options that will enable many different choices for the final hex file. It is important that microcontroller linkers look for library (and object file) functions that are never called by mainline functions and eliminate them from the final hex file. You will probably find that the library consists of many different functions and if you were to load them all into the microcontroller, there wouldn't be any space for the application code.

Compiling or assembling a single-source file application to a hex file is generally a fast, simple process. There is little danger of incorrectly accessing global resources, and the application is very easy to look through and read. With source code simulators and debug tools, I find that a single application is generally easier to debug. The downside to having a single-source file process is that it is difficult for multiple designers to work on an application as well as share different parts of the application with other people. The creation of user-developed libraries that are linked in at compile time will lessen the impact of the latter point.

Multiple source file applications are better suited to multiple developer applications, in which code is strictly controlled. They also enable applications to use predefined and debugged functions. Multiple source file applications can be somewhat difficult for building applications into hex files using linkers, and the debug process can be quite complex if predefined functions are available without source code. The most significant concern of sharing object files without the source, is that linked functions can be updated without the knowledge of the application developer, leading to the chance that the latest function is not included in the application or an updated, not debugged version is linked in.

▨▨▨ Assemblers

Although perceived as the method of producing the most efficient code, assembly language is the most difficult way of writing an application and does not necessarily produce the most efficient code possible. Despite this, assembly language programming is a favorite amongst many robot developers because assemblers for different microcontroller processors can usually be downloaded free from the Internet and there is a lot of satisfaction in developing an assembly language application. The latter point may be hard to understand, but as I describe how I believe successful assembler language programs are written, there is a certain amount of artistry involved that makes the development effort quite a bit of fun.

Before you can start programming an application in assembly language, you must understand the operation of the processor intimately as well as what you want the application to accomplish. It is usually not acceptable to just understand how the different processor instructions operate; there are often pitfalls and tricks that are not obvious when you are looking at a list of instructions.

When I say that you should be familiar with the processor, you must be familiar with how the processor executes code. Personally, I like to take a block diagram of the microcontroller's processor and mark on it how the instruction executes. Many instructions, which seem to be common to all processors, can execute differently in different devices and have advantages that will not be obvious when you first look at the instruction set.

For example, let's look at implementing code that would test the contents of a variable to see if it was zero and if another variable was incremented. As high-level C code, the statements would be as follows:

```
if (A == 0)        //  If Variable "A" is equal to Zero
     B++;          //  Increment Variable "B"
```

In a traditional eight-bit processor, the code that would execute this would be as follows:

```
move Acc, A      ;  Load Accumulator with the contents of "A"
add  Acc, #0     ;  Add Zero from it to see if result is zero
jnz  Skip        ;  If it isn't, skip over
move Acc, B      ;  Else, increment "B"
add  Acc, #1
move B, Acc
Skip:
```

Chances are that the microcontroller you are working with has a single increment instruction. If this is the case, then the three instructions used to increment the variable B could be changed to a single increment instruction.

In some microcontroller processors (such as the PICmicro MCUs), the action of moving the contents of a variable into the accumulator updates the status flags, just as adding something to the accumulator updates the status flags. This feature will eliminate the Add Zero instruction in the previous code.

Along with the capability to test registers, the PICmicro microcontroller processor has the capability to skip over the next instruction if specific status flags are set. Using the "skip if not zero" instruction, along with PICmicro MCU's ability to set the zero flag based on moving the contents of a register and the single instruction increment, the six insructions given above become:

```
movf   A, f    ;  Set STATUS Zero Flag if contents of "A" is 0
skpnz          ;  Skip next instruction if Zero flag is reset
incf   B, f    ;  Increment "B"
```

These three instructions are very simple and have a number of potential benefits that are not obvious:

■ The accumulator is never changed. A is tested as being zero or not, without saving its contents in the accumulator. This can be advantageous in situations when the contents of the accumulator cannot be changed.

■ The use of the "Skip next instruction if not zero" instruction eliminates the need for an additional, arbitrary skip label.

■ The code executes in the same number of cycles whether the increment B instruction is executed or not.

Despite these advantages, if you were to compare these three instructions to the two previous C source lines, you would see very little correlation. Even with comments, what is happening in the instructions is quite hard to follow.

If, after looking at this description on how to convert two simple lines of C code into PICmicro MCU assembler, you think that I am advocating taking each instruction and drawing the data flow along with the affected registers, to get the most efficient assembly code, you would be right. This is the only real way that you can understand how a processor works, and it gives you a way of visualizing how data flows for each instruction. For most simple processors, this is not an insignificant amount of work; although, the efficiency in which you will be able to develop assembler code and find problems in it will be worth the effort.

When you are familiar with the processor, you should be able to visualize the program as a "movie" and you will find that as you single step through the code in a simulator, the data flow should be obvious and expected. If there are any errors in the application, they will become very obvious because the code does not end up where you would expect. This is the artistry that I mentioned at the start of this section; when you can visualize the flow of the code, then you are on your way to being an expert assembly language developer.

Even if you go through the work and become very proficient in assembly language programming, you will have to be aware of a few issues. The first is obvious, but it should be mentioned that any code written for one microcontroller will not work with another—directly porting software is impossible. It is very difficult to explain the movie in your head to another programmer. Second, it can be difficult to link assembly language code with high-level languages. With high-level languages, functions and subroutines have parameters passed to them. The passing mechanism must be understood by the assembly language developer, and testing will have to be put in place to make sure data is being transferred to and from assembly language functions correctly.

The biggest disadvantage of using assembly language programming is that it is very hard to see the structure of the application in the source code. Assembler programming does not enable structured programming practices, and the need for low-level interfaces obfuscates the purpose and function of the code.

For high-level languages, I tend to think of the operation of the code as a series of English language statements. Most high-level languages are syntactically similar to the English language. For example, the English statement

```
"if the contents of variable A are equal to zero, then increment the
contents of variable B"
```

can be converted to the *Beginner's All-purpose Symbolic Instruction Code* (BASIC) statement

```
    if (A = 0), then B = B + 1
```

or the C statement

```
    if (A == 0)
        B++;
```

The original English sentence should be fairly obvious in the BASIC and C example code, but when you go back to the PICmicro microcontroller assembler, the meaning is completely lost.

The most compelling reason for writing in assembler is cost. For many simple robot applications, the availability of free development tools and sample software to build from makes assembler very attractive.

Despite all the negative comments I have given about assembly language programming, it is useful in many cases. Tightly timed loops for device drivers or interfaces often can only be implemented in assembler when having direct control over the instructions that are executing, and this is a distinct advantage. Some microcontrollers can be used to implement simple stand-alone sensors or output devices, and the expense of a compiler and other development tools may not be warranted.

In this book, I will avoid assembler almost exclusively because, as I noted previously, the function that the code is performing is hidden by the assembly language statements themselves. This is in keeping with the theme of this book, which is being device independent rather than specific to the Microchip PIC16F627 that is used to demonstrate the different interfaces used in this book.

Interpreters

When PCs first became available, virtually all software development and learning about programming took place in an interpreter (usually BASIC) that was automatically loaded into the computer when it first booted up. If you are young, this statement is probably surprising, as you will wonder what was the operating system of the computer? The interpreter performed many of the basic functions of what is considered normal for an operating system. The IBM PC changed this, with the introduction of a simple operating system that could either execute binary applications or execute the built-in BASIC interpreter.

An interpreter is a program that runs in the computer and reads and executes another program without the need to convert it into a format specific to the computer. It also provides a simple user interface into the computer. A simple demonstration of an interpreter would consist of a user entering a BASIC application like the following:

```
for i = 1 to 4
      print i
next i
end
```

The user would then command the interpreter to execute it. The interpreter would read through each statement (starting with "for i = 1 to 4") and execute it. When you execute the preceding application, you would get the following list of numbers written on the display:

```
1
2
3
4
```

and then the interpreter would prompt you for another command. PC interpreters would include the capability to save and load applications to and from disk or, for the very early computers, to and from audiotape.

Providing a computer with an interpreter allowed companies to provide computers to customers that were ready to go. Instead of having a computer that would have to be programmed in machine instructions (entered in via switches on the front of the computer), the customer would be given an interface that enabled them to explore the features of their computers as well as run some applications.

When working with a true microcontroller interpreter, you will be accessing the interpreter using something like an RS-232 terminal connection. A microcontroller interpreter often does not easily give you the ability to save and load application source code from a host computer, which can be a drawback in many situations. Like the early PC, most microcontroller interpreters are based on the BASIC language.

There are quite a few reasons for wanting to use an interpreter with robots. Interpreters are great what-if tools, giving you the opportunity to make simple changes to an application to see what happens. Changing an interpreter-based robot application can be as simple as hooking up an RS-232 cable from a laptop to the robot and making the changes, avoiding the need to compile simple changes and download a new application into the robot. Of course, this ability to easily change an application has the potential for losing effective changes because they are being made without being properly saved or documented.

Interpreted code is generally very portable between different devices. This gives you the advantage of being able to test your application on different devices (including a PC) without having to load a robot and potentially chase it around a room if something doesn't work the way you expect. As well as being portable to other devices, you will probably also find that compilers will enable you to build the application as a hex file that can be loaded into a microcontroller.

Interpreters often have complex statements or functions built into them that will enable you to directly interface with hardware devices at full speed, rather than having to develop low-level (elelogic) interfaces. For robot applications, chances are you will still have to create the mechalogic interfaces, but there will probably be built-in functions (such as a switch debounce or *pulse width modulation* [PWM] output) that will lessen the amount of work that you have to do.

The most popular interpreted language is BASIC, which is very easy to learn and is probably the most widely known programming language in the world. This popularity has resulted in there being many different books written explaining how to program in BASIC, as well as lots of source code that you can look at to learn more about programming and use in your own applications.

Lastly, most interpreters are quite inexpensive. A number of microcontroller interpreters can be downloaded free over the Internet. (I have written one myself for the Microchip PIC16F87x.) The source code for these interpreters is usually available, so there is the opportunity to modify them to provide basic robot mechalogic and elelogic functions.

Although strong reasons exist for using an interpreter in your robot applications, there are some downsides to consider when choosing a development tool. The biggest one is the speed at which code in an interpreter executes. Although assembly language will normally execute on the order of millions of instructions per second and compiled code will execute at the rate of tens of thousands of statements per second, you will find that interpreters will execute at hundreds or thousands of statements per second. An interpreter has to read a complete statement, break it down into the tasks it has to perform, and then execute them. For biologic code, this is generally not an issue, but for mechalogic and elelogic, there will be some major concerns and the need for developing workarounds to potential problems.

You should be aware of a couple of other speed issues regarding interpreters as well. The first is how the inclusion of comments slows down the execution of an application. Each character of a comment must be read in and discarded, which slows down the actual execution of the application. Second, anytime a goto is executed, the interpreter has to search through the entire application for the destination label. In many early BASIC interpreters, this issue was mitigated by the use of line numbers and specifying code that goes to specific line numbers that could be easily searched rather than doing a text search for a label.

Along with comments and gotos slowing down an application's execution, precisely timing an interpreted application can be very difficult. You can poll internal resources of the microcontroller (such as timers), but you will find that these will just give you a minimum timing; depending on how the interpreter and the microcontroller operate, you have no guarantee of what the maximum timing will be.

With an interpreter, you can very quickly run into problems by running out of application *random access memory* (RAM). This is due to the limited amount of space available for applications in the microcontroller. Instead of making things worse for the overall readability of the application, you have to remember that each comment takes up space that could be used for code. It isn't unusual to see large, complex applications for interpreters with no comments.

The amount of memory required by the interpreter can also be very surprising. An interpreter is a sophisticated application and will require a large amount of *read-only memory* (ROM) as well as quite a bit of RAM to execute different functions. You will find that interpreters will only be possible in fairly large microcontrollers or ones that are able to access a considerable amount of external memory. Very few microcontrollers can implement an interpreter without any external chips, and in these cases, you will probably find that the list of available built-in functions is quite short.

I have found that with most interpreters, the built-in application debug capabilities are generally quite poor. It is unusual to find interpreters with the capability to stop or "break" applications at specific link numbers. I also find the line-by-line operation of interpreters makes it difficult to follow the application's execution flow.

For hardware support, the interpreter should be able to read and write hardware interface registers built into the microcontroller. Due to the speed of the interpreter, these reads and writes can be quite slow and interrupts cannot be processed in an interpreter. This will attenuate the sophistication of the application that can be run in the interpreter and could make getting the application to work much more difficult.

The last disadvantage of using an interpreter is something of an irony. I stated earlier that the advantage of using an interpreter is that changes to an application could be made without the need of a programmer, but a programmer is needed to load the interpreter into a microcontroller. Some chips can be purchased with an interpreter built in, but these are generally quite expensive and rarely have features built into them that are designed for robot programming.

Finally, the last topic to mention is the Parallax Basic Stamp, which is a microcontroller module that is often confused with an interpreted device. The Basic Stamp software development tools compile PBASIC source code into a series of tokens that are downloaded into the Basic Stamp module. These tokens are bytes that represent different functions built into the PBASIC language. Because the Basic Stamp does not take raw source code, it cannot be considered to be a true interpreter. The Basic Stamp is, however, a popular device for controlling robots because of its reasonable cost, ease of programming, sophisticated built-in functions, and large amounts of sample source code that can be found on the Internet.

When choosing interpreters and the hardware they run on for your robot application, you should only use ones that have the following built-in features:

- Nonvolatile source code storage of the application
- A simple serial communications interface
- PWM generation
- *Inter-Intercomputer Communications* (I2C), SPI, CAN, or other high-speed synchronous serial interfaces
- Button debounce capabilities
- *Analog-to-digital converter* (ADC) interfaces

Compilers

Compilers work through a series of reasonably simple rules to convert high-level language statements into assembly language. There is no magic involved, even though the results sometimes are pretty amazing. Modern compilers also look for opportunities to simplify assembly code, further resulting in smaller and more efficient applications. If you are interested in learning assembly language for a specific processor, you should not be surprised to discover that the code that modern compilers produce is an excellent learning example.

All the code for the sample applications presented in this book is in the high-level C programming language. The C statements are converted into processor instructions (compiled) before being programmed into the PICmicro MCU.

The C compiler used with the examples provided in this book is PICC Lite by HI-Tech Software and is restricted to lower-capability, mid-range, Flash-based PICmicro microcontrollers. The language is based on the ANSI C standard and specifications are given in the appendices. The code produced is quite efficient and, as an added bonus, the PICC Lite compiler integrates very well into the Microchip MPLAB-integrated development environment.

Throughout this book, you will see statements like

```
A = B + (C * D);
```

A number of steps have to be carried out to convert this high-level statement into a series of assembly language instructions. Most compilers take advantage of a processor data stack for working through these types of instructions and keeping track of temporary values. When processing the statements, the values are pushed onto the stack in the reverse order. When the statement is executed, this data is "popped" off the stack and processed.

The first thing a compiler does is to determine the type of statement it has to work on next. In most high-level languages, there are five types of statements:

```
Statement                    Type
variable = ...               Assignment Statement
if (...)                     Conditional Execution Statement
label(...)                   Subroutine/Function Call
type variable [= constant]   Variable Declaration
[type] label(variable,....)  Subroutine/Function Declaration
```

In the first type of statement (Assignment Statement, Conditional Execution Statement, and Subroutine/Function Call), the part of the statement shown as "..." can be considered as an expression. An expression can be a constant or a variable, or a series of constants, variables, along with some operators that perform some kind of function. Examples of arithmetic operators include addition, division, bitwise AND, and logical OR. In the appendices, I have listed the arithmetic operators available to PICC Lite (which consists of the basic operators available to C and many other languages).

The variables, constants, and interim values are the data that will be put on the data stack and executed. Putting the statement in a "heap" stored in a postfix order (the least significant operations given the highest priority) does this.

For the operation in the assignment statement

```
A = B + (C * D);
```

the postfix heap is shown in Figure 2-1.

Figure 2-1
Compiler
postfix heap

A = B + (C * D)

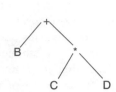

Next, the order of operations is determined by pulling the data from the bottom up, right-to-left order. In this operation, the lowest operation on the right is pushed onto the stack followed by the lowest on the left and the operator is pushed onto the stack. When the operator executes, it pops the previous two (or one) stack elements, and then pushes the result onto the stack. To save the result, the current stack value is "popped." The stack operations for

```
A = B + (C * D);
```

are

```
Push        D
Push        C
Execute     *
Push        B
Execute     +
Pop         A
```

In this sequence of stack operations, D, followed by C, is pushed onto the stack. Next, they are popped off the stack and multiplied together with the result of C * D pushed back onto the stack. The result of C * D is popped off the stack along with B and added together with the result pushed onto the stack. To finish off the instruction, the final result (B + (C * D) is popped from the stack and stored in A.

Depending on your age, you may remember Hewlett-Packard calculators that worked this way. Data entry took the form of the stack instructions listed previously and was known as *Reverse Polish Notation* (RPN). It took a bit of getting used to, but once you were able to think in RPN it actually was easier working through complex problems because you didn't have to remember how many parenthesis were active. In universities and colleges all over the world, the truly cool people could think in RPN and not use a pencil and paper to plan out how they were going to enter statements into their calculators.

Leaving a result on the stack is important for the other two types of statements. In the if statement, if the value on the top of the stack is not equal to zero, then the condition is determined to be true. As will be discussed, for the Subroutine/Function Call statement, the parameters passed to the Subroutine/Function are accessed from the stack by the Subroutine/Function code according to their position relative to the top of the stack. Array elements are also stack values that are popped off when an element is to be accessed.

Another way that a compiler can implement a data stack is to use temporary variables and access them directly. For the assignment statement example, the operations would be as follows:

```
Temp1 = C;
Temp2 = D;
Temp3 = B;

Temp1 = Temp1 * Temp2;
Temp1 = Temp1 + Temp3;

A = temp 1
```

This method gets very complex when statements have more than one stack entry left on the stack during execution. The two types of statements that come to mind are ones that use array elements and those that call subroutines or functions that require more than one input parameter. In these cases, the data is evaluated and left on the stack for later operation.

Adding an array element to the example statement

```
A = B + (C[4] * D);
```

would result in the postfix heap shown in Figure 2-2 and the stack operations

```
Push      D
Push      4
Push      C[
Execute   *
Push      B
Execute   +
Pop       A
```

In this example, the order of operations would be to push D onto the stack, followed by 4. When the Push C[operation is encountered, the previous element (the 4) would be popped off the stack and used as the index into array C. Once C[4] is evaluated, it would be pushed onto the stack and the operation would continue as before with the result left on the stack.

Figure 2-2
Compiler
postfix heap for
array statement

A = B + (C[4] * D)

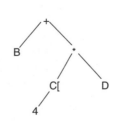

Many subroutines and functions have multiple parameters passed to them. For the function

```
int Func(int varA, int varB)
```

a calling assignment statement could be

```
A = B + Func(C[4], D);
```

which would require the postfix heap shown in Figure 2-3.

The statement would also require the stack loading the following operations:

```
Push        D
Push        4
Push        C [
Call        Func
Push        B
Execute     +
Pop         A
```

The stack is executed similarly to the previous example, but when the Func call is encountered, the two previous values are left on the stack, and then Func is called with them as a local variable parameter.

In Func, the two parameters are referenced according to their position relative to the top of the stack; the first parameter will be one position below the stack top, while the second parameter is the stack top.

If Func was

```
Int Func(int varA, int varB)
{

  varA = varA + 1;

  return varA * varB;

} //  end Func
```

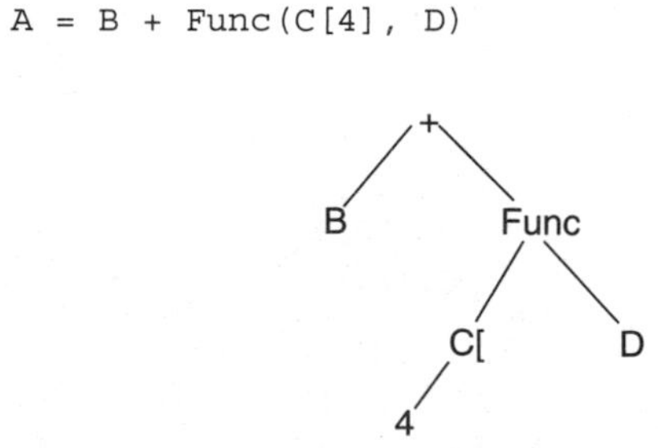

Figure 2-3
Compiler postfix heap for a function call

A = B + Func(C[4], D)

the actual compiled code for the function could be

```
Func:
;   varA = varA + 1
    Push        StackTop - 1      ;  Push "varA" as the New Stack Top
    Push        1                 ;  Push 1 onto the Stack
    Execute     +                 ;  Pop "varA" & "1", Add, Push Result
    Pop         StackTop - 1      ;  Pop Stack Top and Store in "varA"
;   return varA * varB
    Push        StackTop - 1      ;  Push "varA" as the New Stack Top
    Push        StackTop          ;  Push "varB" as the New Stack Top
    Execute     *                 ;  Pop "varA" & "varB", Multiply, Push
                                  ;    Result
    return
```

When the Func has finished executing and control has returned to the statement after the subroutine call instruction, the first stack item would be popped off and saved, and the next two would be popped off and discarded. The saved value would then be pushed onto the stack and execution would continue.

Another way of doing this would be to pop the new top value off the stack and place it as the first stack element before the call. Once this is done, any other values could be popped off the stack and discarded. The code making this call and popping off the unneeded values would look like the following:

```
Push        varA
Push        varB
Call        Func
Pop         StackTop - 1      ;  Put Result in the First Parameter
                              ;    Position ("varA")
Pop         BitBucket         ;  Get Rid of Second Parameter ("varB")
;  Result of "Func" is the Top Stack Element
```

When you look at the actual code produced by a compiler, you will probably see deviations from the strict stack operations outlined in this section. Optimizing code in the compiler causes these deviations. For example, statements like

```
A = B + (C * (4 * 2));
```

should be executed as

```
A = B + (C * 8);
```

with the stack operations looking like

```
Push        8
Push        C
Execute     *
Push        B
Execute     +
Pop         A
```

Linking applications does make a lot of sense when multiple languages are involved (each object file is created from one language type), multiple people are involved with the application development, or the source application is very long (longer than 10,000 lines). In these cases, linking object files together will make the application easier to understand and probably faster to create and debug.

You should also be aware of some concerns about compilers. A poor compiler can produce code that is as slow as an interpreter and be reasonably large. Compilers (at any quality level) can be quite expensive, while assemblers are available free of charge for most processors. When languages are defined, they are often intellectual exercises. Hardware interfaces (including interrupt handlers necessary for robot applications) are not specified and they are often added as poorly thought-out *kludges* afterward. Errors, warnings, and messages produced by a compiler can be quite cryptic and difficult to understand (and to figure out what the actual problem with the code is).

When you are creating your own applications, I strongly recommend that you *never* attempt to program MCU unless the compilation is clean. A warning or message that seems innocuous can result in your application not working properly (if at all).

Simulators and Emulators

It can be very difficult to debug running applications, especially in robots. Later in the book, I will show how different displays on your robot can be added to notify you of the application's operating status, but before you try out the application, I strongly suggest that you test your application on the PC that you are developing the code on. A few minutes spent with a simulator can save many frustrating hours trying to find a problem based on the operation of the robot.

A simulator consists of a software model of the microcontroller processor, which can be controlled along with the ability to pass test I/O signals back and forth to a virtual microcontroller. It is important to remember that the processor and I/O in a simulator are software models, and some situations that can come up in simulator hardware cannot be properly simulated. For example, the Microchip MPLAB IDE is an excellent tool for testing application code, but it is somewhat limited in its capability to model advanced peripheral I/O ports.

Some features that you should always look for in the simulator software you are using include the following:

- Source-code-level debugging is a feature I consider mandatory, as it enables you to see exactly what is happening in the code, make changes, and make sure the application works correctly before programming a microcontroller.

- Execution timing is accurately modeled. This is less of a concern for the high-level language programming presented in this book, but it is a major concern when you are programming in assembler and interfacing with critically timed parts.

■ Some kind of preprogrammed stimulus is needed that will provide you with the ability to repeatedly test the application code with the same inputs to allow you to accurately model what is happening with the robot hardware.

■ The stimulus enables margin testing of the interfaces to make sure they will work, even if timings are out by some margin. This is known as *guardband testing*.

I tend to think of a simulator as a collection of *black boxes* that are controlled by the simulator's User Interface software module as shown in Figure 2-4. I drew the simulator this way because it enables different boxes to be swapped in and out to make up various part numbers without a significant effort required to create simulators for different microcontrollers. Probably more modules are used in a real simulator implementation, but for the purposes of this discussion, the five presented in Figure 2-4 are adequate.

The program memory block is loaded with the hex file that will be programmed into the microcontroller. The processor model will pull data from this simulated program memory as required. The file registers are similar, and the processor model can read and write them as can the real processor.

The I/O and hardware registers block provides some kind of model of the I/O pins. For basic digital I/O, most simulators provide a good model of the I/O hardware, but for advanced peripherals (such as serial and analog I/O), simulator models usually do little more than just accept input from the processor as if the peripheral control registers were simple flip flops.

Typically, the I/O pins have the capability of being driven by external inputs. This is the stimulus driver box in Figure 2-4 and can either be directly used in changing I/O register bits or can run a stimulus file. A stimulus file is created

Figure 2-4
Simulator
software
architecture

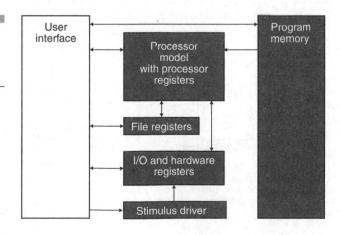

with I/O information that can be processed by the simulator processor without much intervention by the user.

The topic of stimulus files brings up the point that when debugging an application, the simulator must be set up in such a way that it will always run the same. This is an important point and one that I want to make sure you follow because if the simulated application runs differently each time, you will have problems trying to identify problems and fix them. When you debug your application using a simulator, you should be focusing on fixing the problem, not learning how the simulator is set up for this application's debug.

The processor block is obviously the heart of the simulator with the user interface commanding it to single step instructions or execute blocks of code. The processor model is an extremely complex piece of software that not only has to fetch and execute instructions and access registers, but also has to manage such peripheral functions as interrupts. Making the creation of this module even more dificult, is the requirement that the execution of the simulated instructions must be as fast as possible.

The user interface is the primary window into the application code. As such, it should be as configurable as possible to enable user preferences as well as to display different variables and I/O registers based on the user's preferences. Along with displaying customizable registers, the simulator should work through the code source, rather than the simple instructions.

The next step up from a simulator is an *emulator*, which is a device that replaces the microcontroller in your application and is connected to your development PC. The purpose of the emulator is to enable you to monitor the execution of the application code in much the same way as can be done with the simulator, except that you are running on actual hardware. A typical emulator is designed from the block diagram shown in Figure 2-5.

The emulator function can be accomplished in a variety of different ways, from I/O port hardware emulators and simulators running on PCs to specially modified microcontrollers running the application code. The ideal is to have a modified microcontroller used as the interface between the PC and the circuit so that the emulator is modeling the actual chip as much as possible. An advantage of the emulator over the simulator is that actual pin I/O signals can be observed, both

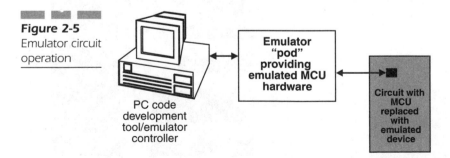

Figure 2-5
Emulator circuit
operation

Emulator
"pod"
providing
emulated MCU
hardware

Circuit with
MCU
replaced
with
emulated
device

PC code
development
tool/emulator
controller

from the processor's perspective as well as from the circuit's. Most emulators are built from the same chips and hardware as the actual device, so there are no missing peripheral interface functions.

The best method of providing an emulator is to use a *bond-out chip*, which is an actual microcontroller with specific access provided to the processor so that it can be started or stopped and provide emulator unique instructions. The Microchip emulators all use actual PICmicro MCU chips, so the hardware interfaces are accurate to actual devices.

Working with an emulator has two disadvantages. The first one is cost. Emulators generally cost $2,000 or more and require separate pods for each device being emulated. This cost is not significant for many companies, but for small companies and individuals, it can be.

To help offset the costs, a number of simplified emulators exist, like Microchip's PIC16F87x-based MPLAB-ICD Debugger. This tool provides many of the same features as a full emulator, but is restricted in its capability to run internal hardware operations while execution is stopped.

The second disadvantage of using an emulator with a robot is connecting it to the software development PC. For most applications, this is not a problem, but it can be a major inconvenience when developing robot applications.

When choosing an emulator, remember the old adage "You get what you pay for and you pay for what you get."

Like the simulator, having a direct connection between the source code and the instructions running in the processor is critical for efficient operation of the emulator. Without this capability, you will be forced to cross-reference a listing file for an absolute address. This is a function that I feel a simulator and an emulator user interface should do for me.

Integrated Development Environments

When I'm developing software, I always find that I'm the happiest (most productive and debugging the fastest) using an IDE. This extends to PC programming, where I always liked the original Borland Turbo Pascal and the modern Microsoft Visual Basic and Visual C++ development tools (part of Microsoft's Visual Development Studio IDE). I don't believe that I am alone in feeling this way, and the Microchip PICmicro MPLAB IDE (shown in Figure 2-6) has done a lot to make the PICmicro MCU one of the most popular microcontroller for application developers.

An IDE integrates all the software development tools I've described in the previous sections. A microcontroller IDE brings the following tools together:

- Editor
- Assembler

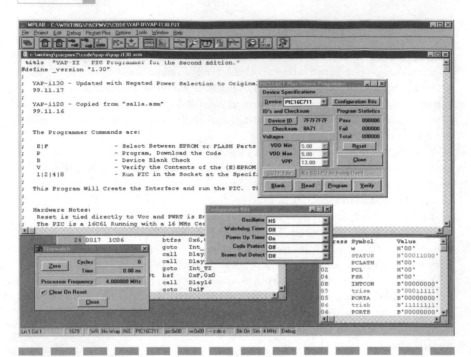

Figure 2-6   MPLAB IDE with PICstart plus the interface

- Compiler
- Linker
- Simulator
- Emulator
- Programmer

The purpose of an IDE is to provide the data in a system that enables it to be shared seamlessly as far as the user is concerned and does not require any special input on his or her part.

When you are looking at different IDEs, I suggest that you make sure you have a tool that is expandable. The IDE should be able to handle

- Adding new microcontroller part numbers
- Inclusion of different vendors' high-level languages
- Connections to different programmers and emulators

Very high-end (read expensive) IDEs are integrated with circuit design systems. The advantages of bringing the two development tools together include a proper electrical connection between the microcontroller and the circuit that it is being built into, and very accurate signal simulations. However, if the IDE is well integrated to the design system, then the whole circuit can be simulated in the circuit design system with the microcontroller behaving as if it were a simple chip in the application, responding to inputs as programmed. This level of integration enables the microcontroller software developer and the circuit designer to be two different people, and they can test the operation of each other's designs without intimately understanding how they work.

Just one last recommendation, make sure that the editing functions are compatible with the standard Microsoft NotePad and WordPad editing commands. These are described in Chapter 3. A tool that uses the Microsoft conventions will enable you to cut and paste code from other files, web pages, or any other source quite effortlessly. Along with making sure the editing functions are standard, also make sure the output (hex) file formats are in a standard format and not in one that will lock you into a certain vendor's equipment (like programmers).

The Microchip PICmicro® Microcontroller

I am probably most known for my books about the Microchip PICmicro® micro-controller. I have been working with the current architecture's *complementary metal-oxide semiconductor* (CMOS) PICmicro microcontroller for well over 10 years (and before that, I had experience with the previous *negative-channel metal-oxide semiconductor* [NMOS] architectures that were built and branded by General Instruments) that the current PICmicro microcontrollers are based on. I've always enjoyed working with these devices because they are easy to work with, reasonably fast, and I find them enjoyable to program. My preference for the PICmicro microcontrollers is not unique; these parts are among the most popular microcontrollers in the world and I'm sure several are used in products you have around the house.

The advantages of the Microchip PICmicro microcontroller include how easy it is to find the parts from numerous distributors around the world. As well as there being many different places to find the parts, there are many different part numbers (or versions) of the PICmicro microcontroller available, one of which will probably have all the features that meet your requirements. Every year, Microchip releases up to a hundred new part numbers, giving customers new internal features as well as packaging options. Part of the strategy behind the multiple product releases is to redesign (or re-spin) existing chips to reduce the amount of silicon they use (resulting in more chips per wafer and lower cost) along with reducing the amount of current the chip requires.

The different features built into different PICmicro MCU part numbers include the following:

- Changeable operations using configuration fuses that are set during programming
- Different clocking options (including internal oscillators and *phased locked loop* [PLL] speed-up circuitry)
- Different reset options (including internally generated reset and brown-out detection)
- Different *input/output* (I/O) pin options including
  - High-current I/O
  - Open collector
  - *Nonreturn to zero* (NRZ) serial I/O
  - Synchronous serial I/O
  - Analog inputs with comparators or *analog-to-digital converters* (ADCs)
  - Bus device emulation
- Flash-based program memory with simple programming for easy and cheap experimentation
- Multiple clock/timer/counter options
- Interrupts from multiple sources (multiple interrupt vectors in some PICmicro MCUs)

Microchip was one of the first manufacturers that provided their chips with a simple, low-cost, and documented programming methodology. Most PICmicro microcontrollers are designed to have their applications programmed in serially using the *In-Circuit Serial Programming* (ICSP) protocol. This synchronous serial programming method enables PICmicro MCUs to be programmed very rudimentary parts, as I will demonstrate with the El Cheapo programmer circuit discussed in this chapter.

Microchip was also one of the first microcontroller manufacturers that built microcontrollers with flash *Electrically Erasable Programmable Read-Only Memory* (EEPROM) for program memory instead of using *ultraviolet* (UV) light *Erasable Programmable Read-Only Memory* (EPROM). Providing devices with flash program memory avoided the need for expensive, "windowed" ceramic packaging for the chips as well as the need for the experimenter to have a UV eraser. Microchip's ICSP, coupled along with Flash-based microcontrollers, enables the experimenter to develop PICmicro MCU applications at a cost that is hundreds of dollars less than what is required for some other microcontrollers.

Many commercial products are based on the PICmicro MCU. In terms of microcontroller sales, Microchip is number two, after the Motorola line of microcontrollers. One of the most popular controllers used in robots is the Parallax Basic Stamp, which is based on the Microchip PICmicro MCU. Several versions of the product provide a complete, single-chip solution using a PICmicro MCU, serial EEPROM memory, and a ceramic resonator as a crystal in a package that is as small or smaller than many chips.

Finally, outstanding support exists for the PICmicro microcontrollers. Microchip provides excellent documentation for the PICmicro MCU product lines as well as the best free application software tools of any microcontroller supplier in the world. Along with Microchip providing outstanding support, you will also find literally thousands of web pages giving information in the form of FAQs, sample applications, lecture notes, explanations, and more. Finally, the many books written about the PICmicro will walk you through the process of developing an application as well as give you some sample applications. I daresay that the PICmicro MCU is the second-best supported electronic device in the world, after the PC.

There are not a lot of downsides to the PICmicro MCU. The most significant issues that people complain about include its "Harvard" processor architecture that can be a challenge for people to learn assembly language for if they have only received training on traditional "Princeton"-based architectures and its price/performance ratio compared to other microcontrollers. Personally, I really enjoy developing assembly language applications on the PICmicro because the architecture enables some very clever things to be written in the software. The price/performance concerns have been addressed as part of the ongoing chip size/cost reductions efforts made by Microchip.

When I originally started this book, I wrote all the example software applications presented in the following chapters for the PIC16F84, which has traditionally been the introductory PICmicro MCU for new users. The reason for choosing the PIC16F84 was the availability of HI-TECH Software's PICC Lite C Compiler, which will enable you to develop C applications for the PIC16F84 simply and without the need for learning assembly language. The PC16F84 has flash program memory and utilizes the ICSP programming protocol, which enables it to be easily (re)programmed. The I/O features of the PIC16F84 can be considered somewhat pedestrian compared to the features of other PICmicro MCU part numbers.

As I was working through the manuscript, HI-TECH Software updated their PICC Lite compiler to support the PIC16F627. This PICmicro MCU, while having the same basic features as the PIC16F84, has greatly enhanced I/O functions, which greatly reduces the amount of effort required to develop robot applications. I'm very happy that I was able to take advantage of this, as using the updated version of PICC Lite with the PIC16F627 simplifies the effort required to develop robot applications, eliminating the requirement for assembly language programming of any type. Along with this, when demonstrating how some peripherals were implemented, I had to come up with circuitry and application code that used multiple PICmicro microcontrollers. I believe that the PIC16F627 and PICC Lite will become a standard for new (and experienced) PICmicro MCU application developers.

As you read through this chapter, you will see that I have described the PIC16F627 in a great deal of detail. This was done to make sure that you understand the different hardware features built into the microcontroller before attempting to work with it in a robot. The obvious question from reading this information is if you should create notes with a similar amount of detail if you are going to use a different controller for your robots. I would say the answer to this question is yes, because you should be as familiar as possible with the hardware you are using to make sure that you fully understand the different features to the controller, along with any potential issues that you will have to overcome.

As a final note, the PICmicro MCU is often known as just the PIC. This was the original name given to the family of microcontrollers by General Instruments, the company Microchip spun off from. Although the term PICmicro MCU is the correct one, you will find that many web sites (and their authors) still use the term PIC. You must take this into account when you are searching for information.

Different PICmicro MCU Devices and Features

Four different PICmicro MCU families exist and the defining feature of the different families is the architecture of the processor built into the microcontroller. The differences in the processor center around how special functions (hardware) and variable registers can be accessed, whether or not interrupts are enabled in the processor, and if any advanced instructions (such as a single-cycle multiplication) are available. Table 3-1 outlines the different families with their capabilities and some important points to note.

All the different PICmicro microcontroller part numbers have common features to the other devices. I find that all the PICmicro microcontrollers are reasonably fast with four clock cycles required for each instruction cycle. Some microcontrollers advertise faster speeds (by the use of PLL clock multipliers and using different chip technologies) and I expect to see more new PICmicro MCU part numbers with these features built in (see Table 3-2).

It is important to note that microchip quotes program memory space size is in instructions, not bytes. If program memory size were quoted in bytes, you would find that the values would be 50 to 100 percent higher. Lastly, the PICmicro processor architecture is designed in such a way that an experienced assembly language programmer can develop applications with as little as 30 percent of the instructions required by a traditional processor to carry out the same task.

Application Development Tools Used in This Book

When I was deciding which microcontroller to present the sample applications in this book, I decided on the Microchip PICmicro microcontroller primarily due to the availability of outstanding free software development tools. Other factors in favor of the PICmicro MCU include the availability of low-cost, Flash-based parts with sophisticated, built-in functions and the ease in which a programmer circuit could be designed. For about $20, you can have a complete robot microcontroller consisting of an *American National Standards Institute* (ANSI)-compliant C compiler, a full-featured *integrated development environment* (IDE), a programmer circuit, and a sample PICmicro microcontroller.

The software application development tools that I will work with in this book include HI-TECH Software's PICC Lite C high-level language compiler (with integrated) assembler and Microchip's MPLAB IDE, integrated with PICC Lite, providing a complete application development system and programmer interface. These tools are all designed for Win32-based operating systems (Windows 95, Windows 98, Windows ME, Windows NT, Windows 2000, and Windows XP) from Microsoft. If you want to use another microcontroller, and if you were to purchase application development tools of similar quality, it could cost you as much as $2,000. If you have never worked with any of these tools before, I know you will be very impressed with them.

PICC Lite is the free version of HI-TECH Software's C programming language compiler for PICmicro microcontrollers. Although the compiler is MS-DOS command-line driven, it integrates seamlessly with Microchip's MPLAB, enabling you to run application code in a PICmicro MCU simulator before

Table 3-1

The PICmicro MCU Families

Family	Instruction Word	Register Layout	Comments
Low-end	12 Bits	32 regs/bank, 4 banks	■ Best suited for small, peripheral functions ■ No interrupts ■ Single timer ■ Up to 20 MHz clock ■ No enhanced I/O pins ■ No flash program memory ■ Parallel programming except for PIC12C5xx and PIC16C505, which use ICSP
Midrange	14 bits	128 regs/bank, 4 banks	■ Good general-purpose microcontrollers ■ Most popular PICmicro MCU family ■ Most part numbers/most diversity ■ Single-vector interrupts ■ Up to three timers ■ Up to 20 MHz clock ■ Many different types of enhanced I/O pins, including analog I/O, LCD drivers, and serial interfaces ■ Flash program memory available for most devices ■ ICSP programming ■ Built-in debugger (ICD) hardware in some devices

Table 3-1 (*continued*)

The PICmicro MCU Families

PIC17Cxx(x)	16 bits	224 regs/bank, 8 banks, 48 SFRs	▓ Bus interface microcontrollers/access parallel devices directly ▓ Architecture quite different from other PICmicro MCU families ▓ Enhanced instruction set and indexing capabilities ▓ Small number of part numbers ▓ Multiple interrupt vectors ▓ Three timers ▓ Up to 33 MHz clock ▓ Some enhanced I/O pins available ▓ No flash program memory ▓ Parallel programming, including self-programming
PIC18C/Fxx2	16 bits	256 regs/bank, 16 banks	▓ Enhanced PICmicro MCU architecture based on midrange ▓ Able to access 2 MB of program memory, up to 4K of variable memory ▓ Plans to replace midrange with PIC18C/Fxx2 ▓ Enhanced instruction set and indexing capabilities ▓ Single interrupt vector with programmable priority for sources ▓ Similar I/O capabilities as midrange PICmicro MCUs ▓ Up to 40 Mhz operation with 10 MHz clock and PLL clock multiplier ▓ Flash program memory versions available for all parts ▓ ICSP with modified data interface with self-programming capability

55

Table 3-2

PICmicro Microcontroller Part Numbers

Part Number	Architecture	Features	Applications
12C5xx	Low-end	Internal Osc/Reset	Simple interfacing
12C6xx	Midrange	ADC/Internal Osc/Reset/Data EERPOM	Simple interfacing
14C000	Midrange	ADC/Vref	Power supply control
16C5x	Low-end		Basic applications
16C505	Low-end		Basic applications
16HV540	Low-end	Voltage regulator	Basic applications
16C55x	Midrange		Basic applications
16C6x	Midrange		Digital applications
16C62x	Midrange	Voltage comparator	Analog monitoring
16F62x	Midrange	Voltage comparator/flash program memory	Analog monitoring
16C7x	Midrange	ADC	Analog interfacing
16x8x	Midrange	Flash program memory	Application development
16F87x	Midrange	ADC/flash program memory	Analog interfacing/application development
16C9xx	Midrange	ADC/I2C	Analog interfacing
17Cxx	High-end	External memory	Advanced applications
18Cxxx	18Cxx	ADC/I2C	Analog/digital interfacing applications

programming a PICmicro MCU and finding (and trying to fix) problems the hard way. As you will see in the following chapters, PICC Lite enables very substantial applications to be created without the need for assembly language. I have included a brief outline of the language in the Appendices, and the HI-TECH Software language manual PDF file is included on the CD-ROM.

The PICC Lite compiler is a subset of the full HI-TECH Software's C compiler. The full version of the compiler can produce code for each of the four PICmicro MCU processor architectures and has include files written for all the different PICmicro microcontroller part numbers. PICC Lite will only produce code for the PIC16C84, PIC16F84, PIC16F84A, and PIC16F627. This may seem like a minimal choice, but the PIC16x84 part numbers have been traditionally used by new PICmicro MCU application developers because of their flash program memory. I am expecting the PIC16F627 (and the PIC16F628, which has more program memory) to become the new standard for developers learning about the PICmicro microcontroller.

PICC Lite is ANSI/*International Organization for Standardization* (ISO) C programming language compliant, although there are a few features that you should be aware of. As I have pointed out, PICC Lite is designed to be used with Microchip's MPLAB IDE and provides source-code-level debugging and access to the Microchip linker, enabling standard assembly language functions to be added to a PICC Lite application. If a variable is given the type int (integer), then the compiler will allocate 16 bits (2 bytes) of space and not 8 bits (1 byte), as might be expected. This is a bit of a departure from the ANSI specification (which expects int to specify a variable with the data word size) but provides the programmer with a very useful default. If you want to declare a variable as being 8 bits in size, then you will have to give it the char type. All the PICmicro MCU's hardware registers can be accessed directly from C statements and they have been defined in the pic.h include file.

You should be aware of a few restrictions of the compiler. Recursive functions are not available in PICC Lite because of the limited stack built into the PICmicro MCU (only eight levels deep). Reentrant functions (ones that can be invoked from the interrupt handler while they have also been invoked from the mainline) cannot be implemented because of the lack of a data stack available to the processor. These limitations are actually quite minor and not a significant issue for microcontroller applications.

In this book, I will be focusing on developing single-file applications rather than multiple-file, linked applications. This will enable the applications to be built faster and will minimize confusion on what module is performing which task. Creating applications from just one source code file will also simplify the task of debugging applications.

PICC Lite has a built-in assembler that enables the inclusion of in-line PICmicro MCU assembly language statements and avoids the need for linking separate assembly-language-based object files. With the exception of the *sleep* and *clrwdt* hardware function instructions, I will not be using any assembly language code in any of the presented applications.

PICmicro microcontroller assembly language, while not terribly difficult to learn, is different from the assembly language used by traditional Princeton-architected processors (like the Motorola 68HC11), and putting it in the sample applications would require me to explain it in some detail. Being able to port the code directly from the PICmicro MCU to another microcontroller without having to translate assembly language is another reason for not taking advantage of the PICC Lite's built-in assembler.

As I will discuss, I developed the sample application code in this book twice (the first time for the PIC16F84 and the second time for the PIC16F627 when an updated version of PICC Lite became available). Even when I was working with just the PIC16F84, which does not have many of the advanced features of the PIC16F627, I was able to create sample code without having to resort to using assembly language. The code generated by PICC Lite is very well optimized and I confirmed this by writing several tests, comparing the PICC-Lite-generated code with how I would perform the same function.

PICC Lite can be somewhat tricky to install in different operating systems. When you are installing PICC Lite on your PC, make sure you follow the instructions on the CD-ROM that comes with this book. The CD-ROM has the latest information concerning how PICC Lite is installed in the different versions of the Windows operating system.

The MPLAB IDE is Microchip's free tool designed to enable developers to start creating applications using Microchip's PICmicro microcontrollers. I cannot say enough good things about MPLAB—I have been working with it for over five years (as I write this) and I have been very impressed with the effort Microchip has put into supporting this product. Not only have numerous small errors been fixed, but the product has been updated to support the numerous new PICmicro MCU part numbers that have become available.

The version included on the CD-ROM that initially came with this book is 5.62, and I expect that in a few months after the book's first printing, updated versions of the tool will be available. I suggest that you check Microchip's web site to make sure you have the latest version of the MPLAB IDE installed on your computer. You should expect the release of a 32-bit version of the MPLAB IDE in 2002, but before you install it, make sure that it is compatible with PICC Lite.

The MPLAB IDE provides a single, consistent interface, giving users a series of tools to help them with their PICmicro MCU application development. Although written originally for Microsoft Windows 3.1(1), the MPLAB IDE has been updated to be Win32 compatible with the following built-in features:

- A Microsoft-compatible editor
- Assembler register definition files for all PICmicro microcontroller part numbers
- Assembler for all PICmicro MCU families and part numbers
- Linker to put together multiple .obj files

■ Project manager to keep track of all files needed for a specific project as well as maintain the appearance and placement of dialog boxes on the MPLAB desktop

■ A simulator limited for some peripheral features, not an issue for the PIC16F84(A) part available for use by PICC Lite

■ Emulator interfaces for the Microchip-designed MPLAB-ICE and PICMASTER PICmicro MCU emulators

■ Debugger interface for the PIC16F87x, which has a built-in hardware debugger known as MPLAB-ICD (In Circuit Debugger) that enables the user to monitor the execution of the application in circuits, like an emulator from the MPLAB IDE

■ Programmer interface for Microchip-designed PICSTART Plus and PROMATE II PICmicro microcontroller programmers

The "what you see is what you get" Microsoft-compatible editor built into the MPLAB IDE uses the standard editing functions listed in Table 3-3.

To delete and move text, I use the cut and paste functions. To select text to cut and paste, I mark the text first pressing the Shift key while moving the cursor to highlight the text to be relocated. If you have marked text incorrectly, then simply move the cursor without the Shift key pressed to turn off the text marking. Pressing your mouse's left button and moving the mouse across the desired text will also mark it. Left-clicking on another part of the screen will move the cursor there and eliminate the highlighting for the text.

Next, the keystroke Ctrl-X is used, which removes the marked text from the file and places it into the window's clipboard. Ctrl-C copies the marked text into the clipboard and doesn't delete it. To put the text at a specific location within a file after the current cursor location, Ctrl-V is used.

Note that I do not use delete or insert. Deleting marked text destroys it completely, whereas Ctrl-X saves it in the clipboard so it can be restored (with Ctrl-Z or Ctrl-C) if you made a mistake. The Insert key toggles between data Insert or Replace mode for Microsoft-keystroke-compatible editors. Normally, when an editor boots up, it is in Insert mode, which means any new keystrokes are placed in front of the current text. This is the preferable mode to be in.

I should point out that if you are looking for a string in a file, instead of pressing Ctrl-F, as you would in many Microsoft-compatible editors, in the MPLAB IDE you will have to press F3. This is the only significant deviation that I have found between the MPLAB IDE's editor and a standard editor like WordPad.

Microchip has a number of different hardware tools that can be connected to your PC and are integrated to the MPLAB IDE. These tools, like the MPLAB IDE, are uniformly excellent and are continually updated by Microchip as problems with the products become known and new PICmicro device part numbers become available. I have had a PICSTART Plus for several years (and many internal firmware revisions). One tool you may be interested in is the MPLAB ICD, which is based on the PIC16F876 and PIC16F877 and it gives you many of

Table 3-3

MPLAB IDE
Editor
Functions

Keystrokes	Operation
Up Arrow	Move cursor up one line
Down Arrow	Move cursor down one line
Left Arrow	Move cursor left on character
Right Arrow	Move cursor right on arrow
Page Up	Move viewed text up
Page Down	Move viewed text down
Ctrl-Left Arrow	Jump to start of word
Ctrl-Right Arrow	Jump to start of next word
Ctrl-Page Up	Move cursor to top of viewed text
Ctrl-Page Down	Move cursor to bottom of viewed text
Home	Move cursor to start of line
End	Move cursor to end of line
Ctrl-Home	Jump to start of file
Ctrl-End	Jump to end of file
Shift-Left Arrow	Increase the marked block by one character to the left
Shift-Right Arrow	Increase the marked block by one character to the right
Shift-Up Arrow	Increase the marked block by one line up
Shift-Down Arrow	Increase the marked block by one line down
Ctrl-Shift-Left Arrow	Increase the marked block by one word to the left
Ctrl-Shift-Right Arrow	Increase the marked block by one word to the right
Ctrl-Z	Undo the last change to the text
Ctrl-X	Delete the marked text and save it in the clipboard
Ctrl-C	Save the marked text in the clipboard
Ctrl-F	Find text in the file from the cursor. Note, this is not available in MPLAB (use F3 instead)
Ctrl-S	Save the current file
Ctrl-O	Open a new file
Ctrl-N	Create a new file
Ctrl-P	Print the current file (or portions of it)

the features of an emulator, including a direct interface to MPLAB to help you debug your applications.

For professional PICmicro MCU application developers, I can recommend the Microchip PROMATE II programmer and MPLAB ICE emulator. The PRO-MATE II can be used to production-program all PICmicro MCUs (rather than development programming provided by most simple programmer circuits) along with different serial EEPROMs. Adapters are available for the PROMATE II, which will enable you to program SMT parts as well.

The MPLAB ICE emulator is probably the best chip emulator that I have ever seen. Not only is it well integrated with the MPLAB ICE, it was also designed for real-world use. The MPLAB ICE emulator hardware is quite small and can be mounted on a small tripod to keep it close to the project board being debugged and minimize the stress on the socket and wiring. All PICmicro MCU chips (which are used in the emulator) are designed with emulator hardware built in and are used in the customization modules package. When you connect the emulator to your project board, you can feel comfortable knowing that when you put in a chip, it will behave exactly the same way as when you were using the emulator.

The MPLAB IDE has the capability to display specific register and bit contents in the PICmicro during simulation. When I created each of the different example applications presented in this book, I will have gone through the process of simulating the application before burning it into a PICmicro MCU. To view the contents of registers and variables, the MPLAB IDE enables you to create *watch windows*. These windows or dialog boxes, such as the one shown in Figure 3-1 enable you to select the registers and PICC Lite variables to monitor.

To define a watch window or add more registers to it, click Window followed by Watch Windows . . . and then New Watch Window . . . . The Register Selection window is brought up for you to select the registers you would like to monitor. The Properties window is selected from the Register Selection window (as is shown in Figure 3-2) to specify the characteristics of the register that is displayed.

Watch windows should be started *after* the application has assembled without any errors, warnings, or messages. If there are errors when the watch window is created, the file register information is not available to the MPLAB IDE and the list of registers available is restricted to the basic set available to the device.

Figure 3-1
Watch window
sample

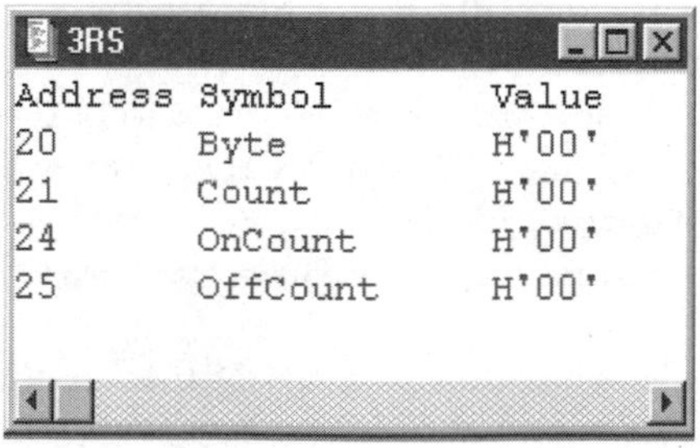

Address	Symbol	Value
20	Byte	H'00'
21	Count	H'00'
24	OnCount	H'00'
25	OffCount	H'00'

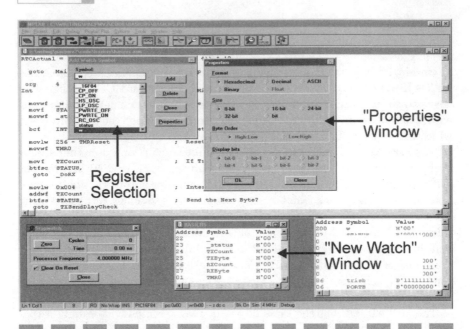

Figure 3-2 MPLAB watch window

Registers can be displayed in decimal, hex, binary, and *American Standard Code for Information Interchange* (ASCII) formats, and they can be one, two, three, or four bytes in size. Multiple-byte data can be selected as being the least significant byte first (my normal default) or the most significant byte first. Along with registers, individual bits within them can also be displayed. This gives you a pretty complete range of options for displaying data.

On the CD-ROM, I have included some standard watch window files for the PIC16F84 and PIC16F627 devices as well as some watch window files for the different advanced I/O functions of the PIC16F627 and the application code. Please feel free to use these files for your own applications or use them as models. If you look at the actual files, you will see that they are quite simple and easy to understand, but I do not recommend editing your own simply because any mistakes made in the file could cause the MPLAB IDE to crash.

I consider the MPLAB IDE's built-in simulator to be probably the most important part of the IDE. The simulator will enable you to work through your PICmicro MCU application in a virtual PICmicro MCU environment, allowing you to monitor the operation of the code as well as see which values are being stored in variables and hardware interface registers. However, there are a few issues that you should be aware of.

The simulator provides an excellent model of the PIC16F84, although for the PIC16F627, there are some built-in functions that are not modeled. The functions that are not modeled are the analog-to-voltage comparators and the *universal asynchronous/synchronous receiver/transmitter* (USART), although I will show some methods that are available for working as though the simulator software had these functions built in.

The other issue with regards to the simulator that you should be aware of is the speed at which it operates. The simulator typically runs at 1,000 to 3,000 instruction cycles per second, which means that it could take up to 5 minutes to simulate operations for 1 second. This will be an issue when you simulate your applications, as the biologic code requires a few seconds of operation.

When the MPLAB IDE software was originally written, the operating system of choice was Microsoft Windows 3.11. This operating system did not provide the full multitasking and hardware virtualization features of the Win32 operating systems that have followed. As a result of writing the software for Windows 3.11, the MPLAB IDE relies on a polling model for checking the keyboard and mouse during operation, rather than the interrupt-like event-driven programming model of most modern applications.

To speed up the operation of the MPLAB IDE simulator, you can either move your mouse around in circles as the simulator is executing or you can run the speed.exe application (which is found on the CD-ROM). In both these cases, the polling interval is decreased, which enables the simulator to run much faster (30,000 instruction cycles or more on a modern PC). If you use speed.exe, please remember that it is shareware and you should send a $5 check to the author (his address is given in the readme file of the speed.exe package).

Five different methodologies exist for passing input to the application code as it executes in the MPLAB IDE simulator. The range of choices is a bit overwhelming (especially since most other simulators only have one or two input options). The five different methodologies enable you to tailor the input to your actual requirements to find and fix your problems. In the later chapters, I will show you how the different simulation methodologies are used.

The most basic input method is the Asynchronous Stimulus window (shown in Figure 3-3), which consists of a set of buttons that can be programmed to drive any of the simulated PICmicro pins and can set the button to change the pin by the following:

```
Pulse
Low
High
Toggle
```

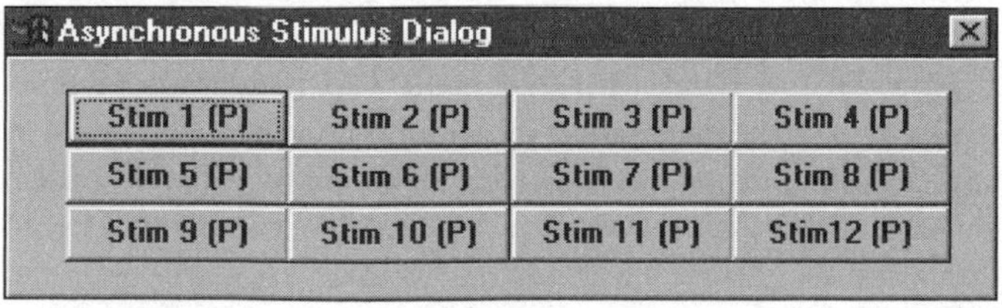

Figure 3-3
Asynchronous
Stimulus
window

Asynchronous Stimulus Dialog			☒
Stim 1 (P)	Stim 2 (P)	Stim 3 (P)	Stim 4 (P)
Stim 5 (P)	Stim 6 (P)	Stim 7 (P)	Stim 8 (P)
Stim 9 (P)	Stim 10 (P)	Stim 11 (P)	Stim12 (P)

The Pulse option changes the input pin to the complemented state and then back to the original state within one instruction cycle. This mode is useful for clocking timers or requesting an external interrupt. Setting the pin High or Low will drive the set value onto the pin. To change the value of the pin between the two states, you can program two buttons in parallel with each other and each button changes the state. This can also be done with a single Toggle button, which changes the input state each time the button is pressed.

The asynchronous inputs are not something I like to use a lot because they are completely user generated and there have been many times when simulating an application that I have forgotten to press the appropriate buttons or have pressed them at the wrong point in time. Another issue with the asynchronous inputs is how much they slow down the operation of the simulator. Depending on the number of input buttons customized for the application, you can see the speed of the simulator decrease to one-twentieth or less of the norm.

To eliminate the problem of not pressing an asynchronous input button at the correct time, I normally develop a stimulus file for the application. This file provides an instruction countervalue (step) along with a pin state value at this point. An example stimulus file showing how a rotation sensor (from a mouse or a robot's motor) could be implemented is shown here:

```
!  Rotation Sensor Example
Step      RB0       RB1
  1        0         0
100        0         1
200        1         1
300        1         0
400        0         0
500        0         1
600        1         1
700        1         0
800        0         0
```

The ! character in the stimulus file is a comment (everything to the right of it is ignored). The first actual line of a stimulus file contains the step directive followed by the pins to be driven. In this example, the stimulus data is passed to the RB0 and RB1 pins.

The input pins can be any of the pins in the simulated device using the Microchip convention of R$# where $ is the port identifier (A, B, C, and so on) and # is the number of the pin in the port. Reset can be driven from a stimulus file as the MCLR pin.

Note that in the previous stimulus file, I included two movement cycles to work the code through, without having to reset the application. Stimulus data will not repeat once the counter has gone beyond 800, and the RB0 and RB1 inputs will stay reset until the simulated PICmicro is reset.

To count the instruction cycles, the StopWatch window (see Figure 3-4) is invoked and put onto the MPLAB IDE desktop. Clicking the Window pull-down menu and then Stopwatch . . . initiates this window. The Stopwatch function

Figure 3-4
MPLAB
Stopwatch

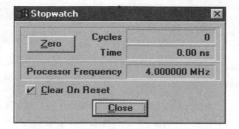

enables you to time applications and I generally keep the Clear on Reset button checked and never click the Zero button unless I know exactly what I am doing.

The reason for this is that resetting the Stopwatch will cause any stimulus input to become reset and not behave as I was expecting. The only time I will click Zero is for manually timing loops. In these cases, the timing of the code execution is critical and I want to avoid having to subtract a previous value that is an opportunity for me to make an error.

The speed of the Stopwatch (and the simulated PICmicro) is set by clicking the Option pull-down menu, selecting Development Mode, and clicking the Clock tab. The processor speed (along with its type) can be specified at any time when you are working with MPLAB.

The task of changing the stimulus file timing or specifying it in the first place is not difficult. The formula that I use is

```
Step = Time Delay * Clock Frequency/4
```

Using this formula in a PICmicro MCU running at 3.58 MHz, a 250-μsec Step point would be calculated as follows:

```
Step = Time Delay * Clock Frequency/4
     = 250 μsec * 3.58 MHz/4
     = 223.75
     = 224 (After Rounding)
```

Simulated clock inputs can be input into the MPLAB IDE simulator by clicking the Debug pull-down menu and the Simulator Stimulus and Clock Stimulus . . . selections. The Clock Stimulus dialog box (see Figure 3-5) can input regular clocks into a PICmicro by selecting the pin and then the high and low times of the clock, along with whether or not the clock is inverted (which means at reset the clock will be low rather than high). The clock counts in Figure 3-5 are in instruction cycles, which are found exactly the same way as shown earlier.

Clock stimulus can be used for simple I/O tests, but it is really best suited for putting in repeating inputs that drive clocks or interrupts. The example of the rotation sensor that I used for the stimulus file could be implemented very easily using the clock stimulus feature of the MPLAB IDE simulator.

As I have indicated elsewhere, the MPLAB IDE simulator does not handle advanced peripheral operations well. This deficiency can be alleviated somewhat

Figure 3-5
MPLAB Clocked
Stimulus dialog
box

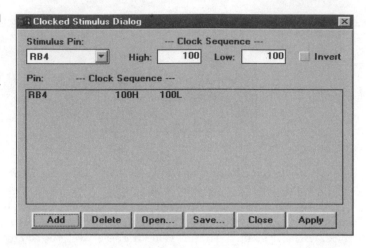

by using the Register Stimulus feature of MPLAB. This feature stores a two-digit hex value in a specified register every time a specific address is encountered in the simulated application execution. To load the operating parameters of the Register Stimulus method, the Debug pull-down menu is clicked, followed by Simulator Stimulus, and then Register Stimulus is enabled. This brings up the small window shown in Figure 3-6 where you will select the address of the register to change as well as the address that this happens at. Once the addresses have been specified, the register stimulus file is selected by clicking Browse . . . .

The register stimulus file is simply a text file consisting of a column of two-digit hex numbers. The following example file lists all the values from 0x000 to 0x0FF:

```
00
01
02
 :
40
41
 :
C0
C1
 :
FF
```

No comments or multibyte values are allowed in this file. The colons (:) are used to indicate that the file continues with the data. The Register Stimulus feature does not allow more than one register to be modified at any given time.

The Register Stimulus is very useful for simulating the operation of a hardware function that is not modeled in the MPLAB IDE simulator. When I discuss using the USART in an application, I will show how a register stimulus file can be created and used, avoiding the possibility that you will program a PICmicro MCU with code that has been tested on the simulator first.

Figure 3-6
MPLAB Register
Stimulus
specification

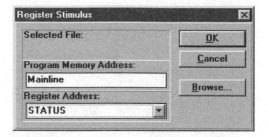

To modify more than one register, I use the Register Modify window (shown in Figure 3-7). This window is available from clicking the Window pull-down menu and then selecting Modify . . . . This window can access any register in the simulated device, including w, which cannot be directly addressed.

I should point out that all five of the simulator pin/register input methodologies I have presented in this section can work together. This means that any two (or more) inputs can be enabled simultaneously to interface with the application and help you to debug it and find any potential problems. Doing this is not always as easy as it sounds because of the difficulty in synchronizing the inputs so that they work together.

When you work with the different methods of specifying the inputs to the PICmicro microcontroller or the registers, you should notice that the MPLAB IDE enables you to use simplified pin names (such as RA0), defined register names, and program memory address labels inside the source code. This makes the input features much easier to work with and avoids your having to read through the listing file to find absolute addresses or values.

The development tool that I want to discuss is the Hi-Tech Software PICC Lite compiler. I was surprised at how quickly I was able to start developing applications using this tool. PICC Lite is very well thought-out and has been thoroughly debugged over years of work. If you are familiar with the C programming language, you will understand what a really nice feature the warning on "if (variable = constant)" instead of "if (variable = = constant)" is. Other built-in features will help you find and fix syntax errors very easily (which is somewhat unusual for a C compiler).

To install PICC Lite, please follow the directions listed on the CD-ROM that comes with this book. PICC Lite installs differently for the different flavors of Microsoft Windows, and you should follow the instructions specific to the operating system you are running on your PC.

Using the following application, I will demonstrate how a project is created in the MPLAB IDE, compiled by PICC Lite, and debugged using the simulator. I created an application called example.c that I put the code example subdirectory of the CD-ROM:

```
#include <pic.h>

//  02.03.31 - Myke Predko
```

Figure 3-7
MPLAB Register
Modify
window

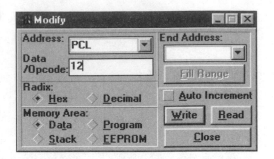

```
//
//   Example of a PICC Lite application
//
//   Demonstrate a simple error.
//
//
//   Hardware Notes:
//   PIC16F84 Running at 4 MHz
//   RB0 will be toggled at full speed
//

//   Global Variables

//   Configuration Fuses
        __CONFIG(0x03FF1);              //   Set Configuration Fuses to:
                                        //   - XT Oscillator
                                        //   - 70 msecs Power Up Timer On
                                        //   - Watchdog Timer Off
                                        //   - Code Protection Off


//   Mainline
void main(void)
{

        while (1 == 1)                  //   Toggle RB0
             RB0 = RB1 ^ 1

}  //   End of Mainline
```

The application itself has two simple errors that I will find and fix using the compiler and simulator.

After starting up the MPLAB IDE, check to make sure that the PICmicro MCU part number used for the application is correct (it is displayed on the bottom line of the MPLAB IDE desktop). If it isn't correct (the PIC16F84 for this example), click Options, followed by Development Mode . . ., and select the correct part number.

With the correct PICmicro MCU selected, click Project followed by New Project . . . as in Figure 3-8. This brings up the New Project dialog box shown in Figure 3-9 and prompts me to enter in the name of the project that I would like to

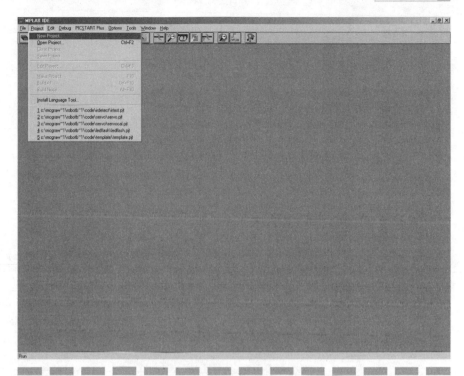

Figure 3-8   Creating a new project in the MPLAB IDE

Figure 3-9
Specifying the
new project
name

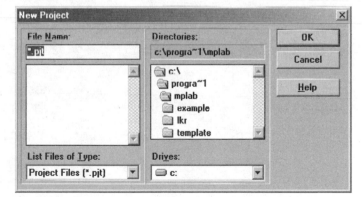

work with. As can be seen in Figure 3-10, I am creating the example project in the example subdirectory.

With the project name selected, click OK. This brings up the Edit Project dialog box, as shown in Figure 3-11. When the Edit Project dialog box first comes up, the MPLAB IDE assumes that your application source code will be written in

Figure 3-10
The new
project name
and directory
selected

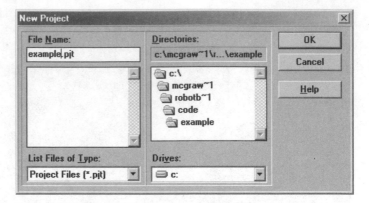

Figure 3-11
Selecting the
project type

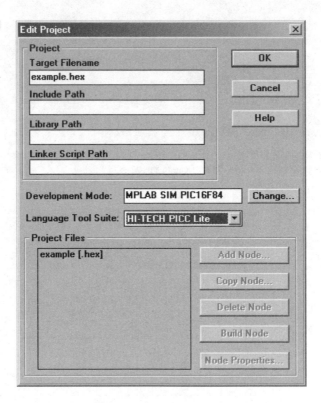

assembler. Change this to PICC Lite, as I have shown in Figure 3-11. Next, click the destination file (example.hex in Figure 3-11) and the Node Properties . . . button will become active.

Click Node Properties . . . and the MPLAB IDE will bring up the Node Properties dialog box shown in Figure 3-12. Make sure that the following options are selected in the Node Properties dialog box:

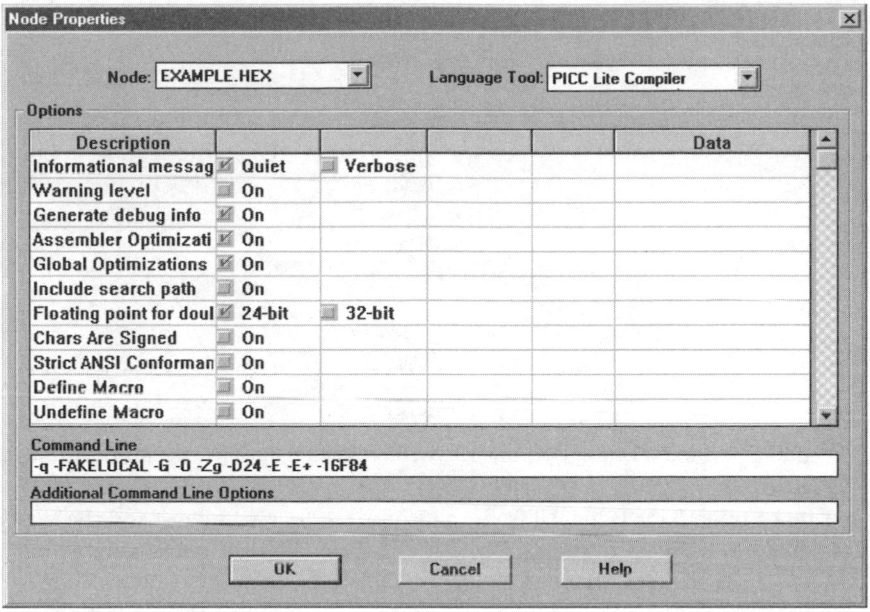

Figure 3-12 Selecting compiler options

- Informational message—Quiet
- Generate debug info—On
- Assembler Optimization—On
- Global Optimizations—On
- Floating point for double—24-bit
- Error file—On
- Append Errors to file—On

These are the basic properties that you will require to provide basic optimizations to your application. Use the smallest amount of space possible for floating point values and provide simple error messages, which can be found in the MPLAB IDE. When the Node Properties dialog box has the correct options, click OK.

Next, click Add Node . . . on the Edit Project dialog box and you will be prompted, as shown in Figure 3-13, with the source files available in the project directory. I place the source code in the project directory so that all the files relevant to the project are all located in the same directory in my PC's hard file.

After clicking example.c and OK, the MPLAB IDE returns to the Edit Project dialog box. Click OK in the Edit Project dialog box and the MPLAB IDE desktop returns to the original, blank state. Click File followed by Open . . . and load example.c into the MPLAB IDE desktop by selecting it from the Open Existing

Figure 3-13
Selecting the
PICC Lite
source file

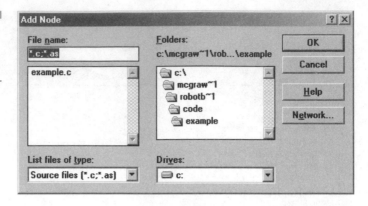

File dialog box and clicking OK. The file will be loaded onto the desktop as shown in Figure 3-14.

Press Ctrl-F10 (or Build All from the Project pull-down menu) and the PICC Lite compiler will be invoked. When it finishes, you will get a new window (labeled Build Results) with the error:

```
Error[000] C:\MCGRAW~1\ROBOTB~1\CODE\EXAMPLE\EXAMPLE.C 32 : ; expected
```

If you place your mouse on this line and double-click, you will find that the window containing the example.c source code will become the primary focus and the cursor will jump to the last line in the application (} // End of Mainline).

The error indicates that a semicolon (;) was expected and not found on this line. To find the error, go back to the previous line of code (RB0 = RB1 ^ 1) and add a semicolon at the end of it so it looks like

```
    RB0 = RB0 ^ 1;
```

That was the error. Now, press Ctrl-F10 again and the PICC Lite Build Results window will become the primary focus and indicate that the application compiled correctly with the following information:

```
Memory Usage Map:

Program ROM    $0000 - $0005  $0006 (      6) words
Program ROM    $03FC - $03FF  $0004 (      4) words
                              $000A (     10) words total Program ROM

Config Data    $2007 - $2007  $0001 (      1) words total Config Data


Program statistics:

Total ROM used       10 words (1.0%)
Total RAM used        0 bytes (0.0%)


Build completed successfully.
```

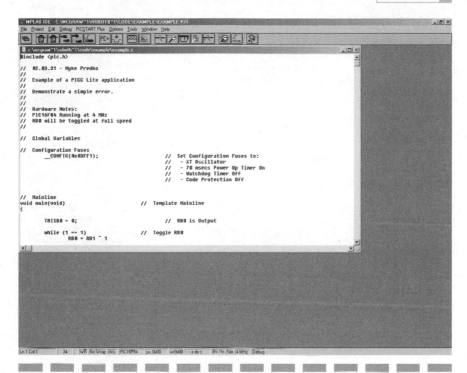

Figure 3-14 The project file loaded in the MPLAB IDE

You can delete the Build Results by clicking the X in the top right-hand corner of the window.

Now the application is ready to simulate to make sure it is running properly. Click Windows followed by Watch Windows ... and click Load Watch Window .... From the file selection dialog box that comes up, go to the procwat directory and select the PIC16F84.WAT file. When this watch file comes up in a window, resize it so that all the registers and their data are displayed, and place it on the MPLAB IDE desktop in a location that won't interfere with other windows or dialog boxes. I normally place this window in the bottom right-hand corner of the desktop as I've shown in Figure 3-15.

The application is now ready for simulation. The toolbar line (the line with the icons) on the MPLAB IDE should be in simulator mode and resemble the line in Figure 3-15. If it doesn't, click the leftmost icon of the line, which will cycle through the different lines available. If you run the mouse over the icons, you will see that the bottom line will indicate what their functions are.

To reset the simulated PICmicro MCU, click the Reset button of the toolbar, or click the Debug pull-down menu and then Run followed by Reset. You can also press F6 for the same function. If you look at the bottom line of the MPLAB IDE, you will see that the Program Counter (PC) will have the value of 0x03FC. This is the start of the application. The RB0 = RB0 ^ 1; line of the application may be

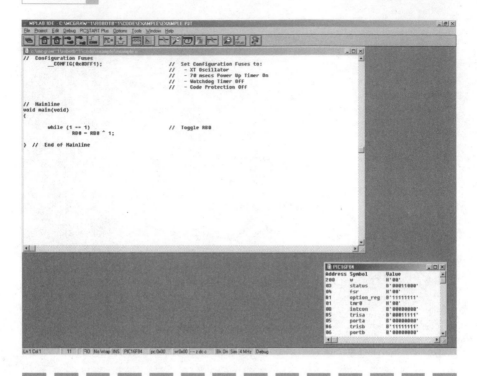

Figure 3-15 The project ready for simulation

highlighted in black. Figure 3-16 shows what the MPLAB IDE desktop should look like.

Now, you can click Step or Step Over to work through the application. As you do this, you will notice that the highlighted line flashes, but nothing seems to be happening. Looking at the PIC16C84.WAT window, none of the registers change. When a register bit is changed, the entire line that the register is on changes from black to red.

As you will read later in this chapter, the *tri-state buffer enable* (TRIS) registers are used to control the operation of the I/O ports, and all the bits that are to be used for output should be loaded with a 0 (not the 1 that is stored at reset). The problem with the application is that the RB0 is in input mode, not output mode. To put RB0 in output mode, add the line

```
TRISB0 = 0;
```

Before the while (1 == 1) statement, rebuild the application (Ctrl-F10). This statement will place RB0 into output mode.

Now, after you reset the application, the highlighted line should be on the TRISB0 = 0; line and when it single steps through it, you should see RB0 toggling with the registers changing, as shown in Figure 3-17.

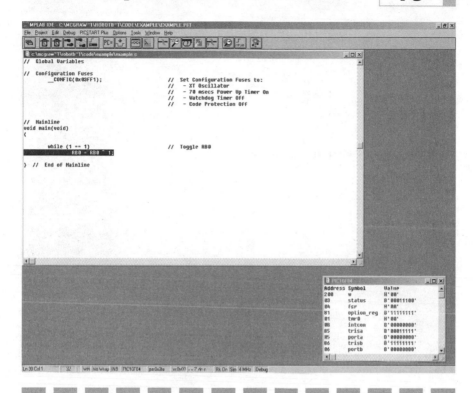

Figure 3-16 Reset simulated project

In Figure 3-17, you will see that the least significant bits of the trisb and portb registers of the PIC16F84 watch window are different than the values of Figure 3-16. This is an indication that the application is running correctly (RB0 is toggling high and low).

This tutorial and application may see somewhat trivial, but I think they illustrate how a PICC Lite C application is built in the MPLAB IDE. They also show some of the steps that are used to simulate the application and debug it before it is burned into a PICmicro MCU. Being able to see the application executing before it is put into a PICmicro MCU avoids the situation when you have built the circuit and you are left shaking your head as to why nothing is working.

Basic Circuit Requirements

When I wrote the *Handbook of Microcontrollers*, I suggested to the reader that if a microcontroller required more than just +5 volts, a reset line, and a crystal, he or she should be looking for another microcontroller. This is a statement that I

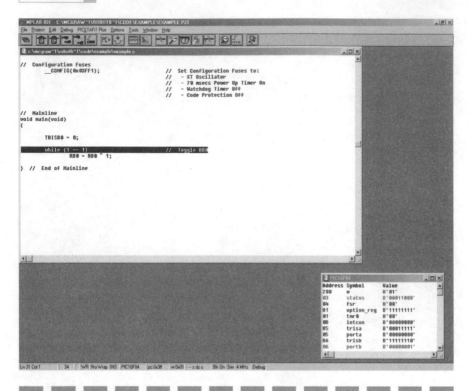

Figure 3-17 A working simulated project

truly believe in—life is too short to try to figure out the best way to implement a microcontroller in a circuit. This is especially true when there are devices like the PIC16F627, which has many built-in features that will make wiring the micro-controller into your robot an absolute pleasure and will enable you to focus how the robot's software works, which is the important aspect of robot design.

The PICmicro MCU is quite robust in terms of power, with virtually all devices either being able to run at voltages from 2.5 to 6 volts DC or having low-power versions that can execute with only 2.5 volts available. I typically design my PICmicro MCU circuits to work at just 5 volts to avoid any problems with finding peripheral chips that run at lower voltage levels.

To provide this regulated power, I typically use 78(L)05 voltage regulators or a 5.1-volt Zener diode. A sample 7805 circuit is shown in Figure 3-18 and can source up to 500 mA. A Zener diode can be used as a voltage regulator using the circuit shown in Figure 3-19. Note that the current-limiting resistor's value and power rating must be calculated using the formulas shown in Figure 3-19.

For the diode and resistor power rating in the Zener diode power regulator circuit, after the power dissipated is calculated, you have to choose a power rating

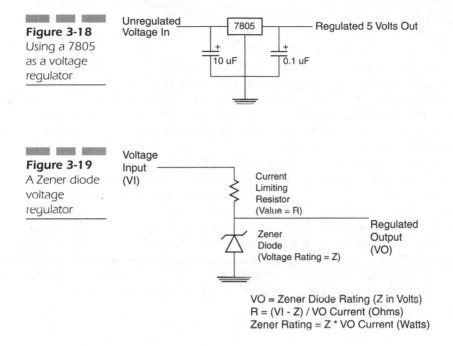

Figure 3-18
Using a 7805
as a voltage
regulator

Figure 3-19
A Zener diode
voltage
regulator

VO = Zener Diode Rating (Z in Volts)
R = (VI - Z) / VO Current (Ohms)
Zener Rating = Z * VO Current (Watts)

greater than the calculated value. For example, if the power calculation through the diode was 0.625 watts, then you should choose a diode with a 1-watt power rating.

Once power is available to the PICmicro microcontroller, the next function to provide is reset (used by the PICmicro MCU to recognize when power is available and the processor can start executing). The reset signal can be independently produced or it can be based on the microcontroller's power supply. When power is at the appropriate level, the microcontroller's processor is allowed to start executing instructions. The PICmicro MCU provides a number of different options that you can choose from for implementing reset.

The PICmicro microcontroller has a negatively active reset pin known as "_MCLR" that can be controlled by three different interrupt sources: external, internal, and brown-out detection. Negative Active means that the PICmicro MCU will be held to reset if _MCLR is held at a low logic level. When the PICmicro microcontroller is reset, the clock is stopped and I/O devices and the program counter are reset.

Many PICmicro microcontrollers provide an internal reset capability that is triggered when power is applied to the device. Along with eliminating the need for the external circuitry shown in Figure 3-20, the internal reset also frees up a pin on the PICmicro MCU that can be used as an input (never as an output). Unless I absolutely need the input pin, I always attach a 10K pull-up on _MCLR.

This enables me to reset the PICmicro microcontroller by shorting _MCLR to ground, which is accomplished in the 18-pin PICMicro MCUs by shorting pin 4 (_MCLR) to pin 5 (Vss or Gnd).

The _MCLR pin is also used as the Vpp pin and will put the PICmicro MCU into programming mode when more than 12 volts are applied to it. In some PICmicro microcontrollers (the PIC12C5xx and PIC16C505 specifically), you will discover that the _MCLR pin, when it is defined as an input pin, defaults to being a clock input. To change the pin to a digital input, you have to write a value to the OPTION register within the chip.

The brown-out detect reset built into some PICmicro MCUs resets the device when Vdd or the voltage at _MCLR falls below a preset value (normally 4.5 volts). Although the PICmicro microcontroller runs without any problems at 4.5 volts, other devices in the circuit may fail at this voltage. A brown-out detect circuit, like the one shown in Figure 3-21, may be built into some PICmicro MCUs, which will pull down _MCLR if Vdd drops below 4.5 volts.

Some PICmicro devices have programmable, voltage-level brown-out detection circuits, while others require a separate, external circuit. A circuit like the one shown in Figure 3-21 could be built, but a number of small chips in transis-

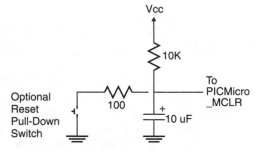

Figure 3-20
External
PICMicro reset
circuit

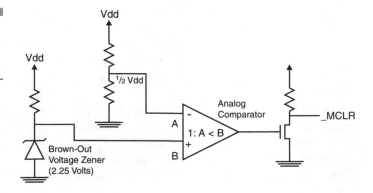

Figure 3-21
PICMicro MCU
brown-out
reset circuit

tor TO-92 and SOT-23 packages provide this same function along with a built-in delay to enable the power supply to stabilize before the processor starts executing instructions. These chips can be purchased at a cost that would be less than the parts needed to implement the circuit in Figure 3-21.

The PICmicro microcontroller's *Power-up Wait Reset Timer* (PWRT) should always be enabled. This timer will delay the startup of the PICmicro microcontroller's processor for 72 milliseconds after _MCLR has been driven high. This additional delay will ensure that the power applied to the PICmicro MCU (and other circuitry) will be stable, as is the clock before the processor starts to execute code.

The last requirement for the PICmicro microcontroller to run is a processor clock (or oscillator). This clock will be used to sequence instructions within the microcontroller's processor as well as provide a timebase for peripheral functions. There are eight different methods of implementing the clock in the PICmicro MCU, each one with its own advantages in different circumstances.

The most basic type of oscillator is the relaxation oscillator built using a simple resistor and capacitor, as shown in Figure 3-22. This type of clock uses the characteristic waveform of an resistor/capacitor (RC) network to supply a specific clock period to the PICmicro MCU. Three types of RC oscillators are built into different PICmicro MCU part numbers.

When the PICmicro microcontroller was first introduced, an external resistor/capacitor oscillator, as shown in Figure 3-22, was implemented. This type of oscillator is very imprecise (although it has the advantage of being very cheap to add to a circuit). In later PICmicro MCUs, the resistor/capacitor network was built into the chip with a variable resistor, enabling the oscillator to be calibrated to a fairly precise frequency. The internal RC oscillators, using a calibration value supplied by Microchip, are usually accurate to within 1.5 percent. Along with

Figure 3-22
PICMicro MCU
RC oscillator

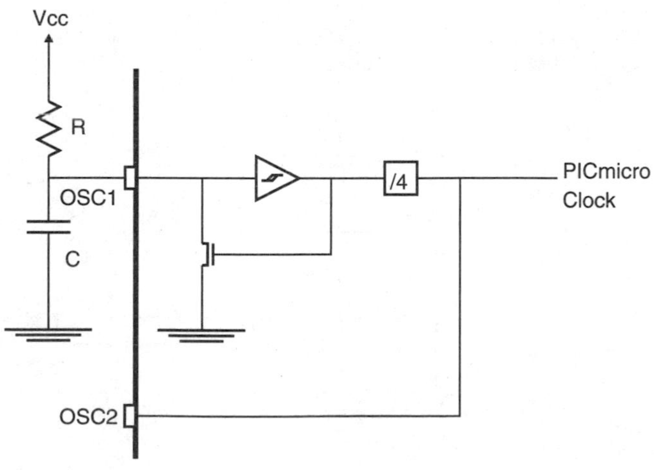

being reasonably precise, the internal RC oscillator enabled the two pins that would normally be used for the oscillator circuit to be used as I/O pins.

Along with the internal RC oscillator, some PICmicro MCU part numbers (like the PIC16F627) have the capacitor built into the microcontroller and are designed to have a resistor connected externally. This oscillator type has the low cost of the external RC oscillator, but it is designed to be tuned to a very precise frequency. It is important to remember that the resistor value that gives a specific frequency is unique to the PICmicro microcontroller that it is running on.

Three different modes exist in which crystals of ceramic resonators can be used with the PICmicro MCU. These devices delay the passage of an electrical signal within them and are wired to the PICmicro microcontroller's OSC pins, as shown in Figure 3-23.

You are probably familiar with crystals, but you may not have heard the term ceramic resonators before. These devices perform like a crystal but are much more robust (a good thing in robots) and often have capacitors built into them, making them quite a bit easier to wire into a circuit.

Three crystal/ceramic resonator operating modes can be selected depending on the speed of operation shown in Table 3-4. When the different modes are selected in the PICmicro MCU's configuration fuses, the oscillator hardware is reconfigured to best suit the needs of running at the different speeds.

The signal at the OSC2 pin can be passed to another CMOS input. If the signal is used to drive more than one CMOS input, then the operation and speed of the oscillator will be changed and there is a good chance that it will stop running altogether.

External (canned) oscillators can be used with the clock signal input into the OSC1 pin of the PICmicro MCU. In some part numbers, the OSC2 pin is made

Figure 3-23
PICmicro MCU
crystal oscillator

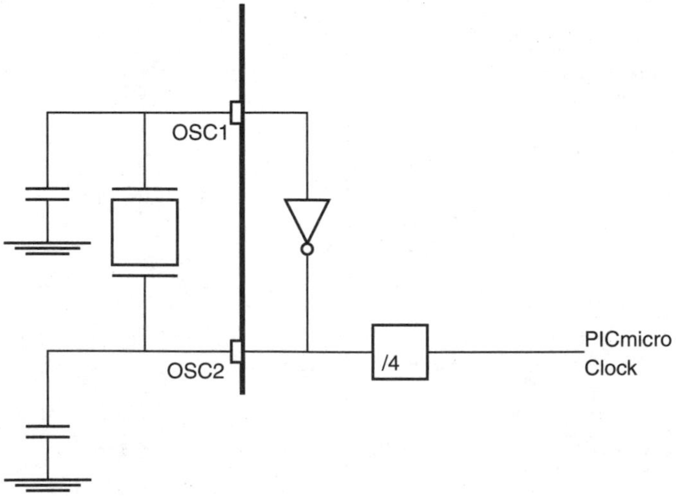

Table 3-4

Crystal/ceramic Resonator Operating Modes

	Speed	Comments
LP	0–200 KHz	Low-power oscillator mode
XT	200 KHz–4 MHz	Normal oscillator operation
HS	4+ Mhz	High-speed operation

available to the application as an I/O pin when OSC1 is driven by an external clock source.

The last clock option that I would like to discuss is the addition of a PLL clock multiplier added to the clock input circuitry. Currently, this feature is only available in the PIC18C/Fxx2 devices. The advantages of the PLL clock multiplier center around using a lower-frequency clock input. The lower-frequency clock input lowers the amount of power required by the PICmicro MCU's oscillator circuitry and lessens the amount of electromagnetic emissions from the complete circuit.

You will probably note that I have not listed any component values for the different clock types. I did not specify any values because each PICmicro microcontroller part number requires different component values for different oscillator types and operating speeds. When you are developing a circuit with a PICmicro MCU built in, you must consult the device's data sheet for the appropriate oscillator component values.

The PIC16F627

To help you go forward with your robot applications using PICC Lite and the PIC16F627 microcontroller, I want to give you a reference to the internal features of the chip. The PIC16F627 actually contains a large number of the available I/O functions available in different PICmicro MCUs. The PIC16F627 contains a superset of the features available in other low-end PICmicro microcontrollers (including the PIC16F84), so the description of the different hardware features within the PIC16F627 can be used for other PICmicro microcontroller part numbers. When I describe how different features within the PIC16F627 are accessed, I will use PICC Lite C as example code.

The pinout of the PIC16F627 is shown in Figure 3-24, and from this diagram, it can be seen that most of the pins can be used for multiple functions. In this section, I will explain the operation of the different I/O pins and the hardware behind them.

All the different PICmicro microcontrollers utilize a Harvard processor architecture, which is similar to the PIC16F617's (see Figure 3-25). The processor is

Figure 3-24
PIC16F627
pinout

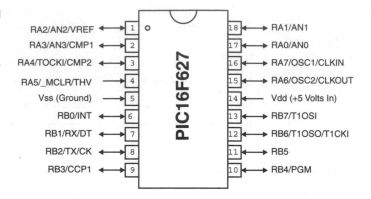

tightly integrated with the different I/O functions of the chip, resulting in a device that can perform different functions very effectively.

By working with the PICC Lite compiler, you will not have to intimately understand the architecture of the PIC16F627. What I want to show is how immediate the hardware interface function control and status registers are to the processor. This point is emphasized even more when you look at the memory map of the PIC16F627, as shown in Figure 3-26. Application variables will be stored in the General Purpose Register areas of the memory map and PICC Lite will be responsible for choosing and allocating where the variable data will be located.

One feature of the PICmicro microcontroller that is somewhat unusual compared to other microcontrollers is its configuration register. This register is a set of programmable bits responsible for specifying the following:

- Oscillator mode used
- Program memory protection
- Reset parameters
- Watchdog timer (WDT)
- PIC16F87X debug mode

The configuration register gives the application developer a great deal of flexibility in how the PICmicro devices are used in an application. I should point out that if the configuration register bits are not specified correctly, the PICmicro MCU will not run the application code properly (if at all).

The configuration register value is written when the device is being programmed and is accessed as part of the microcontroller's bootup process to select the hardware options required for the application. The configuration value cannot be accessed by the application in the low-end and midrange parts. During programming, the PICmicro's program counter is used to access the configuration register's address. In the midrange PICmicros, the configuration register is always at 0x02007. The register bits are used to enable hardware or set hardware

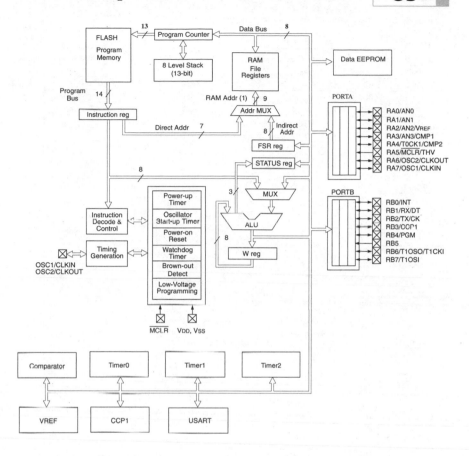

Figure 3-25 PIC16F627 block diagram

Device	Memory		
	FLASH Program	RAM Data	EEPROM Data
PIC16F627	1024 x 14	224 x 8	128 x 8
PIC16F628	2048 x 14	224 x 8	128 x 8
PIC16LF627	1024 x 14	224 x 8	128 x 8
PIC16LF628	2048 x 14	224 x 8	128 x 8

states but cannot be read by the program, and their states can only be indirectly determined within software by the operation of the application.

In each PICmicro microcontroller's MPASM device .inc files, there is a list of parameters for the different options. These parameters are used with the "__CONFIG" statement of an assembler file. The __CONFIG statement is used to AND together all the constants in the .inc file. This is because some bits will be

File Address

Bank 0		Bank 1		Bank 2		Bank 3	
Indirect addr.(*)	00h	Indirect addr.(*)	80h	Indirect addr.(*)	100h	Indirect addr.(*)	180h
TMR0	01h	OPTION	81h	TMR0	101h	OPTION	181h
PCL	02h	PCL	82h	PCL	102h	PCL	182h
STATUS	03h	STATUS	83h	STATUS	103h	STATUS	183h
FSR	04h	FSR	84h	FSR	104h	FSR	184h
PORTA	05h	TRISA	85h		105h		185h
PORTB	06h	TRISB	86h	PORTB	106h	TRISB	186h
	07h		87h		107h		187h
	08h		88h		108h		188h
	09h		89h		109h		189h
PCLATH	0Ah	PCLATH	8Ah	PCLATH	10Ah	PCLATH	18Ah
INTCON	0Bh	INTCON	8Bh	INTCON	10Bh	INTCON	18Bh
PIR1	0Ch	PIE1	8Ch		10Ch		18Ch
	0Dh		8Dh		10Dh		18Dh
TMR1L	0Eh	PCON	8Eh		10Eh		18Eh
TMR1H	0Fh		8Fh		10Fh		18Fh
T1CON	10h		90h				
TMR2	11h		91h				
T2CON	12h	PR2	92h				
	13h		93h				
	14h		94h				
CCPR1L	15h		95h				
CCPR1H	16h		96h				
CCP1CON	17h		97h				
RCSTA	18h	TXSTA	98h				
TXREG	19h	SPBRG	99h				
RCREG	1Ah	EEDATA	9Ah				
	1Bh	EEADR	9Bh				
	1Ch	EECON1	9Ch				
	1Dh	EECON2*	9Dh				
	1Eh		9Eh				
CMCON	1Fh	VRCON	9Fh		11Fh		
	20h	General Purpose Register 80 Bytes	A0h	General Purpose Register 48 Bytes	120h		
					14Fh		
General Purpose Register 96 Bytes					150h		
			EFh		16Fh		1EFh
			F0h		170h		1F0h
		accesses 70h–7Fh		accesses 70h–7Fh		accesses 70h–7Fh	
	7Fh		FFh		17Fh		1FFh

▨ Unimplemented data memory locations, read as 0.
* Not a physical register.

Figure 3-26 PIC16F627 register map

left unprogrammed (set) and others will be programmed (reset). To simplify putting them all together, when creating the configuration constants, they are bitwise ANDed together.

PICC Lite also provides the pseudofunction "__CONFIG(value);" statement for specifying the value to be stored in the configuration register. The configuration register value is stored in the .hex file and can be programmed into the PICmicro MCU as if it were an instruction. I highly recommend that you only use a PICmicro microcontroller programmer circuit (like the El Cheapo described later) that picks up the configuration register values from your .hex file and programs them in automatically. Some programmer circuits provide the capability of programming these options manually, which can lead to problems with the application if you forget to change them from the default (all bits set) value.

When you read through the previous section, I explained that the electrical properties of the PICmicro MCU's built-in oscillator are changed according to the oscillator type and speed range, which is specified in the configuration registers. If you are using a crystal or ceramic resonator and you have selected the wrong operating speed in the configuration fuses, you will find that the PICmicro microcontroller's oscillator will not start up or run reliably.

Before starting any application, you should make sure that you have decided whether or not the WDT is going to be used in your application. In many PICmicro MCUs, the WDT enable bit in the configuration register is positive active. This means that if the WDT bit is set, then the WDT will be enabled, resetting your application anywhere from every 18 to 2.3 milliseconds without you really understanding why. Always make sure that you either disable the WDT or put in a single clrwdt instruction that executes in half the time it takes the WDT to time out.

Along with this, as I will discuss, there is the opportunity for ruining the PICmicro microcontroller, which is exacerbated by the differences in the configuration register bits implemented in the different PICmicro MCUs. In EPROM-based PICmicro MCUs, the *code protection* (CP) bit or bits are usually protected by a metal shield that prevents them from being erased. The purpose of the metal shield is to prevent software pirates from selectively erasing the chip (that is, just the CP bits) and reading back the application. If these bits are ever inadvertently programmed, you will not be able to reuse the PICmicro MCU. To be on the safe side, always specify "CP_OFF" in the configuration register values to make sure that you never change the part so that it can never be programmed again.

Make sure that you understand whether or not *low voltage programming* (LVP) is to be enabled. Incorrectly enabling this function will prevent the PICmicro from executing, unless RB4 is driven to a specific value.

The PICC Lite .h files contain specifications for some configuration bits but are not easy to use and require that you understand how the configuration fuse bits are implemented in the specific PICmicro microcontroller that you are working with. The PIC16F627's configuration register is defined in Table 3-5.

Table 3-5

PIC16F627's
Configuration
Register

Bit	Name	Function
13–12	CP1:CP0	These bits are "shadowed" as bits 11 and 10 (and should have the same values as 13 and 12). 1x, code protection off. 01, instruction addresses 0x00200 to 0x03FF protected. 00, code protection for entire instruction address space.
9	Not Used	Ignored, leave programmed.
8	CPD	1, data memory code protection off. 0, data memory code protection enabled.
7	LVP	LVP (5-volt Vpp) enabled. 1, enabled. RB4 is now a programming control bit. 0, disabled. RB4 is digital I/O.
6	BODEN	1, brown-out detect enabled. 0, brown-out detect disabled.
5	MCLRE	1, RA5/_MCLR pin works as _MCLR. 0, RA5/_MCLR is a digital I/O.
3	_PWRTE	72-millisecond power-up timer. 1, disabled. 0, enabled.
2	WDTE	WDT enabled. 1, enabled. 0, disabled.
4, 1–0	FOSC2: FOSC0	Oscillator selection bits. 111, external resistor on RA7. RA6 drives the instruction clock. 110, external resistor on RA7. RA6 is digital I/O. 101, internal RC oscillator. RA6 drives the instruction clock. 100, internal RC oscillator. RA6 is digital I/O. 011, external clock into RA7. RA6 is digital I/O. 010, HS oscillator. 001, XT oscillator. 000, LP oscillator.

Table 3-6

Configuration
Register
Values

Value	Functions
0x03F70	Internal oscillator, RA6/RA7 digital I/O, BODEN-enabled, WDT-disabled, external reset.
0x03F74	Internal oscillator, RA6/RA7 digital I/O, BODEN- and WDT-enabled, external reset.
0x03F61	4 MHz XT oscillator, BODEN-enabled, WDT disabled, external reset.
0x03F65	4 MHz XT oscillator, BODEN- and WDT-enabled, external reset.

I recommend that the list of sample configuration register values in Table 3-6 be used with the __CONFIG(. . .) statement in PICC Lite. These values are designed for a 4 MHz internal or external oscillator with no code protection, and they are programmed using the El Cheapo programmer circuit and various special functions that should be useful for robot applications.

The I/O hardware registers consist of the OPTION, TMR0, PORT, I/O PINS, and enable registers. They also include INTCON as well as other interrupt control and flag registers, along with any other hardware features built into the PICmicro microcontroller. The important difference between these hardware registers and processor registers is that except for INTCON, these registers are bank specific and although some conventions are used for the placement of these functions, for part numbers and specific functions, the registers are located in different addresses.

Although it may not qualify as a feature, unused registers in a PICmicro's register map will return zero (0x000) when they are read. This capability can be useful in some applications.

Zero registers (or undefined registers that return zero when read) are normally defined in the Microchip documentation as "shaded" addresses in the device register map documentation. In Figure 3-26, the 16F627's register map is shown with grayed-out addresses, indicating that they return zero when read. Of course, when these registers are written to, their values are lost and not stored in the register. (One might say the information has gone to the great bit bucket in the sky.)

The most basic way of getting data in and out of the PICmicro is via the parallel I/O bits that are located in the ports. In many PICmicros, these pins have peripherals behind them to provide advanced I/O capabilities. Despite this capability, in virtually every PICmicro application that you create, the straight I/O port functions will be required.

The PICmicro's typical I/O pin is capable of being either an input pin or an output pin. When in output mode, the pins are able to source or sink roughly 20 mA of current.

The block diagram of a PICmicro I/O pin is shown in Figure 3-27. Each register port is made up of a number of these circuits, one for each I/O bit.

Depending on your previous experiences, this I/O pin can look as if it is very complex, needlessly complex, or pretty basic. Regardless of your own feelings about the I/O pins, you should be aware of a few of their aspects.

I/O pins are associated with the bit number of the port they belong to. The minimal size for an I/O port is eight bits (or pins) for one byte. The convention used by Microchip (and used in PICC Lite) is to label the pins according to their bit number and port they're associated with.

The convention is as follows:

    R%#

where % is the port letter (port A, port B, and so on) and # is the bit number of the port.

Using this convention, RB3 is port B, pin 3, and I will use it throughout the book to identify specific pins. In some places, you will see the convention

    PORT%.#

or

    PORT%#

which uses the same values for % and # as the R%# method of labeling pins.

The TRIS register is used to control the output capabilities of the I/O pin. When the register is loaded with a 1 (which is the power-up default), the pin is input only (or in input mode) with the tri-state buffer disabled and not driving the pin. When a 0 is loaded into a pin's TRIS bit, the tri-state buffer is enabled (output mode), and the value that is in the data-out register is driven onto the pin. I always remember which is which by remembering that the digit is similar to the first letter of the resulting pin function (0 = Output and 1 = Input).

Figure 3-27
Standard
PICmicro MCU
I/O pin block
diagram

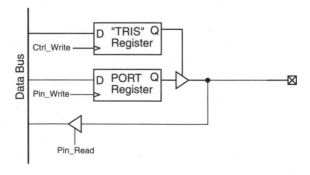

The TRIS bits follow the convention of

```
TRIS%#
```

which uses the same values for % and # as the convention used by R%# and PORT%#. PICC Lite uses the convention of R%# and TRIS%# when defining the I/O pin bits in the pic.h include file.

To make the RB2 I/O pin a digital output and drive a high value, the following PICC code is used:

```
TRISB2 = 0;        //  RB2 is an output bit
RB2 = 1;           //  RB2 drives a "high" value out
```

Along with accessing individual bits, the entire TRIS and PORT registers can be written to explicitly.

Going back to Figure 3-27, pin input is read from the I/O pin and not from output of the data-out register. This is important to remember because there are times where this arrangement will cause problems. This issue seems to be unique to the PICmicro; other microcontrollers either have a separate input address or have a method of selecting which to read (the data-out register or the pin) based on whether or not the pin is in input or output mode.

I wish I kept track of problems that people report on the PICList. I would bet the number one or two problem encountered by people is with RA4 or the PORTA.4 pin (see Figure 3-28). This pin is an *open drain*, which can be used on a pulled-up, dotted AND bus that has multiple dotted AND outputs on it, each able to pull the bus down. When RA4 is used by most people for the first time (I'm guilty of this as well), they forget or don't realize that the pin does not have the capability to drive a high voltage and can't understand why this pin seems to be broken.

This PORTA.4 pin cannot drive a positive voltage out unless it is pulled up (I normally use a 1k to 10k resistor depending on the input capacitance of what is being driven).

Figure 3-28
PORTA.4 I/O
pin block
diagram

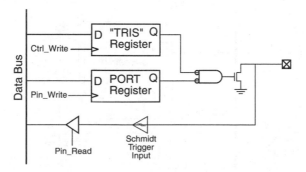

Also note that this pin has a *Schmitt Trigger* input, which has a different threshold voltage for signals going low to high than the threshold for going high to low, as is shown in Figure 3-29.

The purpose of the Schmitt Trigger is to provide "hysteresis" for an input signal and to eliminate some incorrect bouncing errors. This input causes RA4 to behave differently than other I/O pins in some circumstances; most notably, this pin should never be used for an RC circuit to read a potentiometer's value. In this case, the Schmidt Trigger input will enable some current flow from the pin into it, which affects the result of the potentiometer read operation.

A feature you should be aware of in the midrange parts is the availability of a controllable pull-up on the PORTB pins. This pull-up is controlled by the _RPBU bit of the OPTION register and is enabled when this bit is reset and the bit itself is set for output. The port B pin block diagram is shown in Figure 3-30.

Changing pin inputs can initiate interrupt requests to the processor. The function that is normally used is the RB0/INT pin, which can request an interrupt if the INTE bit is set in the INTCON register. An interrupt request can be made on rising or falling edges as selected by the INTEDG bit of the OPTION register. If INTEDG is set, interrupts can be requested on the rising edge of a signal into RB0/INT. INTEDG reset will cause an interrupt on a falling edge. I like to think of the RB0/INT interrupt request hardware as shown in Figure 3-31.

Once the interrupt is acknowledged by the processor, the INTF bit of the INTCON register has to be reset to enable another interruption RB0/INT pin transition. It should be noted that the interrupt request goes through a Schmidt

Figure 3-29
A Schmitt
Trigger data
signal's "edges"

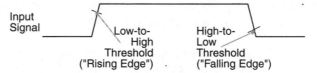

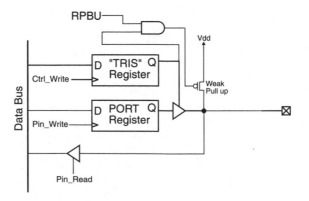

Figure 3-30
PORTB I/O pin
with pull-up

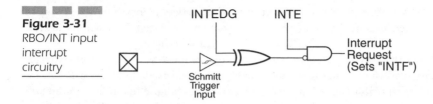

Figure 3-31
RBO/INT input
interrupt
circuitry

Trigger buffer, while the standard RB0 interface pin does not. This results in a somewhat different loading on RB0 compared to other PORTB I/O pins.

The other type of interrupt that can be requested from the I/O pins is the port B "Change on interrupt." If the RBIE bit of the INTCON register is set, then any changes to the RB4 through RB7 pins while they are in input mode will request an interrupt on "port change" and set the RBIF flag in the INTCON register. To clear this interrupt request, PORTB has to be first read to set the current value that is followed by resetting the RBIF flag.

The port change on interrupt is only available on pins RB4 to RB7 when they are in input mode. Changing the state of any of these pins while they are in output mode will not cause a port change interrupt request.

This interrupt is a bit tricky to use, but remember that PORTB should never be polled to eliminate any possibilities of there being an unexpected interrupt request reset before it can be acknowledged. To avoid this problem, when I use the port change interrupt feature, I do not make any other PORTB pins inputs.

In this book, I use interrupts as a base for the mechalogic and elelogic code because they simplify the effort required to write the application code, keeping the different spectrums of interfaces separate. In the next chapter, I will show how an interrupt handler is written in PICC Lite and what are the issues to be considered.

In the PIC16F627, the INTCON register is the central focus point for interrupts. This register is used to globally enable interrupts. The register consists of four different bit types and is located at address 0x0B in all active register banks. The bit usage is similar for all midrange devices. All the bits are positive (Set) and active.

The GIE bit must be set for interrupt requests to be passed to the processor. This bit can globally mask or allow (unmask) interrupt requests going to the processor. If a critical section of code is being entered, by resetting this bit the interrupt request will be shown. This bit is reset upon acceptance of the interrupt request and then set upon exit from the interrupt handler.

The INTCON bit names that end in E are the interrupt enable flags. When these bits are set, any incoming interrupt requests will set the corresponding interrupt request active flags, which have a bit name that ends in F. The request active flag must be reset in hardware and is not reset automatically by the operation of the interrupt acknowledgement. As well, the requesting hardware may have to be reset before the F (request active flag) can be reset.

As shown in Figure 3-31, if the E bit for a particular interrupt source is set, then interrupt requests, which have set their appropriate F bits, will request an interrupt of the PICmicro MCU processor if the GIE bit is set.

When the processor receives the interrupt request, it completes the current instruction before jumping to the interrupt vector. Instruction execution in the PICmicro microcontroller can be one or two cycles long. When added to the two-instruction delay for calling the interrupt handler, the total delay (which is known as *interrupt latency*) is three or four instruction cycles. In the PIC16F627 (as well as other midrange PICmicro microcontroller part numbers), the interrupt vector's address is 0x00004 for all interrupt sources. What happens during execution is shown in Figure 3-32.

In the PIC16F627, three interrupt sources are handled from within the INTCON register (see Table 3-7). The TOIF and TOIE bits enable an interrupt request when TMRO overflows (when it equals 255 or 0x0100). The second responds to an input on the RB0 (usually marked RB0/INT by Microchip) when the input on the pin goes high or low (depending on the status register state).

Along with these three interrupt sources, the PIE1 and PIR1 registers contain the E and F bits, respectively, used for the other built-in interrupt functions. The PIC16F627 PIE1 is defined in Table 3-8. The PIC16F627 PIR1 is defined in Table 3-9.

Table 3-7

INTCON
Register Bit
Defintions

Bit	Name	Function
7	GIE	Global interrupt enable
6	PEIE	Enable peripheral interrupts
5	T0IE	TMRO overflow interrupt enable
4	INTE	RB0/INT pin interrupt enable
3	RBIE	PORTB input change interrupt enable
2	T0IF	TMRO overflow interrupt request active
1	INTF	RB0/INT pin interrupt request active
0	RBIF	PORTB input change interrupt request active

Figure 3-32
Interrupt
operation

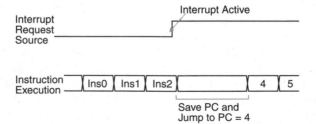

Table 3-8

The
PIC16F627
PIE1 Register

Bit	Name	Function
7	EEIE	Data EEPROM write interrupt enable
6	CMIE	Comparator interrupt enable
5	RCIE	USART receive interrupt enable
4	TXIE	USART transmit interrupt enable
3	N/A	Not implemented, read 0
2	CCP1IE	CCP module interrupt enable
1	TMR2IE	TMR2 overflow interrupt enable
0	TMR1IE	TMR1 overflow interrupt enable

Table 3-9

The
PIC16F627
PIR Registers

Bit	Name	Function
7	EEIF	Data EEPROM write completed interrupt pending
6	CMIF	Comparator interrupt pending
5	RCIF	USART character received pending interrupt
4	TXIF	USART transmitter holding register empty
3	N/A	Not implemented, read 0
2	CCP1IF	CCP module interrupt pending
1	TMR2IF	TMR2 overflow interrupt pending
0	TMR1IF	TMR1 overflow interrupt pending

Note that PIR1 and PIE1 will be defined differently for different PICmicro microcontroller part numbers. To find out what is the exact bit operation of these parts, you will have to consult their datasheets.

When the PICmicro MCU starts executing from reset, the PCON register can be polled and written to in order to specify at what speed the microcontroller is running as well as what was the reason for the PICmicro MCU starting up. The bit definitions for the register are shown in Table 3-10.

It is useful to poll the _POR bit in robots because if the microcontroller has reset itself because of losing voltage the batteries are losing their charge. It could also mean there is a potential problem in the circuit with the motors taking too much current from the batteries (such as if the robot is stalled against an obstacle).

Table 3-10

PCON
Register Bit
Definitions

Bit	Name	Function
7-4	N/A	Not implemented, read 0.
3	OSCF	INTRC/ER oscillator speed. If set, it runs at 4 MHz for internal RC oscillator, or if ER, then the external resistor is active. If reset, it runs at 37 KHz (for low-power operation).
2	N/A	Not implemented, read 0.
1	_POR	Reset when "power on reset." Must be set in software after checking value at boot.
0	_BOD	Reset on "brown out reset." Must be set in software after checking value at boot.

Robots with noisy power or that have large electrical fields can cause the circuits within the PICmicro microcontroller to stop executing properly. These events can cause the processor program counter to change to an invalid address or the instruction decoder to process an instruction improperly. Often when this happens, the PICmicro microcontroller locks up and will stop executing the application. To help counter this problem, Microchip has designed a Watchdog Timer (WDT) built into the PICmicro MCU that resets the chip if normal application execution is lost and the processor starts executing incorrectly or locks up.

The WDT has a nominal 18-millisecond clock delay, which will reset the PICmicro if it times out. Normally in an application, it is reset before timing out by executing a clrwdt instruction. The WDT is shown in Figure 3-33.

The WDT Oscillator in Figure 3-33 is an RC oscillator, which drives a counter. The *overflow* (O/F) output of the counter is optionally passed to a prescaler before going to the PICmicro microcontroller's reset circuit. The prescaler will be discussed in more detail later in this section, but its basic purpose is to count clock inputs, either to the PICmicro's TMR0 or the WDT. The overflow is passed to the PICmicro MCU's reset circuit when the specified number of WDT overflows has occurred. The prescaler enables WDT reset delays from 18 milliseconds to 2.3 seconds.

When the WDT causes a reset in the PICmicro microcontroller, the _TO bit of the status register is reset. In the initial code of your application, if WDT resets are used, then the _TO bit should be checked because the current file register settings will probably be set from a previous application execution.

It is recommended that the WDT be reset by the clrwdt instruction after half of the reset period has passed. The nominal error in the RC oscillator used for the WDT function is 20 percent, which means that WDT timeouts can take place anywhere from 14 to 22 milliseconds when no prescaler is used. To be on the safe

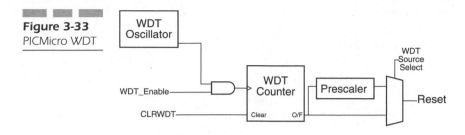

Figure 3-33
PICMicro WDT

side, executing clrwdt every 9 milliseconds in this situation will avoid any potential invalid WDT resets.

To reset the WDT in PICC Lite, you should put in the statement:

```
asm("clrwdt");
```

The WDT is enabled from within the configuration word and cannot be disabled within the application. This means you have to be very careful to avoid enabling the WDT unless you have provided support for it in the application code. Providing support for the WDT means that the clrwdt instruction is executed repeatedly to avoid the WDT from resetting the PICmicro microcontroller unexpectedly.

In the PIC16F627, the WDT enable bit of the configuration word is positive and active and is enabled when the bit is set (unprogrammed). This can cause some problems with new application developers that forget to disable the WDT explicitly. If you are going to create an application that uses the WDT, it is recommended that it only be enabled when you are about to release the application. It can be a problem during debug and can cause the PICmicro to reset itself when you least expect it, making it difficult to debug the application.

TMR0 is the basic eight-bit timer available to all PICmicros and can be used to request an interrupt in the PIC16F627 (as well as most PICmicro MCU part numbers). TMR0 has a few peculiarities with regards to its operation that you should be aware of, but for the most part it is quite straightforward to use and I will present some example applications later in the book that use it.

TMR0 is an eight-bit incrementing counter that can be preset (loaded) by application code with a special value. The counter can either be clocked by an external source or by the instruction clock. Each TMR0 input can be matched to two instruction clocks for synchronization. This feature limits the maximum speed of the timer to half the instruction clock speed. The TMR0 block diagram is shown in Figure 3-34.

The TOCS and TOCE bits are used to select the clock source and the clock edge that increments TMR0 (rising or falling edge). These bits are located in the option register that is discussed later.

The synchronizer is a glitch elimination feature of the PICmicro MCU. This circuit only updates TMR0 when two instruction cycles have passed without a

■■ ■■ ■■

Figure 3-34
TMR0 block
diagram

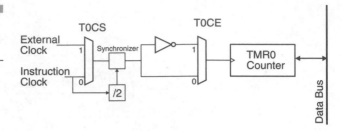

change to the input. For most applications, this circuit means that TMRO is updated after two instruction cycles have passed.

TMRO can be driven by external devices through the T0CKI pin. In the PIC16F627, the pin can also be used to provide digital I/O. When a clock is driven into the TMRO input, the input is buffered by an internal Schmidt Trigger to help minimize noise-related problems with the input.

Input to TMRO can be made with and without the prescaler, which provides a power of two "divide-by" feature to the TMRO or WDT counter inputs. Its purpose is to divide the incoming clock signals by a software-selectable power of two value to enable the eight-bit TMRO to time longer events or increase the watchdog delay from 18 milliseconds to 2.3 seconds.

The prescaler's operation is controlled by four bits contained within the OPTION register. PSA selects whether the WDT uses the prescaler (when PSA is set) or TMR uses the prescaler (when PSA is reset). Note that the prescaler has to be assigned to either the WDT or TMRO.

The divide by prescaler value is selected by the PSO, PS1, and PS2 bits of the OPTION register. I call prescaler operation a power of two because the number of cycles' delay is a function of the PSA bit value. Table 3-11 shows the prescaler cycle delay for varying values of the PS# bits.

The WDT and TMR0's block diagrams can be combined with the prescaler to show how the functions work. This is shown in Figure 3-35.

TMRO is located at register address 0x001 in the PIC16F627 (along with the other low-end and midrange PICmicro MCU part numbers). The contents of TMRO can be read from and written to directly. One thing to always remember when TMR0 is being updated is that the synchronizer (and prescaler, if connected) will be reset.

In the midrange PICmicro microcontrollers (like the PIC16F627), when TMR0 transitions from 0x0FF to 0x000, an interrupt can be requested. This is first accomplished by loading TMR0 to the specified delay value and then setting the T0IE bit. When TMR0 overflows to 0x000, T0IF will be set and if GIE is set, the interrupt request will be acknowledged.

The T0IF flag must be reset explicitly before enabling TMR0 interrupts to prevent the chance for a spurious interrupt when the TMR0 interrupt is enabled

Table 3-11

The Prescaler
Cycle Delay
for the PS Bits

PS2–PS0	Prescaler Delay
000	1 cycle
001	2 cycles
010	4 cycles
011	8 cycles
100	16 cycles
101	32 cycles
110	64 cycles
111	128 cycles

Figure 3-35
PICMicro
prescalar circuit

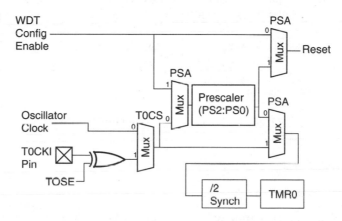

and TMR0 has already overflowed. T0IF is not reset automatically when the interrupt request is acknowledged.

One of the basic operations of a timer is to provide a set time delay. In many applications, I have used TMR0 to request an interrupt after a specific time delay. It isn't difficult to use a timer to calculate a delay, although there are a few things you should be aware of before you attempt it.

The first thing you should recognize is that the interrupt is requested when TMR0 overflows or equals 256. Two hundred fifty-five or 0x0FF is the largest value that can be saved in the bit timer register. 0x0FF does not cause an interrupt, although an interrupt will be requested when the timer increments or overflows to 256 or 0x0100. Therefore, the delay is calculated for the time when the timer equals 256 and not 255.

The initial timer value must be the number of clock increments (or ticks) required to get to 256 within a specific period of time because the timer can only count up. To calculate the initial value (with no prescaler, which will be described later), the following formula is used:

```
TMR0 initial = 256 - delay cycles
```

The delay cycles value is found by taking the delay time, dividing by the frequency, and multiplying by 4. The result is divided by two because the input to TMR0 divides the cycles by two:

```
delay cycles = (delay time * frequency/4)/2
             = delay time * frequency/8
```

These formulas can be used to calculate a 160-μsec delay in a 4 MHz PICmicro MCU application. Starting with the second formula, the number of delay cycles timed by the is calculated as

```
delay cycles = delay time * frequency/8
             = 160 μsecs * 4 MHz/8
             = 160(10**-6) secs * 4(10**6)/secs/8
             = 160 * 4/8
             = 80
```

This value is then used to calculate the initial timer value using the first formula:

```
initial = 256 - delay cycles
        = 256 - 80
        = 176
```

So, in order to wait 160 microseconds before a TMR0 overflow interrupt in a PICmicro MCU running at 4 MHz, an initial value of 176 must be loaded into TMR0.

For delays longer than 512 cycles, the prescaler can be used to divide the number of cycles input into the timer. When calculating the delay, the delay cycles are continually halved until the value is less than 256. For example, if a delay of 5 microseconds is required, the delay cycles would be calculated in a 4 MHz PICmicro microcontroller as follows:

```
delay cycles = delay time * frequency/8
             = 5 msec * 4 MHz/8
             = 5(10**-3) secs *  4(10**6)/secs/8
             = 5(10**3)/2
             = 2.5(10**3)
             = 2,500
```

Since the calculated delay cycles are greater than 256, it is continually divided by 2 to get an appropriate prescaler divisor. Dividing by 2, the delay count would be 1,250, which is still greater than 256. Divided by 2 again, the delay count would be 625, which is also greater than 256. Dividing the original value by 8 yields a delay count of 312.5. Finally, dividing the original by 16 yields a delay count of 156.25. Rounding off this value, the original TMR0 delay formula can be used:

```
initial = 256 - 156
        = 100
```

Now, the rounding that I did (taking off 0.25) will result in a difference of 4 cycles between the delay of 5 milliseconds and the delay provided by TMR0 and the interrupt handler. Actually, you will find that there are multiple delay latencies for which you will have to be prepared for. The instructions produced by the PICC Lite compiler are generally unknown in length, as is the number of cycles executed before the interrupt is acknowledged. For these reasons, I should point out that the TMR0 delay presented previously is only a minimum delay; the actual maximum delay is probably longer.

If your application requires a precise delay, then there are two things that you can do. The first is to use the CCP module (described later) in the PIC16F627 in PWM mode to output a precisely timed pulse. The second thing that you can do is build an external timer (single shot or oscillator) into your application circuit that will provide this precise timing. I am not being facetious when I am making these suggestions; there are times when a precisely timed signal is required in an application. What I am saying is that you cannot rely on the standard TMR0 interrupt in PICC Lite.

The OPTION register is a cornerstone register to the operation of the PICmicro. The register controls the operation and delay of the prescaler, selects the clock source, and specifies the operation of the interrupt source pin. It is a very useful register and one that you should always keep in the back of your mind and make sure it is set up properly for your application.

If you look in the Microchip documentation, you will see that this register is labeled as OPTION_REG. In most texts and the PICC Lite compiler, this register is given the label OPTION. The midrange device's OPTION register does not have any device-specific bits, as shown in Table 3-12.

For very low power operation, where the PICmicro microcontroller is waiting for an event, you can consider using the sleep instruction and mode of the PIC16F627. Sleep will reduce the power required by the PICmicro MCU to 10 uAmps or less, while leaving the external interrupts enabled and waiting for an external event. When the PICmicro microcontroller is in sleep mode, the timers will not operate.

Before using the sleep instruction, you must know if the minimized PICmicro MCU power drain will affect the entire system. Often in robot applications, having a microcontroller that is only consuming 10 µA is irrelevant due to the large battery capability built into the system.

Sleep mode is invoked by the sleep instruction and can be "woken up" by a system power-down/power-up, a _MCLR reset, an interrupt request, or a WDT reset.

Table 3-12

OPTION
Register Bits

Bit	Name	Function
7	_RBPU	Enable PORTB weak pull-ups 1, pull-ups disabled 0, pull-ups enabled
6	INTEDG	Interrupt request on: 1, low to high on RBO/INT 0, high to low on RBO/INT
5	TOCS	TMRO clock source select 1, Tock1 pin 0, instruction clock
4	TOSE	TMRO update edge select 1, increment on high to low 0, increment on low to high
3	PSA	Prescaler assignment bit 1, prescaler assigned to WDT 0, prescaler assigned to TMRO
2–0	PS2–PS0	Prescaler rate select 000, 1:1 001, 1:2 010, 1:4 011, 1:8 100, 1:16 101, 1:32 110, 1:64 111, 1:128

If the PICmicro MCU is taken out of sleep mode by an interrupt, then the next instruction will be executed, even if the interrupt handler is not invoked.

To jump to the interrupt vector upon wake-up, the appropriate E bit of the INTCON or PIE registers is set along with the GIE bit of INTCON. When the PICmicro MCU starts executing again, the instruction after sleep is executed and then the program counter is loaded with the interrupt vector (address 0x004). If GIE is reset when the sleep instruction is executed, then the PICmicro MCU continues executing at the address following the sleep instruction. The execution of the instruction after sleep no matter what is why it is suggested that a *nop* instruction should be placed after the sleep instruction. Using PICC Lite, this sleep/nop instruction sequence is implemented as

```
#asm
  sleep
  nop
#endasm
```

or

```
asm("sleep");
asm("nop");
```

An increasingly popular feature in PICmicro devices is the availability of built-in EEPROM memory that can be used to store configuration, calibration, or software data. In the PIC16F627, 128 bytes of EEPROM data can be accessed either using read and writes to registers or using built-in PICC Lite functions.

Along with accessing data, some Flash devices (the 16F62x and 16F87x) can also read and write flash program memory from within the application. This is useful for storing larger amounts of data in nonvolatile memory as well as changing application code, enabling the implementation of monitor-like applications for the PICmicro MCU.

For EEPROM I/O, you should be aware of four registers: EECON1, EECON2, EEADR, and EEDATA. These registers are used to control access to the EEP-ROM. As you would expect, EEADR and EEDATA are used to provide the address and data interface into the 256-byte EEPROM data. EECON and EECON2 are used to initiate the type of access as well as indicate that the operation has completed. EECON2 is a pseudoregister that cannot be read from but is written to with the data, 0x055/0 x0AA, to indicate the write is valid. EECON1 contains the bits shown in Table 3-13 for controlling the access.

Using these bits, a read can be initiated as the following:

```
EEADR = Address;                // Load Data Address
  RD = 1;                       // Start Data Read
  EEPROM_data = EEDATA;         // Read the Data Byte
```

or the built-in function can be used as follows:

```
    EEPROM_data = eeprom_read(Address);
```

Table 3-13

EECON1 Bits

Bit	Name	Function
7-4	N/A	Not implemented, it is read as 0.
3	WRERR	Set if a write error is terminated early to indicate that a data write may not have been successful.
2	WREN	When set, a write to EEPROM begins.
1	WR	Set to indicate an upcoming write operation. Cleared when the write operation is complete.
0	RD	Set to indicate a read operation. Cleared by the next instruction automatically.

Code to do an EPROM write consists of the following:

```
EEADR = Address;                      //  Load the Data Address
EEDATA = EEPROM_data;                 //  Load Data to Store
GIE = 0;                              //  Disable interrupts
WREN = 1;                             //  Enable Data EEPROM Write
EECON2 = 0x055; EECON2 = 0x0AA;       //  Critical Timing Specific code
WR = 1;                               //  Start Write Operation
GIE = 1;                              //  Can have Interrupts Enabled
while (WR == 1);                      //  Wait for Interrupt to complete
WREN = 0;                             //  Disable Data EEPROM Write
```

or the built-in function can be used like this:

```
eeprom_write(Address, EEPROM_data);
```

The critically timed code (writing 0x055 and 0x0AA to EECON2) is used to indicate to the EEPROM access control hardware that the application is under control and a write is desired. Any deviation in these instructions (including interrupts during the sequence) will cause the write request to be ignored by the EEPROM access control hardware.

Instead of polling, after the WR bit is set, the EEIE interrupt request bit can be set. Once the EEPROM write has completed, then the EEIF file is set and the hardware interrupt is requested.

Along with TMR0, the PIC16F627 has additional 16-bit (Timer 1 [TMR1]) and 8-bit (Timer 2 [TMR2]) timers. These timers are designed to work with the compare/capture program hardware feature. Along with enhancing this module, they can also be used as straight timers within the application.

TMR1 (see Figure 3-36) is a 16-bit timer that has 4 possible inputs. What is most interesting about TMR1 is that its own crystal can be used to clock it. This enables TMR1 to run while the PICmicro's processor is asleep.

To access TMR1 data, the TMR1L and TMR1H registers are read and written. Just like in TMR0, if the TMR1 value registers are written, the TMR1 prescaler is reset. A TMR1 interrupt request (TMR1IF) is made when TMR1 overflows.

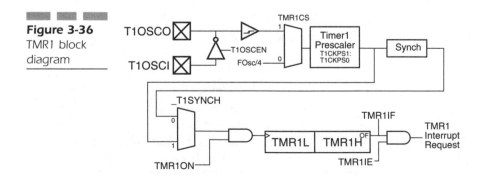

Figure 3-36
TMR1 block
diagram

TMR1 interrupt requests are passed to the PICmicro's processor when the TMR1IE bit is set.

TMR1IF and TMR1IE are normally located in the PIR and PIE registers. To request an interrupt, along with TMR1IE and GIE being set, the INTCON PIE bit must also be set.

To control the operation of TMR1, the T1CON register is accessed with its bits defined as shown in Table 3-14.

The external oscillator option is designed for fairly low-speed, real-time clock applications. Normally, a 32.768 kHz watch crystal is used along with two 33pF capacitors. 100 kHz or 200 kHz crystals could be used with TMR1, but the capacitance required for the circuit changes to 15 pF. The TMR1 oscillator circuit is shown in Figure 3-37.

Table 3-14

T1CON Register Bits

Bit	Name	Function
7-6	N/A	Not implemented, it is read as 0.
5-4	T1CPS1:T1CPS0	Select TMR1 prescaler value: 11, 1:8 prescaler 10, 1:4 prescaler 01, 1:2 prescaler 00, 1:1 prescaler
3	T1OSLEN	Set to enable TMR1's built in oscillator.
2	T1SYNCH	When TMR1CS is reset, the TMR1 clock is synchronized to the instruction clock.
1	TMR1CS	When set, the external TMR1 clock is used (crystal/ceramic resonator on RB6/RB7 or external oscillator on RB6).
0	TMR1ON	When set, TMR1 is enabled.

Figure 3-37
TMR1 oscillator circuit

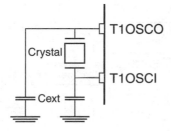

When TMR1 is running at the same time as the processor, the T1SYNCH bit should be reset. This bit will cause TMR1 to be synchronized with the instruction clock. If the TMR1 registers are to be accessed during processor execution, resetting T1SYNCH will make sure there are no clock transitions during TMR1 access. T1SYNCH must be set (no synchronized input) when the PICmicro microcontroller is in sleep mode. In sleep mode, the main oscillator is stopped, halting the synchronization clock to TMR1.

The TMR1 prescaler enables 24-bit instruction cycle delay values to be used with TMR1. These delays can either be a constant value or an overflow, similar to TMR0. To calculate a delay, the following formula is used:

```
Delay = (65,536 - TMR1Init) x prescaler/T1frequency
```

The T1frequency can be the instruction clock, the TMR1 oscillator, or an external clock driving TMR1. Rearranging the formula, the TMR1init initial value can be calculated as

```
TMR1Init = 65,536 - (Delay x T1frequency/prescaler)
```

When calculating delays, the prescaler will have to be increased until the calculated TMR1Init is positive. This is similar to how the TMR0 prescaler and initial value are calculated for TMR0.

TMR2 (see Figure 3-38) can be used as a recurring delay timer, almost exactly the same as TMR0 running an internal clock. When it is used with the CCP module, it is used to provide a *pulse-width-modulated* (PWM) timebase frequency. In normal operations, it can be used to create a 16-bit instruction cycle delay, similar to TMR0.

TMR2 is continually compared against the value in PR2. When the contents of TMR2 and PR2 match, TMR2 is reset, and the event is passed to the CCP as TMR2 Reset. If the TMR2 is to be used to produce a delay within the application, a postscaler is incremented when TMR2 overflows and eventually passes an interrupt request to the processor.

TMR2 is controlled by the T2CON register, which is defined as shown in Table 3-15.

The TMR2 register can be read or written at any time with the usual note that writes cause the prescaler and postscaler to be zeroed.

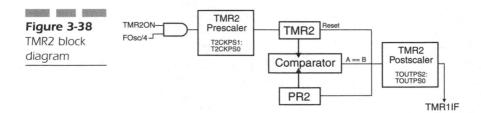

Figure 3-38
TMR2 block diagram

Table 3-15

T2CON
Register Bits

Bit	Name	Function
7	N/A	Not implemented, it is read as 0.
6-5	TOUTPS3:TOUTPS0	TMR2 postscaler select: 1111, 16:1 postscaler 1110, 15:1 postscaler : 0000, 1:1 postscaler
2	TMR2ON	When set, TMR2 is enabled.
1-0	T2CKPS1:T2CKPS0	TMR2 prescaler select: 1x, 16:1 prescaler 01, 4:1 prescaler 00, 1:1 prescaler

The timer itself is not synchronized with the instruction clock because it can only be used with the instruction clock. This means that TMR2 can be incremented on a one-to-one instruction clock ratio.

PR2 contains the reset or count-up-to value. The delay before the reset is defined as

```
Delay = prescaler x (PR2 + 1)/(Fosc/4)
```

If PR2 is equal to zero, the delay is

```
Delay = (prescaler x 256)/(Fosc/4)
```

I do not usually calculate TMR2 delays with an initial TMR2INIT value. Instead I take advantage of the PR2 register to provide a repeating delay and just reset TMR2 before starting the delay.

To calculate the delay between TMR2 overflows (and interrupt requests), the following formula is used:

```
Delay = (prescaler x [PR2 + 1|256])/((Fosc/4) x postscaler)
```

Interrupts use the TMR2IE and TMR2IF bits that are similar to the corresponding bits in TMR1. These bits are located in the PIR1 and PIE1 registers. Because of the exact interrupt frequency, TMR2 is well suited for applications that provide "bit-banging" functions like asynchronous serial communications or PWM signal outputs.

Included with TMR1 and TMR2 is a control register and a set of logic functions (known as the Capture/Compare/PWM or CCP module) that enhance the operation of the timers and can simplify your applications. Because there are some PICmicro MCUs with two CCP modules built in, the Capture/Compare/PWM module in the PIC16F627 is called CCP1.

The CCP hardware is controlled by the CCP1CON register, which is defined as shown in Table 3-16.

The most basic CCP mode is capture, which loads the CCPR registers (CCPR14, CCPR1C, CCPR2H, and CCPR2L) according to the mode the CCP register is set in. This function is shown in Figure 3-39 and shows that the current TMR1 value is saved when the specified compare condition is met.

Before enabling the capture mode, TMR1 must be enabled (usually running with the PICmicro MCU clock). The edge detect circuit in Figure 3-39 is a four-to-one multiplexor, which chooses between the prescaled rising edge input or a falling edge input, and passes the selected edge to latch the current TMR1 value and optionally request an interrupt.

In capture mode, TMR1 is running continuously and is loaded when the condition on the CCPx pin matches the condition specified by the CCP1M3:CCP1M0 bits. When a capture occurs, an interrupt request is made. This interrupt request

Table 3-16

CCP1CON
Register Bits

Bit	Name	Function
7-6	N/A	Not implemented, it is read as 0.
5-4	CCP1X: CCP1Y	Two least significant CEPST bits of the PWM compare value.
3-0	CCP1M3: CCP1M0	CCP module operating mode: 11xx, PWM mode 1011, compare mode—trigger special event 1010, compare mode—generate software interrupt 1001, compare mode—on match CCP pin low 1000, compare mode—on match CCP pin high 0111, capture on every sixteenth rising edge 0110, capture on every fourth rising edge 0101, capture on every rising edge 0100, capture on every falling edge 00xx, CCP off

Figure 3-39

CCP capture
module

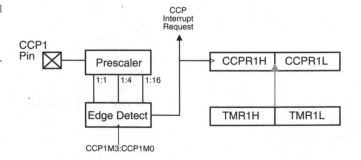

should be acknowledged and the contents of CCPR1H and CCPR1L should be saved to avoid having them written over and the value in them lost.

Capture mode is used to time repeating functions or determine the length of a PWM pulse. If a PWM pulse is to be timed, when the start value is loaded, the polarity is reversed to get to the end of the pulse. When timing a PWM pulse, the TMR1 clock must be fast enough to get a meaningful value with a high enough resolution so that there will be an accurate representation of the timing.

Compare mode changes the state of the CCP1 pin of the PICmicro microcontroller when the contents of TMR1 match the value in the CCPR1H and CCPR1L registers, as shown in Figure 3-40. This mode is used to trigger or control external hardware after a specific delay.

Of the three CCP modes, I find the PWM signal generator to be the most useful. This mode outputs a PWM signal using the TMR2 reset at a specific value capability. The block diagram of PWM mode is shown in Figure 3-41. The mode is a combination of the normal execution of TMR2 and capture mode; the standard TMR2 provides the PWM period while the compare control provides the "on" time specification.

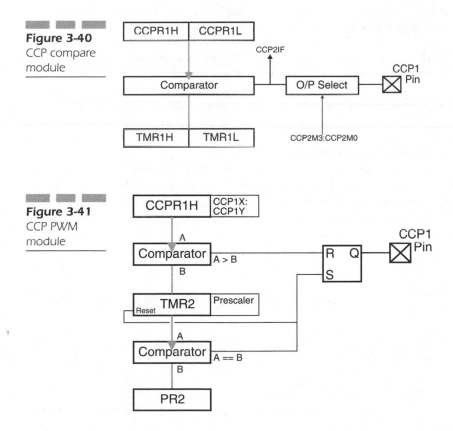

Figure 3-40
CCP compare module

Figure 3-41
CCP PWM module

When the PWM circuit executes, TMR1 counts until its most significant eight bits are equal to the contents of PR2. When TMR2 equals PR2, TMR2 is reset to zero and the CCP1 pin is set high. TMR2 is run in a 10-bit mode (the 4:1 prescaler is enabled before PWM operation). This 10-bit value is then compared to a program value in CCPR1L (along with the two CCP1x bits in CCP1CON) and when they match, the CCP1 output pin is reset low.

To set up a 65 percent duty cycle in a 20 KHz PWM executing in a PIC16F617 clocked at 4 MHz, the following steps are taken. First, the CCPR1L and PR2 values are calculated for TMR2, and the 4:1 prescaler must be enabled, resulting in a delay of

```
Delay = (PR2 + 1) * 4/(frequency/4)
PR2 = delay * frequency - 1
    = 50 msec * 4 MHz - 1
    = 200 - 1
    = 199
```

Sixty-five percent of 200 is 130, which is then loaded into CCPR1L.

For a period of 200 TMR2 counts with a prescaler of 4, the CCPR1L value becomes 131.75. To operate the PWM, I would load 130 into CCPR1L (subtracting 1 to match TMR2's zero start) and then the fractional value 0.75 into CCP1X and CCP1Y bits. I assume that CCP1X has a value of 0.50 and CCP1Y has a fractional value of 0.25. So to get a PWM in this case, CCPR1L is loaded with 130, and CCP1X and CCP1Y are both set. Table 3-17 gives the fractional CCP1x bit values.

The least significant two bits of the PWM are obviously not that important unless a very small PWM "on" period is used in an application. A good example of this is using the PWM module for an R/C servo. In this case, the PWM period is 20 milliseconds with an on time of 1 to 2 milliseconds. This gives a PWM on range of 5 to 10 percent, which makes the CCP1X and CCP1Y bits important in accurately positioning the servo.

Like many microcontrollers, the PIC16F627 has optional, built-in, serial I/O interfacing hardware. This interface enables the PICmicro MCU to work with external memory and enhanced I/O functions (such as ADCs) or communicate with a PC using RS-232.

One thing you will find with the serial I/O hardware is that it is not as flexible as you may want. The lack of flexibility is a concern with regards to interfac-

Table 3-17

CCP1x Bit
Values

Fraction	CCP1X:CCP1Y
0.00	00
0.25	01
0.50	10
0.75	11

ing RS-232 cheaply with PCs or using the synchronous serial interfaces in different situations.

The PIC16F627's USART hardware enables you to interface with serial devices like a PC using RS-232 or synchronous serial devices with the PICmicro MCU providing the clock or having an external clock drive the data rate. The USART module is best suited for asynchronous serial data transmission and in this section I will be concentrating on its capabilities.

The PICmicro transmits and receives NRZ asynchronous data. Synchronous data is sent with a clock and is in the format shown in Figure 3-42.

Synchronous data is latched into the destination on the failing edge of the clock. Although I have discussed packet decoding in detail elsewhere in this book, in this section I'll tend to treat the packet encoding and decoding as a black-box part of the USART and deal with how the data bytes are transmitted and received.

There are three modules to the USART: the clock generator, the serial data transmission unit, and the serial data reception unit. The two serial I/O units require the clock generator for shifting data out at the write interval. The clock generator's block diagram is shown in Figure 4-43.

In the clock generator circuit, the SPBRG register is used as a comparison value for the counter. When the counter is equal to the SPBRG register's value, a clock tick output is made and the counter is reset. The counter operation is gated and controlled by the *serial port enable* (SPEN) bit along with the synch, which selects whether the port is in synchronous or asynchronous mode, and BRGH, which selects the data rate.

Unfortunately, in the PICmicro MCU's USART, the bits used to control the operation of the clock generator, transmit unit, and receive unit are spread between the TXSTA and RCSTA registers along with the interrupt enable and

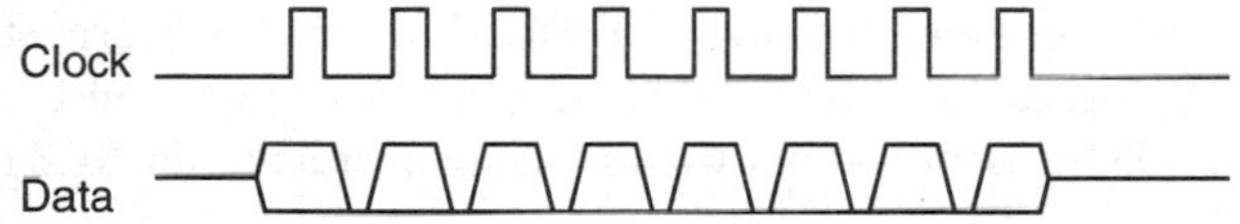

Figure 3-42
Synchronous
data waveform

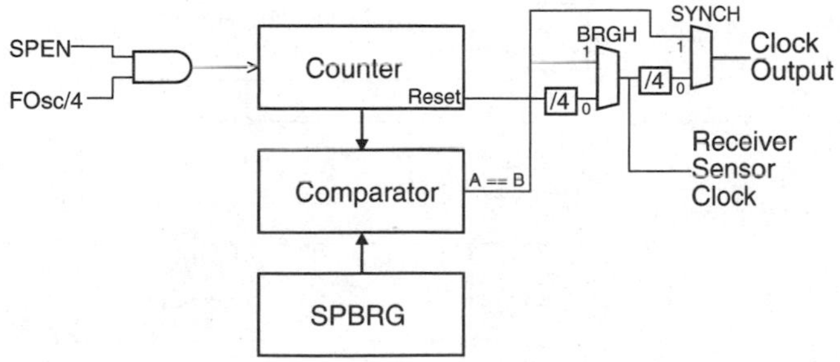

Figure 4-43
USART clock
block diagram

acknowledge registers. The individual bits will be defined at the end of this section, after the three functions of the USART are explained.

For asynchronous operation, the data speed is specified by the following formula:

```
Data Rate = Fosc/(16 x (4 ** (1 - BRGH)) x (SPBRG + 1))
```

This formula can be rearranged so that the SPBRG value can be derived from the desired data rate:

```
SPBRG = Fosc/(Data rate x 16 x (4 ** (1 - BRGH))) - 1
```

So, for a PIC16F627 running at 4 MHz, the SPBRG value for a 1200 BPS data rate with BRGH reset is calculated as follows:

```
SPBRG = Fosc/(Data rate x 16 x (4 ** (1 - BRGH))) - 1
      = 4 MHz/(1200/sec x 16 (4 ** (1 - 0))) - 1
      = 4 (10**6)/(1200 x 16 x 4) - 1
      = 52.0833 - 1
      = 51.0833
```

With 51 stored in SPBRG, there will be an actual data rate of 1,201.9 bps, which has an error of 0.16 percent to the target data rate of 1200 bps. This error is well within limits to prevent any bits being read in error.

The transmission unit of the USART can send eight or nine bits in a clocked (synchronous) or unclocked (asynchronous) manner. The block diagram of the hardware is shown in Figure 3-44.

If the synch bit is set, then data is driven out on the PICmicro microcontroller's RX pin with the data clock being either driven into or out of the TX pin. When data is loaded into the TXREG, if CSRC is reset, then an external device will clock it out. If CSRC can be shifted eight or nine bits at a time with the operation stopping when the data has been shifted out, an interrupt can be requested when the operation is complete. In asynchronous mode, once data is loaded into the TXREG, it is shifted out with a 0 leading start bit in NRZ format.

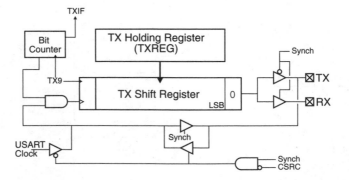

Figure 3-44
USART transmit
hardware block
diagram

The transmit hold register can be loaded with a new value to be sent immediately following the passing of the byte in the transmit shift register. This single buffering of the data enables data to be sent continuously without the software polling the TXREG to find out when is the correct time to send out another byte. USART transmit interrupt requests are made when the TX holding register is empty. This feature is available for both synchronous and asynchronous transmission modes.

The USART receive unit is the most complex of the USART's three parts. This complexity comes from the need for it to determine whether or not the incoming asynchronous data is valid or not using the Pin Buffer and Control unit built into the USART receive pin. The block diagram for the USART's receiver is shown in Figure 3-45.

If the port is in synchronous mode, data is shifted either according to the USART's clock or an external device's clock.

For asynchronous data, the receiver sensor clock is used to provide a polling clock for the incoming data. This sixteen time data rate clock's input into the pin buffer and control unit provides a polling clock for the hardware. When the input data line is low for three receive sensor clock periods, data is then read in from the middle of the next expected bit, as shown in Figure 3-46. When data is being received, the line is polled three times and the majority states read are determined to be the correct data value. This repeats for the eight or nine bits of data with the stop bit being the final check.

Like the TX unit, the RX unit has a holding register, so if data is not immediately processed and an incoming byte is received, the data will not be lost. But if the data is not picked up by the time the next byte has been received, then an overrun error will occur. Another type of error is the framing error, which is set if the stop bit of the incoming NRZ packet is not zero. These errors are recorded in the *receiver status* (RCSTA) register and have to be reset by software.

To control the USART, two registers are used explicitly. The *transmitter status* (TXSTA) register is defined with the bits outlined in Table 3-18.

Figure 3-45
USART receive
hardware block
diagram

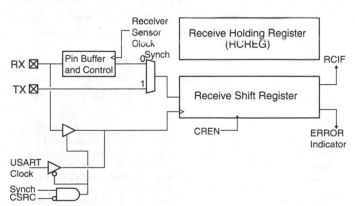

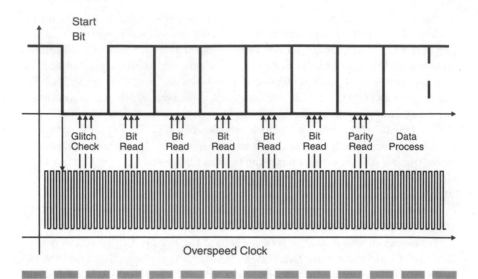

Figure 3-46 Reading an asynch data packet

Table 3-18

TXSTA Bits

Bit	Name	Function
7	CSRC	Clock source select used in synchronous mode. When set, the USART clock generator is used.
6	TX9	Value of the ninth bit.
5	TXEN	Set to enable data transmission.
4	SYNC	Set to enable synchronous transmission.
3	N/A	Not implemented, it is read as 0.
2	BRGH	Used in Asynchronous Mode to enable fast data transmission.
1	TRMT	Set if the transmission shift register is empty.
0	TXD	Bit nine of transmitted data (if TX9 set).

The *receiver status* (RCSTA) register is defined as shown in Table 3-19.

To set up asynchronous serial communication transmit, the following code is used:

```
SYNCH = 0;              //  Not in Synchronous Mode
BRGH = 0;               //  Use Low Speed Clock
SPBRG = DataRate;       //  Load USART data rate divisor
SPEN = 1;               //  Enable the Serial Port
TX9 = RX9 = 0;          //  Only 8 Bits of data
SPEN = 1;               //  Enable the serial port receiver
```

Table 3-19

RCSTA
Register Bits

Bit	Name	Function
7	SPEN	Set to enable the USART.
6	RX9	Set to enable a nine-bit USART receive.
5	SREN	Set to enable a single-byte synchronous data receive. Reset when data has been received.
4	CREN	Set to enable a continuous receive.
3	ADDEN	Set to receive data address information.
2	FERR	Framing error bit.
1	OERR	Overrun error bit.
0	RX9D	Received ninth bit. Valid if RX9 set.

To send a byte, use the following code:

```
while (TMRT == 0);      //  Wait for Holding Register to become Free
TXREG = SendData;       //  Load the USART with Data to Transmit
```

In the data send code, the TRMT bit, which indicates when the TX holding register is empty, is polled. When the register is empty, the next byte to send is put into the transmit shift register. This polling loop can be eliminated by setting the TXIE bit in the interrupt control register and then in your interrupt handler, checking to see if the TXIF flag is set before putting a byte into TXREG.

To wait for data to receive, you can use the code:

```
while (RXIF == 0);      //  Wait for data to become available
ReceiveData = RCREG;    //  Get the received byte
RXIF = 0;               //  Reset the interrupt request flag
```

Analog voltages can be processed by the use of comparators built into PORTA of the PIC16F627 that indicate when a voltage is greater than another voltage. The inputs compared can be switched between different I/O pins as well as ground or a reference voltage that can be generated inside the PICmicro chip.

Enabling comparators is a straightforward operation with the only prerequisite being that the pins used for the analog compare must be in input mode. Comparator response is virtually instantaneous, which enables an alarm or other fast responses from changes in the comparator inputs.

The comparator works very conventionally, as shown in Figure 3-47. If the value of the +input is greater than the -input, the output is high. Two comparators in the PIC16F627 are controlled by the CMCON register, which is defined in Table 3-20.

The CIS and CM2:CM0 bits work together to select the operation of the comparators (see Table 3-21).

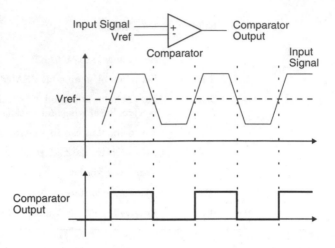

Figure 3-47
Comparator
response

Table 3-20

CMON
Register Bits

Bit	Name	Function
7	C2OUT	Comparator 2 output (high if + greater than -)
6	C1OUT	Comparator 1 output (high if + greater than -)
5	C2VINV	When set, comparator 2 output inverted
4	C1VINV	When set, comparator 1 output inverted
3	CIS	Comparator input switch
2–0	CM2:CM0	Comparator mode

From these selections, some points should be made:

- For CM2:CM0 equal to 000, RA3 through RA0 cannot be used for digital I/O.
- For CM2:CM0 equal to 000, RA2 and RA1 cannot be used for digital I/O.
- RA3 can be used for digital I/O.
- RAO and RA3 can be used for digital I/O.
- RA3 is digital output, the same as comparator 1 output.
- RA4 is the open drain output of comparator 2.
- RA0 and RA3, as well as RA1 and RA2, can be used for digital I/O.

Upon power-up, the comparator CM bits are all reset, which means RA0 to RA3 are in analog input mode. If you want to disable analog input, the CM bits must be set (write 0x007 to CMCOM).

Interrupts can be enabled that will interrupt the processor when one of the comparator outputs changes. This is enabled differently for each PICmicro with

Table 3-21

CIS and CM2:CM0 Bits

CM	CIS	Comp 1 + input	- input	Comp 2 + input	- input
000	X	RA0	RA3 (1)	RA2	RA1 (4)
001	0	RA2	RA0	RA2	RA1
001	1	RA2	RA3	RA2	RA1
010	0	Vref	RA3	Vref	RA1
010	1	Vref	RA3	Vref	RA2
011	X	RA2	RA0 (3)	RA2	RA1
100	X	RA3	RA0 (4)	RA2	RA1
101	X	DON'T	CARE	RA2	RA1
110	X	RA2	RA0 (5)	RA2	RA1 (6)
111	X	RA3	RA0 (7)	RA2	RA1 (8)

built-in comparators. Like the PORTB change on interrupt, after a comparator change interrupt request has been received, the CMCOM register must be read to reset the interrupt handler.

Along with comparing to external values, the PIC16F627 can also generate a reference voltage (Vref in Table 3-21) using its own built- in, four-bit digital-to-analog converter.

The Vref control bits are found in the VRCON register and are defined as shown in Table 3-22.

The Vref output is dependent on the state of the VRR bit. The Vref voltage output can be expressed mathematically if VRR is set as

```
Vref = Vdd * (Vfcon & 0x00F)/24
```

Or, if it is reset as

```
Vref = Vdd*(8 + (Vrcon & 0x00F))/32
```

Note that when VRR is set, the maximum voltage of Vref is $^{15}/_{24}$ of Vdd, or less than two-thirds Vdd. When VRR is reset, Vref can be almost three-quarters of Vdd.

El Cheapo PICmicro MCU Programmer Circuit

Many PICmicro MCU programmer designs are available due to the ease with which most PICmicro microcontrollers can be programmed using the ICSP protocol. These programmer circuits range from just a few simple components to commercial programmer circuits costing thousands of dollars. To enable you to

Table 3-22

Vref Control Bits

Bit	Name	Function
7	VREN	Vref enabled when set.
6	VROE	Vref output enabled. When set, it is RA2 Vref.
5	VRR	Range select: 1 = low range. 0 = high range.
3	N/A	Not implemented, it is read 0.
3–0	VR3:VR0	Voltage selection bits.

work through the sample applications that I have included in the next two chapters, I would like to introduce you to my El Cheapo programmer circuit that you can build for PIC16F627s in just an hour or so for a few dollars.

ICSP is one of the big advantages of the PICmicro microcontrollers. This feature enables you to program PICmicro MCUs after they have been assembled into the application circuit, which eliminates one manufacturing step or eliminates the need for buying specialized sockets and handling equipment for different devices. The ICSP features of the PICmicro MCU enable simple programmer circuits to be built.

The first feature of ICSP, enabling the PICmicro microcontroller to be programmed in circuit, is something only a "card stuffer" would appreciate. Being able to program parts after they have been soldered to a circuit board eliminates the need for costly (and unreliable) programmer sockets and extra device handling, which can result in bent pins or reversed components being put into the placement equipment and onto the boards. This gives the card manufacturer an opportunity to provide a cheaper service to their customers and provide the flexibility of shipping a product with the latest possible software "burned" into the product.

The pins that are affected by ICSP for the different PICmicro MCU pin count devices are shown in Table 3-23.

To program and read data, the PICmicro microcontroller must be put into programming mode by applying 12 to 14 volts to the _MCLR pin; then pull the data and clock lines low for several milliseconds. Once the PICmicro microcontroller is in programming mode, data can then be shifted in and out using the clock line.

Vdd is at 5 volts and requires 20 to 50 mA, which means either a 78L05 regulator or a Zener diode regulator, as discussed earlier in this chapter.

A transistor's physical switches can be used for turning on and off the Vpp and Vdd voltages. If Vpp is not being driven, internal pull-downs in the PICmicro will pull its _MCLR pin to ground, which eliminates the need for a ground driver on the reset line. Putting the PICmicro into programming mode requires the data waveform shown in Figure 3-48.

Table 3-23

ICSP-Affected Pins

Pin	12C5xx	16C50x	18-Pin Mid	28-Pin Mid	40-Pin Mid
1 Vpp	4 _MCLR	4 _MCLR	4 _MCLR	1 _MCLR	1 _MCLR
2 Vdd	1 Vdd	1 Vdd	14 Vdd	26 Vdd	11,32 Vdd
3 GND	8 Vss	14 Vss	5 Vss	8,21 Vss	12,31 Vss
4 DATA	7 GPO	13 RB	13 RB7	28 RB7	40 RB7
5 CLOCK	6 GP1	12 RB1	12 RB6	27 RB6	39 RB6

Figure 3-48
Programmer
initialization

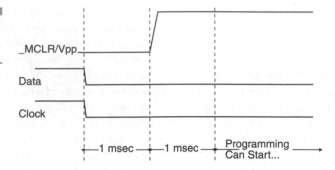

When _MCLR is driven to Vpp, the internal program counter of the PICmicro microcontroller is reset. The PICmicro MCU's program counter is used to keep track of the current program memory address in the EPROM that is being programmed. When it is set to 0x02000, the ID locations of the PICmicro microcontroller will be programmed, and address 0x02007 points to the configuration register.

Data is passed to and from the PICmicro MCU using a synchronous data protocol. A six-bit command is always sent before data is transferred. The commands (and their bit values and data) are listed in Table 3-24.

Data is shifted in and out of the PICmicro microcontroller using a synchronous protocol. Data is shifted out by the least significant bits on the falling edge of the clock line. The minimum period for the clock is 200 nanoseconds with the data bit centered as shown in Figure 3-49, which is sending an increment address command.

When data is to be transferred, the same protocol is used, but a 16-bit transfer (least significant bits first) follows after 1 microsecond has passed since the transmission of the command. The 16 bits consist of the instruction word shifted to the left by 1. This means the first and last bits of the data transfer are always zero.

The programming cycle for the Flash-based PICmicro devices is as follows:

1. Load data command (000010 + data word × 2).

2. Begin programming command (001000).

3. Wait 10 milliseconds.

Configuration memory is programmed the same way with the load configuration command executed (with 0x07FFE as data to avoid incorrect data writes taking place) before the memory writes. The ID locations are set by convention, at 0x02000 to 0x02003. The 16-bit __IDLOCS value is written to each of the 4 ID Locations with the least significant 4 bits written to each location for a 16-bit ID value. The configuration register is at address 0x02007, like the EPROM

Table 3-24

The PIC16F627 ICD

Command	Bits	Data
Load configuration	000000	07FFE in
Load data for program memory	000010	Word × 2 going in
Load data for data memory	000011	Byte × 2 going in
Read data from program memory	000100	Word × 2 going out
Read data from data memory	000101	Byte × 2 going out
Increment PICmicro's PC	000110	None
Begin programming	001000	None
Bulk erase program memory	001001	None
Bulk erase data memory	001011	None

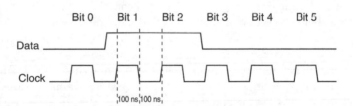

Figure 3-49
Programmer command—six bits

midrange parts with the register set using the same programming cycle as given previously.

The only outstanding issue is how to erase the program memory before programming. The Flash-based PICmicro microcontroller program in operation converts specific unprogrammed 1's in memory to programmed 0's. The Flash-based PICmicro MCU erase step loads 1's in all the memory locations.

This can be done using the bulk erase commands, but I prefer to use the Microchip-specified erase command for code-protected devices. This operation will erase all flash and EEPROM memory in the PICmicro device, even if code protection is enabled, through these steps:

1. Apply Vpp.
2. Execute load configuration (0b0000000 + 0x07FFE).
3. Increment the PC to the configuration register word (send 0b0000110 seven times).

4. Send command 0b0000001 to the PICmicro.

5. Send command 0b0000111 to the PICmicro.

6. Send begin programming (0b0001000) to the PICmicro.

7. Wait 10 milliseconds.

8. Send command 0b0000001.

9. Send command 0b0000111.

Note that two undocumented commands (0b0000001 and 0b0000111) are in this sequence.

When this sequence is followed, all data within the PICmicro MCU is erased and the device is ready for a new application to be burned into it.

Once a memory location has been correctly programmed, the PICmicro microcontroller's program counter can be incremented. If there is nothing to program at a memory location, or the value is 0x03FFF, then you can simply send an increment to skip to the next address and ignore programming the instruction completely.

After the program memory has been loaded with the application code, Vpp should be cycled off and on, and the PICmicro MCU's program memory is read out and compared against the expected contents. When this verify is executed, Vpp should be cycled again with Vdd at a minimum voltage (4.5 volts) and then repeated again with Vdd at a maximum voltage (5.0 volts) value.

When this verify is executed at voltage margins, the PICmicro microcontroller is said to be production programmed. If the margins are not checked, then programming operation is said to be prototype programmed. Most hobbyist programmer circuits (including the El Cheapo and the Microchip PICStart Plus) are prototype programmer circuits because they cannot margin Vdd when the value is checked.

The El Cheapo PICmicro MCU programmer circuit is a PC parallel-port-based interface with a Microsoft Windows interface. The design presented here (see Figure 3-50) is actually the sixth version of a basic programmer circuit that I first published four years ago. The circuit (and driver software) has been updated to allow it to program virtually any ICSP-programmed PICmicro MCU. The reason for the many upgrades to the circuit and software is to make the programmer circuit work without modification for as many PCs as possible, as well as support all the new PICmicro microcontroller part numbers that are being released by Microchip.

The bill of materials for the programmer circuit is given in Table 3-25. Note that for the 2.5 mm power connector (J1), I have included a Digi-Key part number to make it easier for you to find.

These parts are quite easy to find in most electronic stores. In fact, you will probably be able to find most of the components in your local Radio Shack. If you have problems finding any of these components, please let me know and I'll see what I can do about pointing you in the right direction.

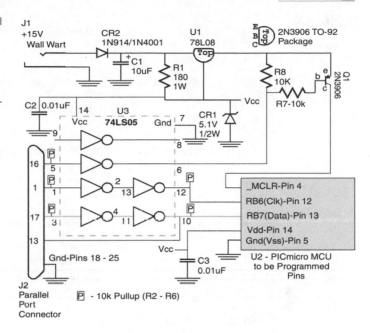

Figure 3-50
El Cheapo
PICMicro
programmer
circuit

The circuit can be broken up into four major subsystems. The first is the power supply. Input DC power is passed to both the 78L12 and 5.1-volt Zener diode regulator circuit to provide the operating power (Vdd) and the programming voltage (Vpp) for the PICmicro microcontroller being programmed. Vpp must be at least 13 volts, and to produce this voltage, I connect the ground input of the 78L08 to the 5.1 volts produced by the 5.1-volt Zener diode (CR1). The input diode (CR2) is used to rectify the incoming voltage and protect the circuit if power is wired incorrectly.

To provide the +15 volts or more DC power to the circuit, I recommend that you use a wall-mounted AC/DC power converter with a 2.5 mm (center positive) connector. These power converters can be very easily found in stores like Radio Shack, Wal-Mart, or Circuit City. Some people call them wall warts.

The Zener diode and resistor circuit are designed to regulate the input power (nominally assumed to be 15 volts) down to 5.1 volts with a capability to drive 60 mAs. When I checked the programming specifications of the various PICmicro devices that I wanted this programmer circuit to handle, I found that the maximum Vdd during programming was 40 mA. This value, along with the current required by the 74LS05 (U3), determined which resistor was going to be used in the circuit. For a voltage drop of 10 volts with 60 mA, a 180 Ohm resistor (R1) was determined to be the best. The 10-volt drop through the 180 Ohm resistor dissipates more than a half-watt of power, which is why I have specified that the 180 Ohm resistor is rated at 1 watt. If there is no load on the 5.1-volt power supply, this 40 mA is passed through the Zener diode. Doing the calculation again,

Table 3-25

Programmer
Circuit Bill of
Materials

Part	Description	Comments
U1	78L08	
U2	18-pin DIP socket	It can be a Zero Insertion Force (ZIF) socket, so use 3M TextTool P/N 218-3341-00—0602R.
U3	74LS05	
Q1	2N3906 PNP bipolar	BC557 can be substituted for this part.
CR1	5.1 V, 1/2 watt Zener	
CR2	1N914/1 N4001 silicon diodes	Virtually any silicon diode can be used here.
C1	10 µF electrolytic	+16-volt rating.
C2, C3	0.01 µF	Any type of capacitor can be used.
R1	180 Ohm, 1 watt	
R2–R8	10K, 1/4 watt	
J1	2.5 mm power socket	Digi-Key part number: SX1152-ND.
J2	DB25-F socket	
PCB	P\N 136437-4	Provided with the book.
Power	+14-volt AC/DC	The power supply must source at least 250 mA. The output must match J1.
PC I/F Cable DB-25M to DB-25M		Straight-through parallel port Switchbox cable. See text.
Misc	Prototyping printed circuit board (PCB), wire	

the total power dissipated by the Zener diode is 0.3 watts. To be on the safe side, this diode should be rated for at least half a watt.

The 5.1-volt power supply is designed so that if it is shorted out, a maximum of 60 mA will be drawn. This enables the PICmicro microcontroller to be plugged in and pulled out of its socket (U2) without switching on and off or disconnecting the power supply. The power supply seems to be a circuit that many people want to change, thinking they know how to design one that is more efficient.

Please do not change or modify the El Cheapo's power supply circuit simply because it is reasonably inexpensive and easy to build; it will protect the circuit against short circuits and other transients caused when the PICmicro MCU is plugged in and out. I know of people that have used bench supplies and one poor soul that used his PC's +5- and +12-volt power supplies and ended up burning them out. The circuit I've shown here can take a lot of abuse and should never burn out or damage any other components.

The 74LS05 (U3) is used to give the programmer circuit access to your PC. I have tried a number of different methods of connecting the programmer circuit to a PC and this is the one that works best for the greatest variety of PCs. A 74LS05 can be purchased for just a few cents and while it seems extravagant, it does not add substantially to cost or complexity for the programmer circuit. Note that I have put 10K pull-up resistors on the inputs connected to the PC's parallel port as well as the outputs connected to other parts of the application. The 74LS05 is an open collector output part, and failure to include them could result in your programmer circuit not working properly.

I suggest that you use a standard 25-pin female D-Shell connector to connect your El Cheapo to your PC. Figure 3-50 specifies the connector pins to be used with this programmer circuit. Note that the ground pins of the parallel port are common with (connected to) the ground of the El Cheapo. If there isn't a ground connection between the two, then the control signals from the PC will not be passed to the El Cheapo programmer circuit.

To connect the El Cheapo to your PC, you should use a switchbox cable, which is sometimes known as a parallel transfer cable. This is *not* the same as a laplink cable. The cable that you use will have two male DB-25 connectors at either end. I do not recommend using a cable longer than 10 feet.

Before attempting to use the cable, check to see that all 25 pins are passed directly from one connector to another. Using an ohmmeter, you can test the cable to make sure that it works properly by making sure that each pin of one connector is connected to (and only to) the same pin on the other connector. You might want to make up a chart like Table 3-26 to make sure that the cable is wired correctly to be used with the EL Cheapo programmer circuit.

The third subsystem in this programmer circuit is the Vpp control circuit consisting of the open collector output of U3 controlling the PNP Vpp transistor switch (Q1). This circuit may seem a bit unwieldy and unnecessarily complex, but

Table 3-26	Connector 1 Pin	Connector 2 Pin
Pin Test Chart	1	1
	2	2
	3	3
	:	:
	25	25

Vpp will require up to a 50 mA source to program EPROM-based PICmicro devices. This circuit will switch Vpp on and off and enable the maximum current supplied by the 78L08 to be passed to the PICmicro MCU being programmed.

The last subsystem in the El Cheapo is the PICmicro socket and programming data pins. In Figure 3-50, I have shown the connections for programming an 18-pin PICmicro MCU (like the PIC16F627).

For this book, I built a number of El Cheapo prototypes using just a simple prototyping board, a standard 18-pin socket, a solder-tail DB-25F D-Shell connector, and the other miscellaneous parts. For my prototypes, I straddled the DB-25F D-Shell connector onto the edge of the PCB for a sturdy mounting with no danger of flexing damaging the connections. I wired the 2.5 mm connector so that it is close to the PCB and I used glue to secure it to the PCB. I generally use Weldbond for this purpose. To wire the programmer circuit, I used 28-gauge wire-wrap wire soldered to the different points of the circuit.

When you are laying out your circuit, make sure that you plan to be able to remove the PICmicro microcontroller from the socket. If you use a standard DIP socket, you will have to provide access to one end of the socket to get a small screwdriver in to pry up the PICmicro MCU. I've heard from a number of people that have forgotten to think about this and have ended up building the programmer circuit twice. The same goes for the DB-25F D-Shell connector and the 2.5 mm power connector; make sure that the mating connectors will go in without any kind of interference *before* you solder or glue them.

The total time for me to assemble a prototype was about a half-hour. When you build your programmer circuit, I suggest that you plan for an afternoon; chances are you may have some problems with laying out and building the circuit, as well as debugging it. I always find that when I'm about to build a new circuit, if I plan for more time than I actually need, then the quality of my work is better, with less chance that I will make mistakes.

Another advantage of setting aside an afternoon to build the programmer circuit is that you can start programming some of the example interface applications into a PIC16F627.

Before you start to build your El Cheapo, I suggest that you install the El Cheapo Windows interface software that is on the CD-ROM and have a PIC16F627 on hand to test out the programmer circuit when you are finished. The CD-ROM has instructions, showing you how the software is to be installed, along with a DLL package that must be installed with the software. The package has suggestions for getting the software running on your PC if you have problems.

As can be seen in Figure 3-51, a Build/Test option will guide you through the task of building your own El Cheapo. Along the PC, the only other tool that you will require is a digital multimeter.

When you have built your El Cheapo and tested it using the Build/Test option of the El Cheapo Windows interface software (and clicked Debug End), the circle in the top right-hand corner of the dialog box should be yellow. If the "sun" isn't

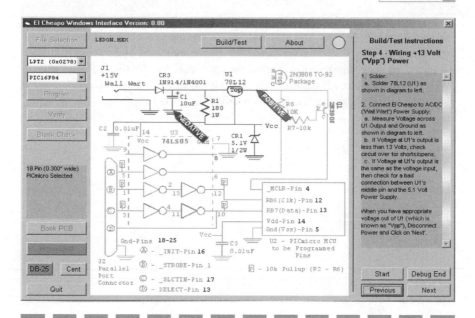

Figure 3-51 El Cheapo Windows interface

out, then the programmer circuit will not attempt to program a part. If the circle stays black, then go through the Test/Build steps to find out what the problem is.

The El Cheapo programmer circuit operation itself is reasonably fast. You will find that it takes about 15 seconds to erase and program a PIC16F627. This is in contrast to other simple programmer circuits that can perform this operation in less than 10 seconds. Although the El Cheapo is a bit slower, it is very reliable and should work on any PC.

You might want to make a couple of enhancements to your El Cheapo or consider them before you build it. The first and most obvious one is to use an 18-pin ZIF socket for the programmer circuit. This will enable you to pop the PICmicro microcontroller in and out easily without having to pry it up with a screwdriver (which can potentially result in a damaged PICmicro MCU).

The second suggestion is a bit more outlandish: why don't you consider building the El Cheapo directly into your circuit? The power supply in the robot dedicated to powering the PICmicro MCU can be used to power it (and the 74LS05) during programming, and a simple 9-volt alkaline radio battery could be connected into the circuit to provide Vpp (eliminating the need for U1). When connecting the 9-volt alkaline battery into the circuit, connect the negative terminal to Vdd and the positive to the emitter of the 2N3906. This will provide a voltage potential of 13 volts or more.

I would suggest that you dedicate the _MCLR (RA5), RB6, and RB7 pins to the programmer function. You will still have as many as 13 I/O pins available for your use, and you will not have to worry about removing the PICmicro MCU from its socket to program. For many robot designs, this will be the simplest way to program the PICmicro MCU and avoid the added time (and potential damage) of removing and reinstalling the PICmicro MCU into the robot.

Micro-
controller
Connections

In this chapter, I will demonstrate how different devices that are external to the microcontroller (peripherals) can be implemented. The code controlling these interfaces consists of sample robot mechalogic and elelogic functions that you can use in your own applications. Although the PIC16F627 is used as the target device for these sample applications, I tried to make the examples microcontroller independent and, in many cases, external component independent.

Before starting this chapter, you should have installed the MPLAB IDE, PICC Lite, and the El Cheapo Windows Interface (and built and tested the programmer circuit) on your PC. You need these three pieces of software (along with the programmer hardware) to reproduce the applications presented in this chapter.

When I originally designed the interfaces presented in this chapter, they were for the PIC16F84, which does not have the same advanced *input/output* (I/O) functions as the PIC16F627. Converting these applications to the PIC16F627 makes some of the code simpler, but the basic premise that the mechalogic and elelogic code would be sequenced and controlled by the PICmicro MCU's TMR0 interrupt does not change. The use of the PICmicro microcontroller's basic timer (TMR0) to interrupt the biologic code execution once every millisecond enables the biologic code to be very independent of the mechalogic and elelogic functions. By basing the peripheral interfaces on the TMR0 1-millisecond interrupt, I can change the low-level mechalogic and elelogic functions when the enhanced built-in features of the PIC16F627 made for better or simpler code without changing the biologic code examples in the next chapter.

In order to select the different interfaces presented in this chapter, I used my survey of robots published on the Internet as a guide for understanding the most important interfaces in hobbyist (and professional) robots. Some devices and implementations may seem surprising, but they are designed to fit into the mechalogic or elelogic spectrums of robot programming. I have based the peripheral interfaces on interrupts as much as possible to eliminate the need for the mainline code to spend a significant number of cycles to support peripheral functions and make high-level decisions. To avoid potential conflicts between interfaces, the mechalogic and elelogic interfaces presented in this chapter are designed to be integrated into a timer interrupt handler without affecting other mechalogic and elelogic interfaces.

The circuits presented here can be used as a cookbook of circuits that you can use in your robot applications.

I know that the mechalogic and elelogic interfaces that I present here are not appropriate for all robot applications. When creating your own applications, you should follow the guidelines I have used for creating these interfaces to make sure there is an abstraction layer between the biologic code and mechalogic/elelogic interfaces.

The application circuits presented here use parts that are easily found and, unless otherwise noted, no surplus parts were chosen. I realize that some parts are more easily found in different parts of the world, but you should be able to find substitutions for all of the parts presented in this chapter.

Hardware Interface Sequencing

When I look at many different robot designs, I find that the designers can create controller software that works with a single input and output very easily. When the interface requirements become more complex and more I/O devices are added to the robot, the code begins to become unwieldy and unexpected failures start occurring. Not only do the new functions not work properly (if at all), but the developer also begins to have problems with the functions that originally worked.

The root cause of this problem is a lack of planning when the software for the robot was being built. These kinds of problems can be avoided with proper planning and by designing the software so that all the necessary interfaces are built in right from the start. Unfortunately, this is not possible in the real world where most robots are constantly redesigned, new interfaces become available after the start of the construction of the robot, or parts used for one robot are not available when somebody wants to reproduce the original. Although I always emphasize the importance of planning the robot project with software being taken into account right from the start, it is necessary to come up with a method to change and add hardware interfaces without affecting the existing code.

The first part of understanding different hardware interfaces is to recognize that there are two types of interfaces for mechanical/magnetic devices (mechalogic) and other electronic devices (elelogic), and their characteristics are different. By separating these interfaces from the high-level robot operating code (biologic), hardware changes can be implemented by keeping the same function interfaces. This enables you to move code easily between robots.

To ensure that the mechalogic and elelogic interface code is added or removed from the robot controller with minimum impact to the software left in the controller, you must create an operating framework for it. Ideally, I would write the code as individual tasks in a *real-time operating system* (RTOS), but this is not possible in many different microcontrollers, including the PIC16F627. By making each interface a separate task, the interfaces can be completely compartmentalized and moved in and out without affecting any other software.

When designing the robot using a controller like the PIC16F627, you should design a method of sequencing the operation of the different mechalogic and elelogic interfaces by using the interrupt subsystem built into the processor. By assigning each interface to a different interrupt function or portion of an interrupt handler, the software for the different interfaces can execute independently without impacting the operation of the different interfaces. After going through the PIC16F627, you should be aware of the different hardware functions that can initiate an interrupt request to the processor. There are not that many of them and the ones that are available are not well suited to the requirements of a robot.

To work within the limitations of the PIC16F627 and enable the different interfaces to execute independently, the TMR0 interrupt of the PICmicro MCU can be used to sequence the different operations. I use a 1-millisecond (actually 1,024 µsecs) nominal interrupt interval. This is fast enough for handling most of the different interfaces used with robots and virtually every microcontroller available can produce it. Although this interval is not fast enough for some pieces of peripheral hardware, I have developed some methods for providing a faster sequencing interrupt while still having the 1-millisecond interrupt for standard functions.

After the microcontroller boots up, I immediately pass the clock oscillator to the timer with a prescaler, resulting in the timer overflowing once every 1,024 clock cycles (or 1,024 µsecs when a 4 MHz clock is used). Next, the hardware bits of the interrupt control register are set, which enables interrupt requests to be passed to the processor each time the timer overflows and rolls over from 0x0FF to 0x000.

The 1,024 µsec interrupt delay was chosen because it provides fast operation for sequencing the different hardware interfaces. A 1-millisecond delay may not seem that fast, but it is much faster than a human can perceive and allows the operation of most mechalogic interfaces without requiring additional peripheral hardware. This interrupt rate helps the biologic code work effectively in real time without the presence of a noticeable delay between the input and output response.

As the timer interrupt vector executes, the requesting hardware is reset and a counter is incremented. Once this is done, the mechalogic and biologic code can execute. Looking at it from the high level, the interrupt response code is:

```
void interrupt timer_int(void)          //  1msec Timer Interrupt
Handler
{

    if (TimerOverflow) {

            TimerInterrupt = Reset;  //  Reset Interrupt Flag

            RTC++;                   //  Increment the Clock

//  Put additional interface code for 1 msec interrupt here

        }  //  endif

//  Put Different Interrupt Handlers here

}  //  End Interrupt Handler
```

When describing and listing the interrupt handler, I have tried to be as device independent as possible to show how the 1-millisecond timer interrupt code is implemented. This is so you don't just think in terms of a PICmicro MCU. The code presented here can be transferred to other controllers with very few changes.

You could use the previous code and provide timings for different interfaces by using code like the following:

```
Final = RTC + 6;          //  Get the RTC value in 5+ msecs
while(Final != RTC);      //  Poll RTC until 5+ msecs has passed
```

This causes a delay of at least 5 milliseconds. By creating this delay, I've added 1 to the number of milliseconds I expect for the delay so that at least 5 milliseconds pass before the while loop is left, ensuring that the interrupt handler executes at least 6 times. The actual delay will be anywhere from 5 to 6 milliseconds in length, depending on when the "Final = RTC + 6;" statement is executed.

Even though the interrupt handler can be ignored completely for the mechalogic and elelogic code, by adding interface software within the interrupt handler (at the lines marked by the comments), you can create sophisticated interfaces that will not impact the biologic code at all. Each invocation of the interrupt handler can be used to execute part of the interface. Later in the chapter, I will show how a 32 Hz *pulse width modulation* (PWM) (30-millisecond period) is implemented using each loop of the 1-millisecond interrup as a step of the output. I will also show how you can create an *infrared* (IR) remote control by adding a separate interrupt handler that polls the timer value. Despite using the same resources, these two interfaces are completely separate and run independently.

I tend to base the design of a successful peripheral interface on the concept of a state machine. Once every millisecond, a branch table or switch statement is used to select the part of the interface code that will be executed. After the interface code is executed, the current input is polled and combined with the current state to determine the next step to execute.

Other interrupt sources (such as a RB0/INT pin, an interrupt on port input change, or an *electrically erasable programmable read-only memory* [EEPROM] write in the PIC16F627) can be added to the interrupt handler code presented previously quite easily after the line marked "// Put Different Interrupt Handlers here." This is why I explicitly check for the timer interrupt—when checking for other interrupt request sources, I will poll their respective flags and execute the code within an if/endif block like I did for the timer interrupt source. Using other interrupt sources does not preclude you from using the timer or its interrupt; in fact, to demonstrate how an IR remote control can be read by the robot software, another interrupt source along with the timer value is used to calculate the delays between data send events.

It should be pointed out that all interrupt operations should execute in as few cycles as possible. This is to provide similar timing in all devices that require the

1-millisecond clock and minimize latency in devices that request other interrupt-request-type responses. Overly long interrupt handlers could result in interrupt requests being ignored or the biologic code being starved for cycles. *Mainline starvation* means that the interrupt handler is executing constantly and few processor cycles are left over for the mainline code.

Robot C Programming Template

Before I started developing the sample interfaces that are presented in the following sections and the next chapter, I created the following template file to use as a basis for robot applications. This template is designed to enable the addition and change of elelogic and mechalogic interface functions with minimal impact on the other interface functions executing in the robot:

```
#include <pic.h>

//   Template for basic PICC Lite Program for Robots
//
//   Setup TMR0 to interrupt Mainline once every 1,024 μsecs
//
//   02.03.28 - Updated to allow PIC16F627/PIC16F84 PICmicro MCUs
//   02.01.23 - Originally created by myke predko
//
//
//   Hardware Notes:
//   PIC16F84/PIC16F627 running at 4 MHz
//   PIC16F627 uses internal 4 MHz oscillator
//   External _MCLR connection required
//


//   Configuration Fuses
#if defined (_16F84)
#warning PIC16F84 selected
      _CONFIG(0x03FF1);               //   PIC16F84 Configuration Fuses:
                                      //   - XT Oscillator
                                      //   - 70 msecs Power Up Timer On
                                      //   - Watchdog Timer Off
                                      //   - Code Protection Off
#elif defined(_16F627)
#warning PIC16F627 with External oscillator selected
      CONFIG(0x03F61);               //   PIC116F627 Configuration Fuses:
                                      //   - External Oscillator
                                      //   - RA6/RA7 Digital I/O
                                      //   - External Reset
```

```
                                        //    - 70 msecs Power Up Timer
                                        //    - Watchdog Timer Off
                                   //    - Code Protection Off
                                   //    - BODEN Enabled
#else
#error Unsupported PICmicro MCU selected
#endif


//  Global Variables
volatile unsigned int RTC = 0;         //  Real Time Clock Counter


//  Interrupt Handler
void interrupt tmr0_int(void)          //  TMR0 Interrupt Handler
{

     if (T0IF) {

          T0IF = 0;                    //  Reset Interrupt Flag

          RTC++;                       //  Increment the Clock

//  Put mechalogic/elelogic interface code for 1 msec interrupt here

     }  //  endif

//  Put interrupt handlers for other mechalogic/elelogic interface code

}  //  End Interrupt Handler


//  Mainline
void main(void)                        //  Template Mainline
{

     TMR0 = 0;                         //  Reset the Timer for Start
     OPTION = 0x0D1;                   //  Assign Prescaler to TMR0
                                       //   Prescaler is /4
     T0IE = 1;                         //  Enable Timer Interrupts
     GIE = 1;                          //  Enable Interrupts

//  Put hardware interface initialization code here

     while (1 == 1) {                  //  Loop forever

//  Put robot biologic code here

     }  //  endwhile

}  //  End of Mainline
```

This file can be found in the code template directory and can be loaded onto your PC's hard file from the CD-ROM.

This file will be explained line by line to help you feel more comfortable with adding your own code to this file and show how a PICC Lite application is created.

The first line, #include <pic.h>, loads in the information that is specific to the PICmicro MCU chosen for the application as well as the common library function prototypes. This file loads the appropriate hardware register and bit definitions for the different PICmicro MCUs.

PICC Lite is limited to the PIC16C84, PIC16F84, PIC16F84A, and PIC16F627; any other part numbers that are selected will cause the application to fail. I'm pointing this out because if you try to compile the application for another PICmicro MCU part number, the compile operation will fail without any kind of error message.

Next, I have put the application, author, and hardware information comment blocks in the file. These statements are not required, but they help you understand what the application does and any special requirements it might have. As I work through the different applications, you will see that I put notes to myself regarding the hardware that must be used with this code.

After the function definitions and application comments are taken care of, check the PICmicro MCU part number using the #if define(. . .) statement. The #if is known as a *directive* and is executed during compilation time—it does not execute during run time (the standard if executes at run time). The #if directive is used to check to see if the PIC16F84 has been selected, and if it has, I load the configuration fuses with a standard value that I use for most of my applications.

Notice that I have also put in the #warning directive, which prints a message on the listing file and compile status window, indicating which PICmicro MCU was selected during the application build. This warning message was put in to provide notification of which part number the application was built for. Most of the example code presented in this book will run under both the PIC16F84 and PIC16F627 without modification. It isn't unusual to have the incorrect PICmicro MCU part number selected in MPLAB, especially if this is a new project or is derived from an earlier one. The warning message helps you figure out what's wrong when you build the application for a PICmicro MCU part number that is different from the one you actually use.

If neither a PIC16F627 nor a PIC16F84 is used (the PIC16C84 is the only remaining possibility), an error is forced and the application stops. If you want to use the PIC16F84 (or another part number), then you can change this code to accept whatever device you want to use.

RTC is an unsigned 16-bit counter that can be used to time application code or indicate how long the application is running. The variable is incremented every time the 1-millisecond interrupt handler executes. It can be polled to wait for a specific delay or its value can be used to help sequence different operations.

RTC is declared as a *global* variable—it is declared outside any functions and is accessible by any function. It should only be updated by the 1-millisecond interrupt handler, although it can be read anywhere within the application. The qualifier volatile was used to indicate that its value must always be retrieved. The compiler cannot optimize the code by using a previously saved version of the value. By defining RTC as unsigned, it can be in the range of 0 to 65,535 (0x0FFFF) instead of the range of –32,768 to 32,767 of a standard signed integer.

The interrupt handler function software is located outside of any other function. In the actual PICmicro MCU, this code is placed at address 0x00004 (the mainline code and functions are placed at the end of program memory).

All interrupt requests execute at tmr0_int, and it is up to the code within the interrupt handler function to determine the source of the interrupt. In the template file, notice that I check to see if the T0IF (TMR0 overflow interrupt request flag) bit is set, and if it is, I execute code that resets the T0IF flag and increments RTC. If there is other code that should execute once every millisecond, it should be placed after the RTC increment (RTC++;) statement as indicated by the comment.

If other interrupt sources will be used in your application, they should be placed after the end of the T0IF interrupt response. They should use the same format as the T0IF interrupt response and the interrupt request flag should be checked and the requesting hardware should be reset before the interrupt response executes.

The template file's mainline is simple. It just initializes the prescaler and interrupt controller to request an interrupt once every 1,024 instruction cycles. Before enabling interrupts, notice that I have reset the timer to zero to make sure that 1,024 instruction cycles pass before the first interrupt executes. TMR0 powers up with an indeterminate value, and once it is allowed to start executing, it can roll over from 0x0FF to 0x000 before the interrupt controller is initialized. After the 1-millisecond TMR0 interrupt request has been set up, other application-specific hardware initializations can be placed in the application code.

After the hardware initializations are taken care of, the code starts an endless loop (the "while(1 == 1) {" statement). The reason for the endless loop is to prevent the mainline from ending and starting all over again. The robot application's biologic code is executed from within this loop.

PICC Lite places the mainline code at the end of the PICmicro MCU's program memory, and when the application mainline finishes, it has executed the last instruction in program memory. When a PICmicro reaches the last instruction in program memory, the program counter rolls over to zero and continues executing. In practical terms, this means that the application will start executing over again. By executing the biologic code in an endless loop, the application never ends and executes through the initialization code again.

Adding elelogic, mechalogic, and biologic code within this framework is quite simple. As shown in the following sections, mechalogic and elelogic interface code can be created that takes advantage of this framework, but does not affect other interface code and can be moved to other robot applications easily. This was the goal of the framework template. I'm sure you will be surprised at how it helps facilitate the creation of complex robot interface functions.

Prototyping with the PICmicro Microcontroller

I chose the Microchip PICmicro PIC16F627 microcontroller to demonstrate the robot software functions for a variety of reasons, but when it comes right down to it, the PIC16F627 is an easy device to work with. By using the PICC Lite compiler and MPLAB IDE, software development is done using tools that are very modern and provide the same capabilities as high-end development systems. Programming the PIC16F627 is simple and fast because the chip requires very little in terms of hardware support.

Before you start working through the sample experiments shown in this chapter, you should have a work area. This can be a desktop with enough space for your PC (including a display, keyboard, and mouse), the programmer circuit you want to use, a static-safe assembly area, and a *digital multimeter* (DMM). If at all possible, avoid putting the PC and programmer circuit on a different bench than the prototyping area as it makes code modifications/reworking somewhat more difficult than if you had the circuitry right in front of you.

The following tools and parts are required to create the experiments in this chapter:

- PC
- PIC16F627 programmer
- DMM
- +5-volt application power supply
- Breadboard
- 22-gauge solid core wire or breadboard interconnect wire kit
- Wire clippers
- Wire strippers
- Needle nose pliers
- Small flat-bladed screwdriver
- Two or more PIC16F627s

The PC must have the following features:

- Intel Pentium II or equivalent processor running at 200 MHz or more
- 64MB or more of free memory
- SuperVGA display capable of displaying at least 1,024-by-768 pixels and at least 2MB of video *random access memory* (RAM)
- CD-ROM reader
- Mouse
- Serial port

- Parallel port

- Microsoft Windows 95/98 /NT/2000/XP installed on the hard drive

- Microsoft Internet Explorer or Netscape Navigator web browsers

- Adobe Acrobat file reader

- Internet connection

A PC with these features is not very expensive (actually, as I write this, a new one with these features can be purchased for less than $500). Once you have the PC, you should install the MPLAB and PICC Lite software and application code from the CD-ROM that comes with the book using the instructions that can be found on the HTML files that come up when the CD-ROM is put into the CD-ROM drive.

Along with the PC, you should have a PICmicro programmer. The El Cheapo programmer circuit and software provided on the CD-ROM can be used to program the PIC16F627. There are a large number of programmer circuits, both home built and commercial, that can program this PICmicro MCU part number. When choosing from different programmer circuit designs, it is recommended that you look for PIC16C84/PIC16F84/PIC16F84A programmer circuits because these part numbers have been the introductory PICmicro MCU for new developers and many home-built programmer circuits available on the Internet were designed for it. The PIC16F627 should work with these circuits without requiring modification.

If you want to experiment with other PICmicro MCU part numbers, I recommend that you buy a Microchip PICSTART Plus programmer circuit. This is a development programmer circuit that can program all the different PICmicro MCU part numbers. The software for the PICSTART Plus programmer circuit is continually updated as new PICmicro microcontroller devices become available.

A breadboard is a simple prototyping technique in which the components/ wires of a circuit are pressed into holes of a plastic board. Multiple pins in the breadboard are wired together, enabling multiple parts and wires to be interconnected. As shown in Figure 4-1, the holes are interconnected—as a rule of thumb, the holes in the center of the breadboard are wired so that a 0.300-inch *dual inline package* (DIP) chip package can be wired efficiently into the breadboard.

Before buying a breadboard and wiring kit, you should look around at different sources to see where you can find them for the least amount of money. Chances are you will find that they are unreasonably expensive at many large electronics retailers. Small breadboards (2-by-3 inches), like the ones that the sample circuits are put on, can be bought for as cheaply as $2 or as expensively as $10. A wiring kit should not cost more than $5—if you have trouble finding an inexpensive wiring kit, you might consider buying a roll of 22-gauge solid core wire and a pair of wire strippers.

If you have experience with high-speed circuits, you know that the breadboard does not work well with high-speed (that is, greater than 10 MHz) signals. All of the circuits that I present in this book run at speeds that are considerably lower

Figure 4-1
The interior
connections of
a breadboard

Interior Connections   Exterior

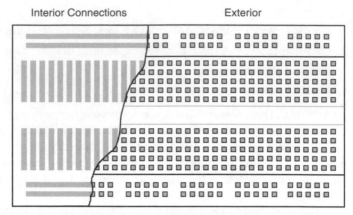

than this, so the parasitic capacitances inherent in the breadboard do not adversely affect the operation of the circuit wired into the breadboard. The breadboard prototyping system is excellent for the sample interface projects shown in this chapter. Wiring diagrams using breadboards will be given everywhere they can be used.

Breadboards are not, however, recommended for use in the final robot. The vibration of the robot moving causes wires (and components) to come free, and the breadboard itself is a magnet for dust and lint that is found on the floor. When you are ready to put the robot together, you should probably consider wiring the circuitry using a wire wrap or some method of soldering to ensure the reliability of the connections.

The flat-bladed screwdriver should be used for prying chips up from the breadboard. A *zero insertion force* (ZIF) socket can be used and eliminates the need for the screwdriver; however, it can be quite expensive and is difficult to plug in properly to a breadboard. Except for adding wires to momentary on switches, RS-232 connectors, and the ultrasonic ranger, a soldering iron is not required.

The screwdriver can also be used to short the _MCLR pin (pin 4) to the PICmicro MCU's Vss (Gnd) pin (pin 5) to reset the PICmicro MCU without having to turn off the breadboard's power supply. To allow this simple reset of the PICmicro MCU, you should wire in a 10K resistor pull-up on the _MCLR line instead of tying it directly to Vcc.

I specified two or more PIC16F627s because in some cases you may want to have two devices programmed with (slightly) different code to compare their operation in a circuit. Not having to reprogram them helps you be more efficient. Along with having extra parts for programming, chances are you are going to damage a PICmicro MCU. Keeping a spare one on hand enables you to keep working without having to wait a day or two for a replacement part.

The sample interfaces are designed to be built and tested using nothing more than a DMM. You may want to get a logic probe, an oscilloscope, or some of the tools listed in the following section to better understand what is happening in the circuit and aid you in debugging your applications:

- Logic probe
- Oscilloscope
- Logic analyzer
- PICmicro emulator

For most of the sample interfaces in this chapter, you just need a 5-volt power supply, a breadboard, and a collection of wires that can be used to wire parts together. I have designed the sample interface circuits to be as simple as possible to wire onto the breadboard. The sample interface power requirements are as modest as possible, and, except for the ultrasonic ranging and the sample robot bases with motor drivers, none of the circuits require more than 100 mA of current. This means that either a 9-volt battery driving a 78L05 or four *nickel metal hydride / nickel cadmium* (NiCad/NiMH) batteries can be used to power the sample circuits.

Always remember to positively turn off all power to the circuit so that you do not wear down the batteries or cause electrical overloads in the PICmicro MCU when you pull it out or put it back into the circuit. If you are using batteries to power your sample interfaces, you could pull the batteries out of the clips, but I recommend adding a switch to the positive wire coming from the battery and going into the interface circuit.

Interconnect wire can be purchased from kits of prestripped wire or made from a roll of 22-gauge solid core wire. I have included suggested wiring diagrams that can be used for wiring the applications together. I suggest that you follow these diagrams as much as possible and work at keeping the wiring as neat as possible (that is, use the shortest possible wire for a particular connection) to minimize the opportunity for incorrect wiring and make it easier to find problems.

Figure 4-2 shows the basic circuit that is used for the demonstration interface circuits presented in this chapter. Figure 4-2 indicates that either a PIC16F627 or PIC16F84 PICmicro microcontroller can be used in the circuit. This is true for many of the applications, but not for all of them. If the PIC16F84 cannot be used in an application because the application requires built-in hardware that is specific to the PIC16F627, it is noted.

The primary reason why I have identified the PIC16F84 as a possible PICmicro MCU to use in the sample applications is because many people have already bought PIC16F84s and used them to learn about the PICmicro MCUs. By designing the applications for both the PIC16F627 and PIC16F84 microcontrollers, developers who already have some parts should be as satisfied as well as these people that cannot find PIC16F627s due to part shortages. To build all the circuits shown in this book, you must have some PIC16F627s.

When creating a breadboard circuit, use the basic layout shown in Figure 4-3.

Any type of 0.1 µF capacitor can be used for decoupling the PICmicro MCU. If at all possible, you should use a tantalum capacitor rated at 16 volts or more;

Figure 4-2
PICmicro
microcontroller
core circuit

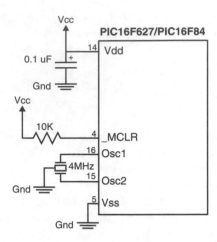

Figure 4-3
PICmicro MCU
breadboard
core circuit

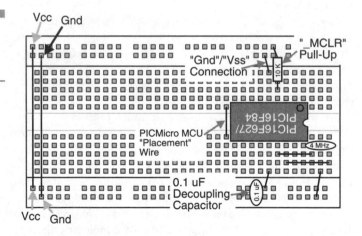

however, as I write this, there is a worldwide shortage of tantalum capacitors. If you look around, you should find other technology 0.1 µF capacitors that will work for this application.

I always use a breadboard with two common rails on the outside of the board that Vcc and Gnd are connected to. Note that I tie the two sets of common rails together at one end of the breadboard. When you are wiring your own applications using a breadboard that is set up like this one, make sure that you pick the most efficient method of connecting to the components for power and make sure there are no opportunities for ground loops.

Lastly, Figure 4-3 includes a PICmicro MCU placement wire that is used to locate the PICmicro microcontroller in the circuit. This wire enables you to easily plug the PICmicro MCU back into where it is supposed to be in the breadboard circuit after you have pulled the part out for reprogramming.

Intercomputer Communications

There are several reasons why robot controllers have to communicate with other processors, including the following:

- Debug software.
- Display/monitor robot operation.
- Interface to peripherals that are controlled by other microcontrollers.

The programming interface may be built into the circuit built for the robot or the microcontroller may be physically removed for programming. Many commercial robot controller boards have (battery-backed) RAM, which is loaded with the application code via RS-232, along with a bootloader/monitor program, which enables you to download the application code. Despite being loaded differently into the robot, software written for a bootloader should follow the same rules that I use in this book.

If you are going to work with the PIC16F84 or PIC16F627 for your own robots, there is no reason why you can't build the robot controller into the robot. If a programmer circuit like the El Cheapo was built into the robot circuitry, you would have to make sure that the _MCLR, RB6, and RB7 pins, which are used for programming, can be disconnected from the robot circuit during programming. Consider placing a DIP switch on your *printed circuit board* (PCB). This isolates these pins from the rest of the board during programming. It is also recommended that you power the programmer circuit functions externally. When programmer power is removed, the 74LS05 and pull-up resistors do not have any voltage applied to them so the programmer circuit does not affect the operation of the robot circuit.

Computer communications for software debugging normally take place using the bootloader/monitor program described earlier or an *in-circuit emulator* (ICE). Normally, computer communications for software debugging use RS-232 because it is a common standard and the relatively high voltage swings along for good noise immunity when the motors are running. Unfortunately, it is awkward to connect a mobile robot to a physical RS-232 link—if you want to use this type of function, you may have to look at wireless (*radio frequency* [RF]) links, which provide an RS-232 interface between the robot and PC controller.

I would discourage using standard intercomputer communications for displaying or monitoring the current status of the robot. I have seen laptops built into the top of large robots so the laptop's display will output the current status of the robot's execution; however, I would tend to discourage displaying the operating data using this method. With the use of a few carefully thought-out and placed *light-emitting diodes* (LEDs), you can often display the current status of the robot and even place a few others to indicate the current state of the biologic

code as well as the current outputs. A laptop's *liquid crystal display* (LCD) will be turned away from you and unreadable at least 50 percent of the time.

Note that monitoring/displaying the current operation of the robot is different from performing a software debug as the operation of the robot cannot be stopped and the manual control of the robot cannot be invoked as is required during a software debug. Monitoring the operation of the robot is a completely passive operation and should not affect the operation of the robot at all.

Lastly, computer communications on a robot can be an interface to different peripherals or controllers built into the circuit. For this type of communication, *Inter-Intercomputer Communications* (I2C) (a synchronous serial protocol) or RS-485 (which has excellent electrical noise immunity) are used. These electrical protocols enable the use of standard data transfer protocols.

When different computers are communicating within the robot, it is important to make sure that there are no communications while a peripheral operation is executing. This is done in order to minimize the possibility that the state of the peripheral will change during the operation and the current information available to the primary controller is incorrect. It also minimizes the opportunity for data to be lost and ignored if the peripheral controller can only perform one task at a time—communicate or operate.

I have a few general recommendations regarding the implementation of computer interconnections in robots. These recommendations are the result of many hours of trying to debug my own robot applications. If you follow these recommendations, your robot's controller will communicate effectively with other devices both within and without the robot:

- Speed isn't critical for most applications. Most microcontrollers (and PCs) are capable of communications of 115.2 Kbps for RS-232 or faster for Ethernet and other network protocols. Just because you can transmit at these speeds doesn't necessarily mean that you should. Lower data rates are generally less susceptible to noise within the robot, require less current, and produce fewer emissions that can cause upsets in an already noisy robot environment.

- Communications within the robot must have priority. As I indicated previously, when the controller(s) within the robot is communicating with other devices, all other operations should stop to make sure that no data is lost or ignored.

- The most important requirement to have when designing the intercomputer communications is that the method must be reliable. This means you must choose electrical protocols and wiring that is noise tolerant. All connections should be soldered or screwed down with strain relief wherever appropriate (my rule of thumb is to have no wires longer than 1 inch without some kind of tie down). Holding down the wiring minimizes the possibility that it will get caught in any of the moving parts of the robot or anything that it

happens to pass by. Unreliable communications just adds a variable when trying to debug problems and fix them.

RS-232

Despite my disparaging comments about RS-232 in the previous section, I must point out that it is a very robust and popular electrical standard and virtually all microcontrollers have the built-in capability of sending *nonreturn to zero* (NRZ) data. Along with available PC software and operating system *application programming interfaces* (APIs), RS-232 communications between robot hardware and other devices are a very attractive method of communicating within the robot and to external devices.

The most common form of asynchronous serial data packets is *8-N-1*, which means 8 data bits, no parity, and 1 stop bit. This reflects modern computers' capability to handle the maximum amount of data with the minimum amount of overhead and still have a very high degree of confidence that the data will be correct.

The RS-232 communications model is shown in Figure 4-4.

In RS-232, different types of equipment are wired according to the functions they perform.

Data Terminal Equipment (DTE) refers to the connector used for computers (the PC uses this type of connection). *Data Communications Equipment* (DCE) refers to modems and terminals that transfer the data.

As mentioned previously, when RS-232 was first developed into a standard, computers and the electronics that drove them were still very primitive and unreliable. Because of that, we have a few legacies to deal with.

The first is the voltage levels of the data. A *mark* (1) is actually –12 volts and a *space* (0) is +12 volts. This is shown in Figure 4-5.

In Figure 4-5, the hardware interface is not simply a TTL or *complementary metal-oxide semiconductor* (CMOS) buffer. Later in this section, I will introduce

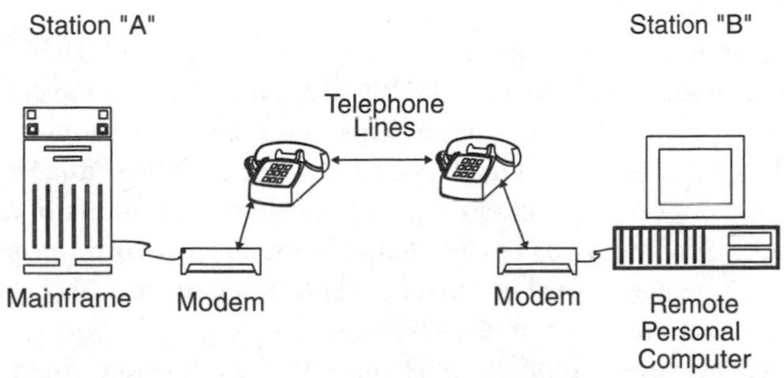

Figure 4-4
Two-computer communication via modem

Station "A"   Station "B"

Mainframe   Modem   Telephone Lines   Modem   Remote Personal Computer

Figure 4-5
RS-232 voltage
levels

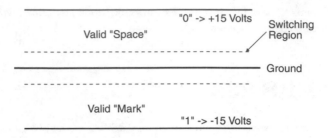

you to some methods of generating and detecting these interface voltages. Voltages in the *switching region* (+/–3 volts) may or may not be read as a 0 or 1 depending on the device. You should always make sure that the signal voltages you are using are in the valid regions. Many computers handle signals that are essentially standard 5-volt TTL/CMOS, but many do not accept these voltage levels and do not respond at all to incoming data.

Either a male 25-pin or male 9-pin D-Shell connector is available on the back of the PC for each serial port. These connectors are shown in Figure 4-6.

When you develop your own RS-232 interfaces, you should not implement any of the handshaking lines that are part of the specification. In most applications, the handshaking lines are almost never used in RS-232 (and not just hobbyist) communications. The handshaking protocols were added to the RS-232 standard when computers were very slow and unreliable. In this environment, data transmission had to be stopped periodically to allow the receiving equipment to catch up.

Today, this is much less of a concern and normally three-wire RS-232 connections are implemented, as shown in Figure 4-7. To make sure these connections work under all circumstances (including those where the device connected to the robot requires the use of handshaking lines), I normally short the *Data Terminal Ready/Data Set Ready* (DTR/DSR) and *Request to Send/Clear to Send* (RTS/CTS) lines together at my hardware. The *Data Carrier Detected* (DCD) and Ring Indicator (RI) lines are left unconnected.

Once the handshaking lines are shorted together, data can be sent and received without having to develop software to handle the different handshaking protocols.

There is a common ground connection between the DCE and DTE devices. This connection is critical for the RS-232-level converters to determine the actual incoming voltages. The ground pin should never be connected to a chassis or shield ground (to avoid large current flows or be shifted and prevent an accurate reading of incoming voltage signals). Incorrect grounding of an application can cause the computer or device that it is interfacing to reset or cause the power supplies to blow a fuse or burn out. The latter consequences are unlikely, but I have seen it happen in a few cases.

To avoid these problems, make sure that chassis and signal grounds are separate or connected by a high-value (hundreds of kilo-ohm) resistor.

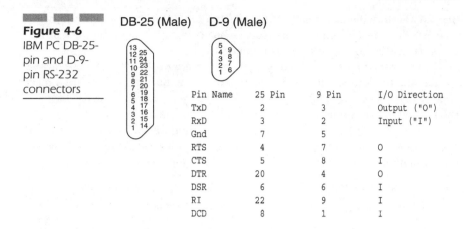

Figure 4-6
IBM PC DB-25-pin and D-9-pin RS-232 connectors

DB-25 (Male)   D-9 (Male)

Pin Name	25 Pin	9 Pin	I/O Direction
TxD	2	3	Output ("O")
RxD	3	2	Input ("I")
Gnd	7	5	
RTS	4	7	O
CTS	5	8	I
DTR	20	4	O
DSR	6	6	I
RI	22	9	I
DCD	8	1	I

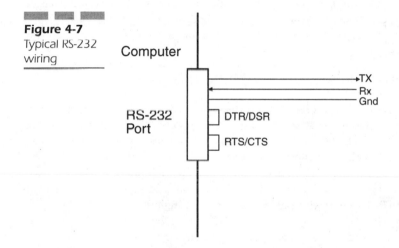

Figure 4-7
Typical RS-232 wiring

The final aspect of the RS-232 to discuss is the speeds in which data is transferred. When you first see the speeds (such as 300, 2,400, and 9,600 bps), they seem rather arbitrary. The original serial data speeds were chosen for teletypes because they gave the mechanical device enough time to print the current character and reset before the next one came in. Over time, these speeds have become standards, and as faster devices have become available, they've been doubled (that is, 9,600 bps is 300 bps doubled five times).

To produce these data rates, *universal asynchronous receiver/transmitters* (UARTs) use a clock divider to produce a clock at 16 times the data rate. If you do not have a device with a built-in UART, you can still communicate via RS-232 to another device by coming up with a *bit-banging* interface (or use a compiler like PICC Lite, which has a software UART built into the libraries). Writing your own RS-232 interface may seem difficult, but it is made easier because of RS-232's strange relationship with the number 13.

If you invert (to get the period of a bit) the standard RS-232 data speeds and convert the units to microseconds, you will discover that the periods are almost exactly divisible by 13. This means that you can use an even MHz oscillator in the hardware to communicate over RS-232 using standard frequencies.

For example, if you had a PICmicro MCU running with a 20 MHz instruction clock and you wanted to communicate with a PC at 9,600 bps, you would determine the number of cycles to delay with the following steps:

1. Find the bit period in microseconds. For 9,600 bps, this is 104 µsecs.

2. Divide this bit period by 13 to get a multiple number. For 104 µsecs, this is 8.

Now, if the external device is running at 20 MHz (which means a 200-nanosecond cycle time), you can figure out the number of cycles as multiples of 8×13 in the number of cycles in 1 µsec. For 20 MHz, 5 cycles execute per microsecond. To get the total number of cycles for the 104 µsec bit period, you evaluate the following:

```
20 cycles/µsec x 13 x 5 µsec/bit = 1,300 cycles/bit
```

When implementing an RS-232 interface, you can make your life easier by doing a few simple things. Whenever I develop an application, I use a standard 9-pin D-Shell with the DTE interface (the one that comes out of the PC) and use standard straight-through cables. By doing this, I always know what my pinout is at the end of the cable when I'm about to hook up a PICmicro to a PC.

By always making the external device DCE and using a standard pinout, I don't have to fool around with null modems or making my own cables.

When creating the external device, I also loop back the DTR/DSR and CTS/RTS data pairs inside the external device rather than at the PC or in the cable. This way I can use a standard PC and cable without having to do any wiring on my own or make any modifications. It looks a lot more professional as well.

Here are a couple of methods that you can choose from to convert RS-232 signal levels to TTL/CMOS (and back again) when you are creating microcontroller-based RS-232 interfaces. These methods do not require ±12 volts; in fact, they just require the +5-volt supply that is available for logic power.

The first method is to use an RS-232 converter that has a built-in charge pump to create the ±12 volts required for the RS-232 signal levels. The MAXIM MAX232 is probably the most well-known chip that is used for this function (see Figure 4-8).

This chip is ideal for implementing three-wire RS-232 interfaces. Ground for the incoming signal is connected to the processor ground.

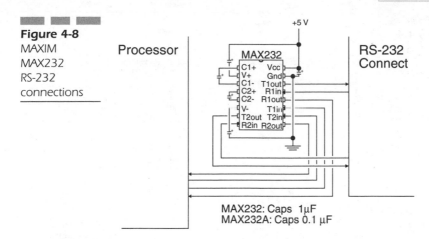

Figure 4-8
MAXIM
MAX232
RS-232
connections

Along with the MAX232, MAXIM and some other chip vendors have a number of other RS-232 charge-pump-equipped devices that enable you to handle more RS-232 lines (to include the handshaking lines). Some charge-pump devices do not require the external capacitors that the MAX232 chip requires, which simplifies the layout of the circuit (although these chips cost quite a bit more).

The other method of translating RS-232 and TTL/CMOS voltage levels is to use the transmitter's negative voltage. The circuit in Figure 4-9 shows how this can be done.

This circuit relies on the RS-232 communications only running in *half-duplex mode* (that is, only one device can transmit at a given time). When the external device wants to transmit to the PC, it sends the data as either a mark (leaving the voltage being returned to the PC as a negative value) or a space by turning on the transistor and enabling the positive voltage output to the PC's receivers. If you refer to Figure 4-5, you'll see that +5 volts is within the valid voltage range for RS-232 spaces.

This method works very well (and consumes almost no power) and is obviously a very cheap way to implement a three-wire RS-232 bidirectional interface.

Unfortunately, this circuit cannot be used unchanged with the built-in serial port for many microcontrollers because they cannot invert the data output as required by the circuit. To get around this, you could use a CMOS inverter between the serial port pins and RS-232 conversion circuit.

To get around this problem, you could use a chip called the Dallas Semiconductor DS275, which basically incorporates the previous circuit (with a built-in inverter) into the single package (see Figure 4-10).

I should point out that there are two part numbers with the same pinout in the DS275: the DS1275 and the DS275. Both work exactly the same way, but the DS275 is a later version of the part.

Figure 4-9
RS-232 to an
external device

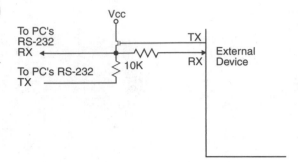

Figure 4-10
Dallas Semi
(1)275 RS-232
interface

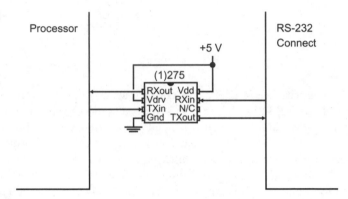

HyperTerminal RS-232 Terminal Emulator

To communicate with your robot project via RS-232, you need a *terminal emulator* that has the following basic features:

- Teletype (TTY) and *American National Standards Institute* (ANSI) terminal emulation
- Varying data rates
- 8-N-1 data format
- Monospace fonts
- User-selectable com port access
- User-selectable handshaking
- Configuration save
- Text file transfer

If you need a PC terminal emulator, I recommend using Hilgraeve Inc's HyperTerminal. This program is bundled with many copies of Microsoft Windows and provides all the basic services you need in a terminal emulator. If you do not have a copy of HyperTerminal, you should upgrade to the latest level (as I write this, 5.0 is available). You can download and install HyperTerminal for free for personal use from the Hilgraeve web site at www.hilgraeve.com.

Once HyperTerminal is installed, you must configure it. To do this, start up HyperTerminal. A dialog box should pop up, as shown in Figure 4-11.

To configure the terminal emulator, first disconnect it (the program connects itself automatically) by clicking the picture of a telephone with the receiver off it, or click Call –> Disconnect. Then select File –> Properties to display the Properties dialog box (which looks like Figure 4-12). This dialog box specifies how you would like HyperTerminal to work for you.

Click the Settings tab and make sure that Terminal Keys, ANSI, Emulator, and 500 backspace buffer are selected, as shown in Figure 4-13.

The Terminal Setup menu enables you to define the cursor settings used in the HyperTerminal dialog box.

The ASCII Setup menu enables you to tailor the data to the application. These parameters are really user and application specific. For the most part, I leave these at the default values.

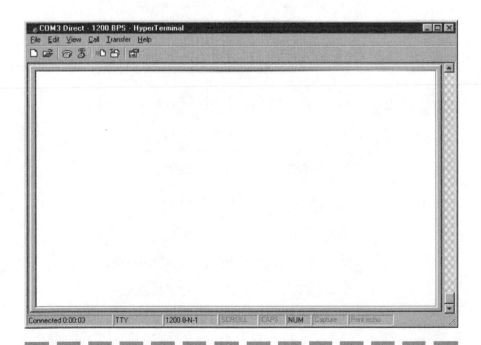

Figure 4-11   HyperTerminal boot screen

Figure 4-12
HyperTerminal
Properties
dialog box

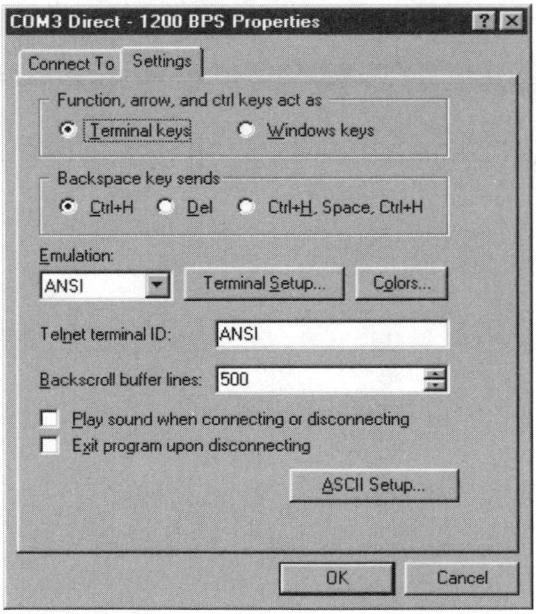

Figure 4-13
HyperTerminal
settings
selection

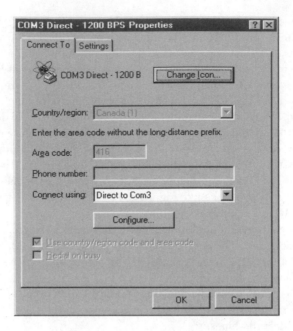

Once the settings are defined, you can click the Phone Number tab and select the appropriate connection in the Connect Using tab using the pull-down menu, as shown in Figure 4-14.

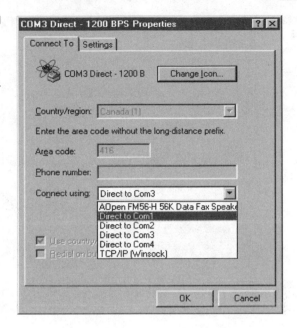

Figure 4-14
HyperTerminal
connection
options

The serial port can be any serial port within the PC that you want to connect to. The modem that comes up is, not surprisingly, your Internet access server and can be used for your terminal emulator if you will be dialing outside.

Next, click Configure and look at the parameters to set up (see Figure 4-15). For a PICmicro application that connects serially to a PC, select 8 for data bits, None for parity, 1 for stop bits, and None for flow control. The data rate (bits per second) is the value you want to use (which is usually 1,200 bps for the projects presented in this book). Selecting None for flow control indicates that a three-wire RS-232 can be used.

Don't worry about the Advanced menu; it is normally used for specifying the serial port's *first-in—first-out* (FIFO) data buffers. To be on the safe side, select No FIFO Operation.

You can now click OK for both the Com Properties and the Properties dialog boxes to save the information.

Next, click View –> Font and select a monospace font like Courier New, as shown in Figure 4-16.

After you click OK to select the font, click File –> Save As . . . so you do not lose these parameters. The default directory should be Desktop. For the filename, enter something descriptive like

```
DIRECT COM 1 - 1200 bps
```

and click Save.

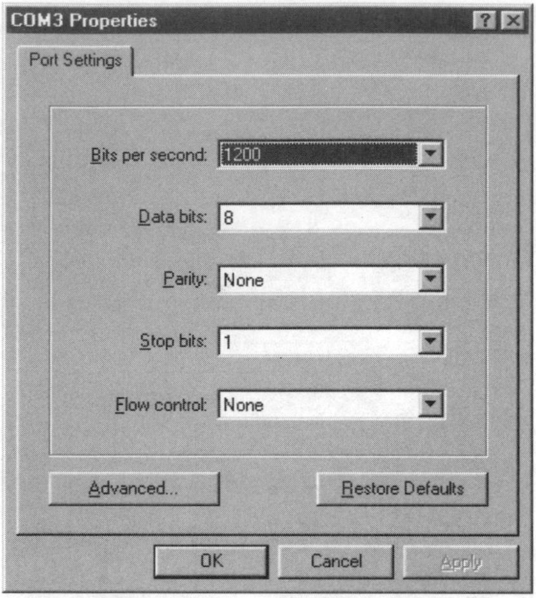

Figure 4-15
HyperTerminal
Port Settings

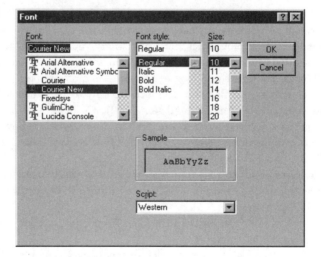

Figure 4-16
HyperTerminal
font selection

If you minimize all the active windows now, you'll see that a new tab is on your desktop that will bring up the HyperTerminal with the parameters that you just entered.

When HyperTerminal has the desktop's focus and is connected, when you press a key, the *American Standard Code for Information Exchange* (ASCII) code for it will be transmitted out of the PC's serial port. Data coming in on the serial port will be directed to HyperTerminal's display window.

Files can also be transferred; this is an excellent way to create test cases instead of repeatedly typing them in manually. Using Send File sends the file exactly as it is saved on disk. Using Send Text File is preferable because it transfers the files of ASCII data in the same format as if it were typed in, including the carriage return/line feed line-end delimiter at the end of each line.

RS-232 Interface Example Between PC and PICmicro MCU

Wiring a PIC16F6287 to a PC is very straightforward and you can use any of the level translation techniques that I presented in the previous section. For this application, I have used the built-in *universal synchronous/asynchronous receiver/transmitter* (USART) hardware of the PICmicro MCU and will not be implementing the interface using bit-banging software. This means that the PIC16F84 cannot be used for this application.

I am assuming that the PICmicro microcontroller will be connected to the PC via a straight-through male-to-female cable. This type of cable is often used to connect PCs to external serial devices.

To wire the PIC16F627's serial I/O bits to the MAXIM MAX232 serial interface, use the circuit I presented in the previous section (see Figure 4-17). It is also very straightforward to wire it into a breadboard (see Figure 4-18), but you must

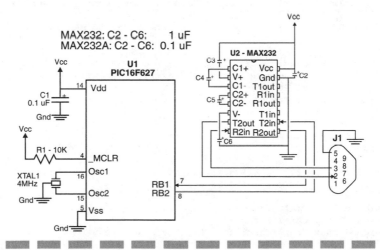

Figure 4-17   RS-232 application circuit

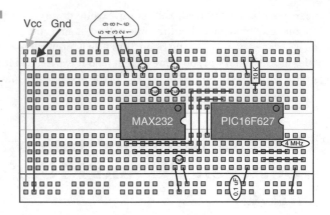

Figure 4-18
RS-232
breadboard
circuit

solder a female 9-pin D-Shell connector with 22-gauge wires on pins 2, 3, and 5 that can be connected into the breadboard. When you wire in the 22-gauge wires, it is recommended that you use different red, green, or blue insulated wires for pins 2 and 3 and a black insulated wire for pin 5. Pin 5 is the ground pin, and this will help you find it when you are wiring the connector into a breadboard.

The bill of materials for the sample interface is listed in Table 4-1.

The software performs the basic tasks used to make sure that I understand how new chips are connected to RS-232—namely, I take streams of text data in and then send it back out capitalized.

If you look at serial.c in robot\serial, you will find the following application:

```
#include <pic.h>

//   02.04.22 - Serial: Echo Back Characters sent to MCU as all caps
//    and when Carriage Return ("Enter") sent, send full string
//
//   PIC16F627's Built in USART used, connected to MAX232
//
//
//   Hardware Notes:
//   PIC16F827 Running at 4 MHz with External Oscillator
//   RB1 - Serial Receive
//   RB2 - Serial Transmit
//

//   Global Variables
unsigned int RTC = 0;              //   Real Time Clock Counter

char StringOut[27] = "\r\nHello\r\n\0              ";
//                    1 2 345678 9 0 1234567890123456

unsigned char CharOutIndex = 0;
unsigned char CharInIndex = 3;
unsigned char SingleCharFlag = 0;    //   Set to Send a Single Byte

//   Configuration Fuses
```

Table 4-1

Materials for
the RS-232
Application
Circuit

Label	Part Number	Comment
U1	PIC16F627	
U2	MAX232	
R1	10K, ¼ watt	Uses _MCLR pull-up
R2	470 ohm, ¼ watt	Uses CR1 current limiting resistor
C1	0.1 µF	Uses any type of U1 decoupling capacitor
C2-C6	1 µF/0.1 µF	Requires 0.1 µF for MAX232A and 1 µF for other versions of the MAX232
XTAL1	4 MHz ceramic resonator with internal capacitors	Required for PIC16F84
Misc.		Requires a breadboard, wiring, and +5-volt power supply

```
#if defined( 16F627)
#warning PIC16F627 with external XT oscillator selected
        CONFIG(0x03F61);        //  PIC116F627 Configuration Fuses:
                                //   - External "XT" Oscillator
                                //   - RA6/RA7 Digital I/O
                                //   - External Reset
                                //   - 70 msecs Power Up Timer On
                                //   - Watchdog Timer Off
                                //   - Code Protection Off
                                //   - BODEN Enabled
#else
#error Unsupported PICmicro MCU selected
#endif



//  Interrupt Handler
void interrupt tmr0_int(void) //  TMR0 Interrupt Handler
{

unsigned char temp;

    if (T0IF) {

            T0IF = 0;           //  Reset Interrupt Flag

            RTC++;              //  Increment the Clock

//  Put additional interface code for 1 msec interrupt here

        } //  endif
```

```
    if (TXIF) {              //   Send out String Character

        TXIF = 0;            //   Reset Interrupt

        if ((temp = StringOut[CharOutIndex]) == '\0')
             if (SingleCharFlag)       //   Just Echoing Back?
                  SingleCharFlag = 0;
             else              //   No, Start new String
                  StringOut[2] = '\"';
        else {
             TXREG = temp;       //   Output the Next Character
             CharOutIndex++;     //    and Point to Next
        }  //  endif

    }  //  endif

    if (RCIF) {                      //   Serial Character Received

        if ((temp = RCREG) < 0x020)
                      //   Control Character received?
             if (temp == 0x008)      //   BackSpace?
                  if (CharInIndex > 3) {
                       //   Yes, and Something to Backspace
                       StringOut[CharOutIndex = CharInIndex]
                       = temp;
                       //   Backspace, Put in Blank, BS
                       StringOut[CharInIndex + 1] = ' ';
                       StringOut[CharInIndex + 2] = temp;
                       StringOut[CharInIndex + 3] = '\0';
                       CharInIndex--;
                       TXIF = 1;
                       SingleCharFlag = 1;
                  } else;
             else if (temp == 0x00D) {
// "Enter", End String and Print
                       StringOut[CharInIndex++] = '\"';
                       StringOut[CharInIndex++] = '\r';
                       StringOut[CharInIndex++] = '\n';
                       StringOut[CharInIndex] = '\0';
                       CharInIndex = 3;
                       CharOutIndex = 0;
                       TXIF = 1;
             } else;
        else {              //   ASCII Char
             if ((temp >= 'a') && (temp <= 'z'))
                  temp -= ('a' - 'A');
             if (CharInIndex < 23) {        //   Echo Back
                  StringOut[CharOutIndex = CharInIndex] =
                  temp;
                  StringOut[++CharInIndex] = '\0';
                       //   Make into String
                  TXIF = 1;
                  SingleCharFlag = 1;
             }  //   endif
        }  //  endif

        RCIF = 0;            //   Reset Interrupt Request
```

```
                } // endif

}  //  End Interrupt Handler


//  Mainline
void main(void)                // Template Mainline
{

unsigned char temp;            // Temporary Storage Value

     OPTION = 0x0D1;           // Assign Prescaler to TMR0
                               // Prescaler is /4
     TMR0 = 0;                 // Reset the Timer for Start
     T0IE = 1;                 // Enable Timer Interrupts
     GIE = 1;                  // Enable Interrupts

//  Put in Interface initialization code here

     SPBRG = 51;               // 1200 bps @ 4 MHz
     TXEN = 1;                 // Enable the USART
     CREN = 1;
     SPEN = 1;

     PEIE = 1;                 // Enable PIE Interrupt Sources

     temp = RCREG;
     RCIF = 0;                 // Enable Receive Interrupt
     RCIE = 1;

     TXIF = 1;                 // Enable Transmit Interrupt
     TXIE = 1;

     while (1 == 1) {          // Loop forever

//  Put in Robot high level operation code here

     }  //  endwhile

}  //  End of Mainline
```

Looking through the code, you might not see anything close to what you were probably expecting. I said that this code took in text data and echoed it back capitalized. You were probably expecting an application that looks somewhat like the following:

```
void main(void)               //   Receive Serial Text
{                             //   Echo it back capitalized

#if defined(_16F627)
#warning PIC16F627 Selected
       CONFIG(0x03F61);
#else
#error Unsupported PICmicro MCU selected
#endif
```

```
unsigned char temp;

        SPBRG = 51;              //  1200 bps @ 4 MHz
        TXEN = 1;                //  Enable the USART
        CREN = 1;
        SPEN = 1;

        RCIF = 0;                //  Clear Received Character Pending
Flag

        while (1 == 1)           //  Loop forever
            if (RCIF) {          //  Character Waiting?
                temp = RCREG;        //  Get the Character Received
                if ((temp >= 'a') && (temp <= 'z'))
                    temp -= ('a' - 'A');    //  Capitalize
                TXREG = temp;
                RCIF = 0;    //  Clear Received Pending Flag

}  //  End of Mainline
```

This code carries out the required task, but it does not work with other interfaces in a robot system. To enable this function to operate without affecting other
interfaces in the robot, you need a method of designing the code so that it does
not affect any other interfaces in the robot system.

I have said that the basis for implementing and sequencing mechalogic and
elelogic interfaces is the 1-millisecond interrupt. I should have qualified this
statement to indicate that you should use the 1-millisecond interrupt for all
mechalogic and elelogic interfaces whenever it makes sense to use it. Most elelogic interfaces cannot take advantage of the 1-millisecond interrupt; instead,
other features of the microcontroller must be used.

In this application's case, the serial receive and transmit interrupt requests
produced by the USART are used to enable the application code to respond to
incoming serial data packets as quickly as possible. This method is somewhat
unexpected, but it can be used for other types of interfaces (I will use it for handling output data with an LCD).

If you were to implement this application using just the USART interrupt
request capability, you might create an application that looks somewhat like the
following:

```
void main(void)               //    Interrupt Receive Serial Text
{                             //    Echo it back capitalized

#if defined(_16F627)
#warning PIC16F627 Selected
        _CONFIG(0x03F61);
#else
#error Unsupported PICmicro MCU selected
#endif


void interrupt serial_int(void)
{
```

```
unsigned char temp;

    if (RCIF) {                    //  Receive Interrupt Request?

        temp = RCREG;        //  Get the Character Received
        if ((temp >= 'a') && (temp <= 'z'))
            temp -= ('a' - 'A');      //  Capitalize
        TXREG = temp;
        RCIF = 0;                //  Clear Received Pending Flag

    } //  endif

} //  End Interrupt Handler

    SPBRG = 51;                 //  1200 bps @ 4 MHz
    TXEN = 1;                   //  Enable the USART
    CREN = 1;
    SPEN = 1;

    RCIF = 0;                   //  Clear Received Character Pending
                                //  Flag

    while (1 == 1);             //  Loop forever

} //  End of Mainline
```

This code is much better—in fact, it implements the most important rule of writing mechalogic and elelogic code. The mainline cannot interface directly with the interface hardware. The problem with this code is that it passes data to the biologic code one byte at a time.

Sending output data to the PC also requires the biologic code to send data one byte at a time and requires the biologic code to poll the status of the serial port and send a byte after the previous one has been sent.

At first glance, this situation is not too bad, but you should consider a real-world situation where the serial port is used to query the command

Send Left Light Sensor Value.

and responds with

251

Writing the biologic code that performs this function becomes very complex, and, because it is waiting for individual bytes to come in or be sent, you will end up polling on data received and data sent completed flags in your application.

The best way to pass data to and from an elelogic interface is to use ASCIIZ strings. An ASCIIZ string consists of a string of ASCII characters that terminate with a NUL (0x000) character. All the characters in the string are processed by the elelogic interface one by one until the NUL character is read by the interface code.

If data is coming from a PC, it is most convenient to end the message by having the user press Enter (which sends a carriage return or 0x00D character). In this situation (the elelogic interface interrupt handler), the carriage return character is replaced with the NUL character. For an RS-232 interface, the interface handler that takes the incoming string could be as follows:

```
if (RCIF) {                   //  Serial Receive Interrupt
     temp = RCREG;
     RCIF = 0;
     if (temp < 0x020) {      //  Control Character?
         if ((temp == 0x008) && (StringInIndex != 0))
             StringInIndex--;  //  Backspace over last
         else if (temp == 0x00D) {    //  CR Character
             StringIn[StringInIndex] = '\0';
             StringInFlag = 1;
                              //  Indicate that string is available
         } else ;    //  No other control characters to
                     //  handle
     else            //  ASCII Character
         StringIn[StringInIndex++] = temp;
}  //  endif
```

This is similar to the code I used in the serial application—the difference is that I echo back the uppercase character each time one is received.

In the previous receive code, you should notice that the ability to backspace over the last character was included. This is a useful function if you are writing code for a human to interface directly to. Humans make mistakes when they type; by providing the backspace function, you are providing a rudimentary form of a user interface to your application.

In addition to converting the received code into a string of characters, you should also send data out of the microcontroller as a string, not as a single character. In the PICmicro MCU, the TXIF flag is set and an interrupt is requested when the transmitter holding register is empty. The following code should be put into the interrupt handler to send the next character in a string when the transmitter holding register is empty:

```
if (TXIF) {                //  TX Holding Register Empty
                           //  Interrupt
     TXIF = 0;
     if ((temp = Message[MessageOffset++]) != 0)
         TXREG = temp;     //  Character to Send?
     else
         TXREGEmptyFlag = 1;
}  //  endif
```

To send a message by using this string serial transmit interrupt handler, the following four statements are used:

```
Message = StringToOutput; // Point to the message to output
MessageOffset = 0;
TXREGEmptyFlag = 0;        // TX is no longer empty
```

```
TXIF = 1;                       // Request interrupt to start data
                                // send
```

The transmit interrupt handler is invoked any time TXIF is set. This can be done manually (by saving a 1 into it) or after a character has finished being sent and the character in the holding register is loaded into it.

The serial receive and transmit interrupt handlers were based on the previous functions. The actual code was complicated by converting all lowercase characters into uppercase and storing the input string in a quoted string that was used for output.

There is one good feature of this code. If you look through it, you will see that the output string is initialized to the greeting (Hello) message and the output index is initialized to zero. To display this message when the PICmicro MCU starts, all I have to do is enable the transmitter holding the register empty interrupt using the following two statements:

```
TXIE = 1;      // Enable Holding Register Empty Interrupts
TXIF = 1;      // Force Holding Register Empty Interrupt Rqst
```

If you look at the actual assembly language code, you will see that I have implemented an initial greeting to the application for the addition of two assembly language statements.

To test out the application, you should load up a HyperTerminal session with the interface running at 1,200 bps. When you have attached the PICmicro MCU and powered it up, you should be greeted with the Hello message. In Figure 4-19, I powered up the application, entered in the string ABCDE, and pressed Enter. The immediate response of the application was to echo back ABCDE and then when the Enter button was pressed, the string ABCDE was displayed on the next line.

I recommend using an ASCIIZ-string-based interface for all elelogic communications. Later in this chapter, I will demonstrate its use for LCD displays and handling data from IR remote controls as complete packets, not individual bits. By passing full strings between the biologic and elelogic code, the biologic code becomes very independent of the intercomputer interfaces.

Before going on to the next sections and their interfaces, three comments must be made regarding the practicalities of RS-232 interfaces. The first should be obvious—if a mobile robot was connected to a computer via RS-232, the cabling required for it would be unmanageable. In some robots, both power and communications cables are suspended from the ceiling. This is not an optimal solution because the wire is going to either be too short for the robot to explore its total environment or become an obstacle for the wheels and maybe get wrapped around the wheel's axles. RF modules are available that provide a wireless RS-232 interface that can be used with robots. I would recommend that you consider using these devices if you want to communicate with your robot during its operation.

For serial.c, I ran the interface at 1,200 bps; that is a very slow communications rate by current standards. I wouldn't be surprised if you had a 100 Mbps

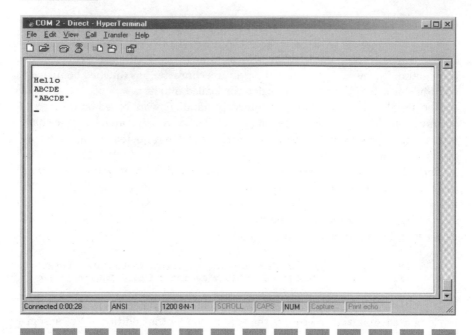

Figure 4-19 HyperTerminal communicating with a serial application

Ethernet network in your home, which is 5 orders of magnitude faster than the 1,200 bps RS-232 used in this application. Before you start looking at increasing the speed of the connection, consider what the application is going to be doing. If it is a user-controlled interface, like serial.c, then the full bandwidth of the connection will not be used if the data rate is faster than 300 bps. Slower data rates generally allow cheaper hardware interfaces, longer cable runs between devices, and more reliable connections.

Faster data rates should only be considered if large amounts of data are transferred. The only time high-speed connections are required for most robot applications is when video is being transmitted from the robot. Robot commands that censor data generally do not require faster data rates than the 1,200 bps used here.

Finally, it is recommended that you always communicate with a robot (or any device over RS-232 for that matter) using human-readable commands. I realize that a few bytes could be saved by sending single-byte commands, but in order to generate and decipher these commands, you must have to have some kind of computer interface. By keeping your commands to simple bytes, even if you want to create a *graphical user interface* (GUI) for the robot for your customers, you can debug the robot application using something like HyperTerminal while the GUI is being developed. In this case, using human-readable commands between a PC and the robot results in a faster, more concurrent development effort.

■ Bidirectional Synchronous Interfaces

RS-232 is called an *asynchronous* data transmission protocol because the receiver accepts data without any kind of indication from the sender. For *synchronous* data communications in a microcontroller, a clock signal is sent along with serial data, as shown in Figure 4-20.

The clock signal strobes the data into the receiver and the transfer can take place on the rising or falling edge of the clock.

A number of synchronous serial bus protocols are available that are quite popular and have a fair number of parts designed for them. The disadvantage that these protocols have compared to a true form of networking is that they need a device to be explicitly addressed by a separate microcontroller I/O pin. Using a networking system, many devices can be connected to a microcontroller without incurring a penalty, such as requiring one pin per device for device selection.

The most popular form of microcontroller network is I2C. This standard was originally developed by Philips in the late 1970s as a method to provide an interface between microprocessors and peripheral devices without wiring a full address, data, and control buses between devices. I2C also enables you to share network resources between processors (which is known as *multimastering*).

The I2C bus consists of two lines: the SCL line and the SDA line. The clock line (SCL) is used to strobe data (from the SDA line) from or to the master that currently has control over the bus. Both of these bus lines are pulled up (to enable multiple devices to drive them).

An I2C-controlled stereo system might be wired as shown in Figure 4-21.

Figure 4-20
Synchronous
data waveform

Figure 4-21
An example of
I2C network
wiring

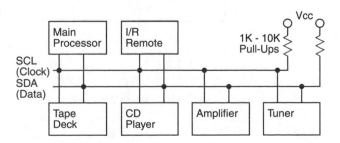

The two bus lines are used to indicate that a data transmission is about to begin and pass the data on the bus.

To begin a data transfer, a master puts a start condition on the bus. Normally, when the bus is in the idle state, both the clock and data lines are not being driven (and are pulled high). To initiate a data transfer, the master requesting the bus pulls down the SDA bus line followed by the SCL bus line. During data transmission, this is an invalid condition (because the data line is changing while the clock line is active/high).

Each bit is then transmitted to or from the slave (the device the master is using to transmit the message) with the negative clock edge being used to latch in the data, as shown in Figure 4-22. To end data transmission, the reverse is executed and the clock line is allowed to go high, which is followed by the data line.

Data is transmitted in a synchronous (clocked) fashion. The most significant bit is sent first, and after 8 bits are sent, the master permits the data line to float (it doesn't drive it low) while strobing the clock so the receiving device can pull the data line low as an acknowledgment that the data was received. After the acknowledge bit, both the clock and data lines are pulled low in preparation for the next byte to be transmitted or a stop/start condition is put on the bus. Figure 4-23 shows the data waveform.

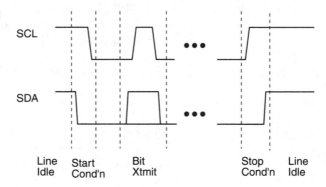

Figure 4-22
I2C signals and waveforms

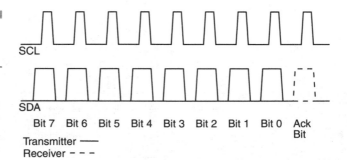

Figure 4-23
I2C data byte transmission

Sometimes the acknowledge bit is allowed to float high, even though the data transfer has completed successfully. This is done to indicate that the data transfer has completed and the receiver (which is usually a slave device or a master that is unable to initiate data transfer) can prepare for the next data request.

There are two maximum speeds for I2C (because the clock is produced by a master, there really is no minimum speed): *standard mode* runs at up to 100 Kbps and *fast mode* transfers data at up to 400 Kbps. Figure 4-24 shows the timing specifications for both the standard (100 KHz data rate) and fast (400 KHz data rate).

A command is sent from the master to the receiver in the format shown in Figure 4-25.

The receiver address is 7 bits long and is the bus address of the receiver. There is a loose standard where the 4 most significant bits are used to identify the type of device, whereas the next 3 bits are used to specify 1 of 8 devices of this type (or further specify the device type).

This is really all there is to I2C communication, except for a few points. In some devices, an initializing packet has to be resent to reset the receiving device for the next command (that is, for a serial EEPROM read, the first command sends the address to read from and the second reads the data at that address).

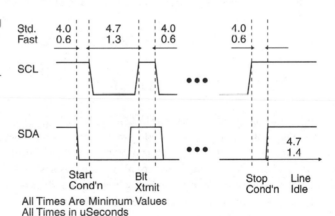

Figure 4-24
I2C signal timing

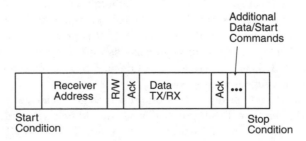

Figure 4-25
I2C data transmission

The last point to note about I2C is that it's multimastering, which is to say that multiple microcontrollers can initiate data transfers on the bus. This obviously results in possible collisions on the bus (which is when two devices attempt to drive the bus at the same time). Obviously, if one microcontroller takes the bus (sends a start condition) before another one attempts to do so, there is no problem. The problem arises when multiple devices initiate the start condition at the same time.

Actually, arbitration in this case is really quite simple. During the data transmission, hardware (or software) in both transmitters synchronizes the clock pulses so they match each other exactly. During the address transmission, if a bit that is expected to be a 1 by a master is actually a 0, then it drops off the bus because another master is on the bus. The master that drops off waits for a stop condition and then reinitiates the message. I realize that this is hard to understand with just a written description.

A bit-banging I2C interface can be implemented in PICmicro MCU software quite easily. However, due to software overhead, the fast mode probably cannot be implemented—even the standard mode's 100 Kbps will be a stretch for most devices. I find that it is best to implement I2C in software when the PICmicro MCU is the single master in a network. That way it doesn't have to be synchronized to any other devices or accept messages from any other devices that are masters and are running a hardware implementation of I2C that may be too fast for the software slave.

▓▓ Output Devices

Software simulators, such as the one in the MPLAB IDE, give you some confidence that your robot application will do exactly what you expect it to do. Unfortunately, a simulator cannot give you complete confidence or accurately predict the operation of an application. There are also some cases where you cannot accurately model the input state of the robot, which makes it impossible to properly simulate what is happening. If you have ever performed any debug of intelligent devices, you may have used an emulator connection between the controller and PC to monitor exactly what is happening. The problem with using an emulator with a mobile robot should be obvious—you might be able to walk with the robot holding a laptop connected to the emulated controller, but 99.9 percent of the time, this will be impractical. The solution to this problem is to come up with a method of displaying the input to the robot as well as its current operating state. I like to use LEDs or buzzers as output devices to give me some feedback as to what is happening with the robot.

It is important to design output into robots to get an idea of what the robot is doing at a given time. Ideally, these devices should help you understand what is happening with the robot from across the room. This is why I focus on LEDs (light) as well as speakers (for a beep). Some other methods of driving out data can be very impractical.

When adding output devices (which might be better described as *feedback devices*), care must be taken to make sure that the current status of the inputs and robot operations are obvious and do not affect the operation of the robot. LEDs of different colors that are placed around the robot (ideally close to the sensor or output being reported on) can be very useful. When doing this, it is important to remember that the output from the LEDs may be received by light receptors on the robot. You can end up in a situation where LEDs cause the problems that you are trying to debug.

Another very obvious feedback method is to use some kind of audible feedback. Personally, I would only use a varying number of beeps from a piezo electric speaker. If you are using audible output, you should be aware of a few rules when you are designing the interfaces. First, the length of time and frequency are subjective—I remember spending several days 10 years ago at a customer's location trying to decode error tones from a PC motherboard that we were trying to debug using a storage oscilloscope. I would recommend just having one frequency and not varying its length. Secondly, the number of beeps can be used to convey information, but you will find that if you send more than three, people will miscount them.

If you feel that adding LED and speaker annunciators makes your robot too toy-like, remember that they can always be disconnected. The alternative of chasing the robot with a laptop to figure out what is happening when the robot is behaving incorrectly is not really a practical approach to determining the problem.

LEDs

LEDs are the most basic output device of modern electronics. They are very inexpensive and trivially easy to control from a microcontroller. They also come in a large number of different forms, including numeric and alphanumeric displays that provide a significant amount of information. They are very useful as tools for outputting status information from the robot to the user. I generally try to put LEDs on my input sensors. They can be used to light when there is a collision or indicate what a light level is.

Electrically, LEDs have some properties that you should be aware of. First and foremost, they are diodes, which means that current in them flows in only one direction. Most modern LEDs require just 5 mAs to light, although there are some (especially high-output) LEDs that require up to 20 mA.

To connect an LED to a microcontroller, it is good practice to connect the anode to +5 volts, the cathode to a current limiting resistor, and the current limiting resistor to the I/O pin of the controller, as shown in Figure 4-26. The reason for

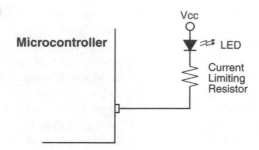

Figure 4-26
LED
connection to a
microcontroller

this configuration is to accommodate microcontrollers (such as the Intel 8051) that cannot source sufficient current to light the LED.

It is important to remember that the voltage drop across them is different from silicon diodes (because they aren't pure silicon diodes). They typically have a 2-volt drop across them, which is needed to take into account when calculating current through current-limiting resistors. I typically use a 470 ohm resistor for this chore when using them in a 5 volt digital logic system.

If you would like to change the brightness of the LED, you should use a PWM (as I demonstrate in a later example application); do not use a higher-resistance current limiting resistor. A different value current-limiting resistor cannot be changed easily, whereas the PWM control can be changed very easily.

Multisegment LEDs are available in a variety of styles. The most common is the seven-segment display used for displaying numbers in digital clocks and other electronic devices. The layout of the display is shown in Figure 4-27.

In virtually all multisegment LED displays, multiple anodes or cathodes are connected to a single pin, which is often referred to as a *commoning bar*. This simplifies the wiring of the display and the bar can connect either multiple cathodes or anodes of the LED. A display with the cathodes connected to the commoning bar is known as a *common cathode*, and, as you would expect, a display with the anodes connected together is known as a *common anode*.

In addition to making the wiring of the display simpler, another advantage of using common anode or common cathode displays is that the multiple displays can be wired in parallel with current through their commoning bar controlled by a transistor. Figure 4-28 shows a sample circuit that uses four common cathode seven-segment LEDs.

Different data is displayed on each of the displays for a very short period before the display is turned off, and the data for the next display is output from the controller and that display is enabled. To avoid perceivable flashing, each display should be turned on and off at least 100 times per second. This means that the four displays are being effectively driven by a PWM—for the sample circuit shown in Figure 4-28, the duty cycle of the PWM is just 25 percent for each display. Adding more displays decreases the duty cycle of each display in the circuit.

The 1-millisecond interrupt handler used for the robot mechalogic/elelogic sequencing is actually perfectly suited for providing control for the scanning soft-

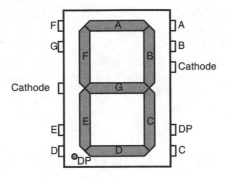

Figure 4-27
Seven-segment
common
cathode LED
display with
pinout

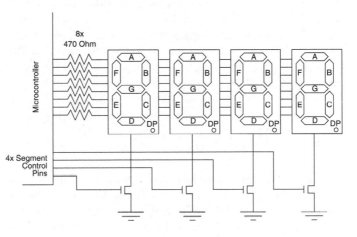

Figure 4-28
Wiring four
seven-segment
LED displays

ware. The following code example shows how a 4-display circuit can be controlled from the 1-millisecond interrupt after the timer is updated:

```
// Put additional interface code for 1 msec interrupt here

// Code for displaying 4 7-Segment LED Displays

    ControlPin[OldRTC] = 0;       // Turn OFF current LED Display

    OldRTC = RTC & 3;             // Use the Least Significant 3
                                  //   Bits of the RTC to select
                                  //   the
                                  //   7 Segment Display
    DataOut = Data[OldRTC];       // Output the 7/8 character
                                  // bits

    ControlPin[OldRTC] = 1;       // Enable the new character
                                  // output
```

This code assumes that DataOut is an I/O port (such as PORTB in the PIC16F627) and ControlPin is a 4-bit array defined over an I/O port (see the appendices to understand how bits are defined in a structure). The biologic code is responsible for loading the 4-byte array "Data" with the segment values for the different values that are to be displayed.

When this code executes, each display is active roughly 250 times a second, which is well above the 100-times-a-second limit that I suggested you follow to make sure that there is no perceivable flashing of the segments.

ledflash—Flashing an LED

After defining the programming template and determining the basic wiring of the PICmicro microcontroller, it is time to start creating some interfaces for the robot. This section introduces you to a simple elelogic interface that flashes an LED approximately once per second. Along with providing the application, I will walk you through the process of simulating the source code before it is burned into the PICmicro MCU.

The purpose of this output elelogic interface is to flash an LED on pin RB1 (pin 7 on the PIC16F627) once per second. If the LED flashing is not enabled, the LED is turned off.

The circuit I used is shown in Figure 4-29 and uses the parts listed in Table 4-2.

To build this circuit on a breadboard, wire it according to Figure 4-30. This circuit uses the original basic circuit and adds the current limiting resistor (R2) and LED (CR1). To light the LED, current must flow through it, and for current to flow through it, the PICmicro I/O pin must be in output mode and driving out a

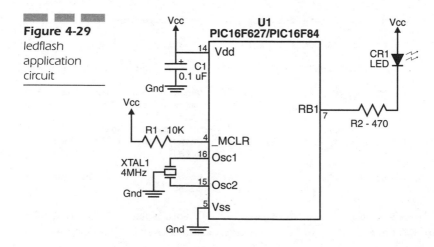

Figure 4-29
*ledflash
application
circuit*

Table 4-2

Parts of the
ledflash
Application
Circuit

Label	Part Number	Comment
U1	PIC16F627	Can use a PIC16F84
CR1	Visible light LED	Uses any type
R1	10K, ¼ watt	Uses _MCLR pull-up
R2	470 ohm, ¼ watt	Uses CR1 current-limiting resistors
C1	0.1 µF	Uses any type of U1 decoupling capacitor
XTAL1	4 MHz ceramic resonator with internal capacitors	Required for PIC16F84
Misc.		Requires a breadboard, wiring, and +5-volt power supply

Figure 4-30
ledflash
breadboard
circuit

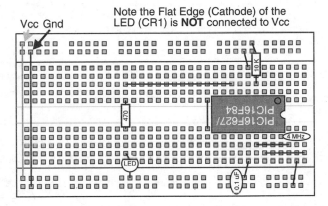

Note the Flat Edge (Cathode) of the
LED (CR1) is **NOT** connected to Vcc

0 value. The polarity of the PIC16F627 (U1) and LED (CR1) should be checked
before applying power to the circuitry.

To create the application software, I modified the template.c file in five places
and renamed the file ledflash.c:

```
#include <pic.h>

//  Flash an LED 1x per second
//
//  Use TMR0, 1 msec interrupt as base
//
//  02.04.09 - Modified for "LEDFlash" Application
//  02.03.28 - Updated to allow PIC16F627/PIC16F84 PICmicro MCUs
//  02.01.23 - Originally created by myke predko
```

```
//
//
//    Hardware Notes:
//    PIC16F84/PIC16F627 running at 4 MHz
//    PIC16F627 uses internal 4 MHz oscillator
//    External _MCLR connection required
//    "LED" - RB1/LED Control Output
//


//   Configuration Fuses
#if defined (_16F84)
#warning PIC16F84 selected
        _CONFIG(0x03FF1);              //   PIC16F84 Configuration
                                       //   Fuses:
                                       //    - XT Oscillator
                                       //    - 70 msecs Power Up Timer
                                       //    - Watchdog Timer Off
                                       //    - Code Protection Off
#elif defined(_16F627)
#warning PIC16F627 with External oscillator selected
        _CONFIG(0x03F61);             //   PIC116F627 Configuration
                                      //   Fuses:
                                      //    - External Oscillator
                                      //    - RA6/RA7 Digital I/O
                                      //    - External Reset
                                      //    - 70 msecs Power Up Timer
                                      //    - Watchdog Timer Off
                                      //    - Code Protection Off
                                      //    - BODEN Enabled
#else
#error Unsupported PICmicro MCU selected
#endif


//   Global Variables
volatile unsigned int RTC = 0;        //   Real Time Clock Counter

static bit trisLED @ (unsigned) &TRISB*8+1;     //   LED Physical
                                                //   Bits
static bit    LED @ (unsigned) &PORTB*8+1;
constant int LEDon  = 0;              //   Declare values for LED
                                      //   ON/off
constant int LEDoff = 1;


//   Interrupt Handler
void interrupt tmr0_int(void)         //   TMR0 Interrupt Handler
{

      if (T0IF) {

            T0IF = 0;                 //   Reset Interrupt Flag

            RTC++;                    //   Increment the Clock

//   Put mechalogic/elelogic interface code for 1 msec interrupt
//   here
```

```
                    if ((RTC % 512) == 0)
                        LED = LED ^ 1;        //  Toggle LED Bit

        }  //  endif

//  Put interrupt handlers for other mechalogic/elelogic interface
//  code

}  //  End Interrupt Handler


void enableLED(int LEDstate)        //  Set "eLED" according to
{                                   //    "LEDstate"

        LED = LEDoff;               //  Start with LED Off
        if (LEDstate)               //  if (LEDstate != 0)
                trisLED = 0;        //    Make LED Bit Output
        else                        //  if (LEDstate == 0)
                trisLED = 1;        //    Make LED Bit Input

}  //  End enableLED


//  Mainline
void main(void)                     //  Template Mainline
{

        TMR0 = 0;                   //  Reset the Timer for Start
        OPTION = 0x0D1;             //  Assign Prescaler to TMR0
                                    //    Prescaler is /4
        T0IE = 1;                   //  Enable Timer Interrupts
        GIE = 1;                    //  Enable Interrupts

//  Put hardware interface initialization code here

        enableLED(1);               //  Start the LED Flashing

        while (1 == 1) {            //  Loop forever

//  Put robot biologic code here

        }  //  endwhile

}  //  End of Mainline
```

The following five areas are changed:

- The program comments are changed. I changed the source code's comments to reflect what ledflash.c does over template.c. This type of change is mandatory so someone who looks at the code later will understand what it does, who wrote it, and what hardware is required for the application.

- I added label declarations for the bit to which the LED is connected. The bit declarations use the standard format used by PICC Lite. After the LED declarations, I have included constant values for turning an LED on or off.

■ If the RTC value is evenly divisible by 512 (using the modulo [%] check), then I toggle the LED pin. The output value of the pin is only changed if the pin is in output mode. If the pin is in input mode, then the value written to the pin output latch is never output from the pin.

■ The function called *enableLED* is used to turn on or off the bit flashing. To enable bit flashing, a nonzero value is passed to it. Inside the function, an LED off value is written to the pin output buffer and the pin is put into output mode. The LED does not turn off until the RTC value becomes evenly divisible by 512. If a zero is passed to the enableLED function, then the LED pin is put into input mode and no changes to the state of the input pin are output.

■ The LED flashing is initiated by the "enableLED(1);" statement.

This code is a good example of what I am advocating with the elelogic, mechalogic, and biologic spectrums. In this code, you can see that no hardware registers are accessed within the biologic code loop (marked with the comment "// Put robot biologic code here"). The LED flashing occurs within the 1-millisecond interrupt handler.

Looking at this code, you might feel like I made it much more complex than it should have been. If you want to minimize the number of programming statements and make the LED flashing code somewhat more obvious, do not modify the interrupt handler; just put in the following code:

```
LED = LEDoff;              //  Start with LED Off
trisLED = 0;               //  LED Bit Output

while (1 == 1) {

        if ((RTC % 512) == 0) { //  1/2 Second passed?
                LED = LED ^ 1;     //    Yes, Toggle LED Pin
                While ((RTC % 512) == 0;
                        //  Wait for RTC to change to prevent
                        //    multiple LED Toggles
        } // endif

} //  endwhile
```

This code requires fewer lines of code and it is probably easier to understand what I've created for this sample interface. The primary concern is what happens later when the robot application has several peripheral interfaces along with the controlling code. I want to be able to update, modify, or change the code without affecting other interfaces or the overall function of the robot. The simple solution to flashing the LED once a second does not give me that ability, whereas the code that I have provided in this example does.

One more question you might have is why I chose to change the LED state every 512 instances of the 1-millisecond interrupt handler instead of 500. PICC Lite, like most compilers, looks at values and chooses the code that executes most effectively. If you change the comparison value to 500 and rebuild the application, you will discover that it requires twice the amount of program memory space as the build comparing to 512. PICC Lite has optimizing code that looks for opportunities where surprisingly intelligent methods of performing operations can be accomplished.

The general case for the statement

```
if ((RTC % n) == 0)
    LED = LED ^ 1;
```

consists of the following operations:

- Temporary = RTC % n. Return the remainder of RTC divided by n.
- If Temporary == 0, then invert LED.

The action of dividing RTC by n requires a separate division function (the PICmicro MCU does not have a built-in division instruction), which must be linked into the application and passed along with the rest of the hex file.

When the PICC Lite compile operation executes, it has built-in rules that enable it to rewrite a statement to make it easier to process. In this case, the rule built into the compiler states that for constants that are powers of two, the modulus statement is equivalent to ANDing the value with the constant minus 1. This could be written out as follows:

```
A % B = A & (B - 1)        When "B" is equal to 2**n
```

The number 512 is actually 2**9, so the two previous statements are converted by the compiler into the following:

```
if ((RTC & 511) == 0)
    LED = LED ^ 1;
```

The second version of the two statements is easy for the PICmicro microcontroller to process and does not require the inclusion of the division function. When the modulus value changes, the division function has to be included. When I tested this out with the constant value 500, I found that the program memory and RAM registers required by the application doubled. For an application of this size, these increases do not affect the overall application, but it could be a problem if you have other functions running in the microcontroller at the same time.

I'm pointing this out because there are many opportunities for decreasing the code size of your application by simply looking at it in a different way or thinking about what is the easiest way for the controller to process the data.

Build the ledflash application and set up the MPLAB IDE desktop (as shown in Figure 4-31) by adding the stopwatch and the P16F627.WAT watch file from robot\procwat, and create a unique watch window to display the 16-bit value RTC.

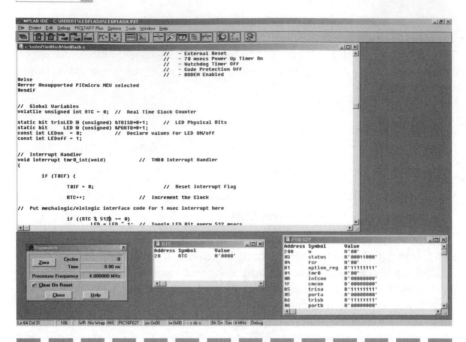

Figure 4-31 The MPLAB IDE ready for simulation

To start single stepping through the application, click the source file to make sure it has the focus of the desktop and then click the Reset PICmicro icon. Click Debug –> Run –> Reset or press F6. This resets the simulated PICmicro MCU to the power-on state. The source window should jump to the first execution line in the main function and highlight it.

Click the Single Step icon (the two-footprint icon), and then click Debug –> Run –> Step or press F7.

As you single step through the application, the highlighted line changes and you may notice some unusual things happening like the line jumping to previous lines of code for no reason. When the source code is compiled, certain statements may be used for multiple operations (to optimize the final application) and the references to the source code lines may no longer be valid. When you single step through the enableLED function, you will see execution jump back and forth in the if statement and the statements that are built into it. You can be quite confident that the code is working correctly in this case.

As you continue this, you will end up in the infinite loop (while(1 == 1) . . .) and single stepping will not do anything except increment trm0 in the processor standard register watch window. To make this less boring, scroll up to the first line in the interrupt handler and right-click the first statement within it (T0IF = 0;). When you do this, a small box (similar to the one shown in Figure 4-32) comes up. Click Break Point(s) and the line will turn red.

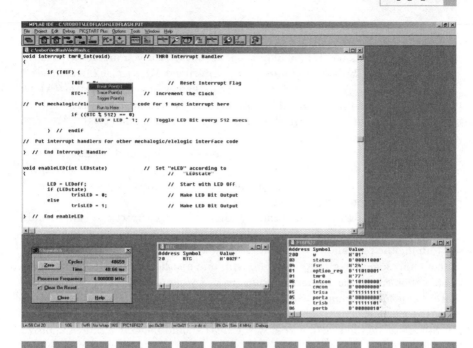

Figure 4-32 Setting a breakpoint

You have just set a breakpoint. Now when the application is allowed to execute freely (and not single stepped), it will stop at this point. To try this out, click the Run icon (the green light), click Debug –> Run –> Run, or press F9. The application executes until it reaches this breakpoint.

When I tried it on my PC, the stopwatch indicated that the breakpoint stopped at 1,087 cycles, or 1.09 milliseconds. This is more than what was expected because of the overhead of the initial setup required for the application. When I clicked the Run icon again, the stopwatch incremented to 2,111 cycles, which is 1,024 more than the last time, indicating that the TMR0 interrupt is working correctly in the simulation.

Disable the interrupt at the "T0IF = 0;" statement by right-clicking the line again and clicking Break Point(s).

Now, set a breakpoint at "LED = LED ^ 1;", which is the statement where the state of the LED is changed. Click the Run icon and wait for the application to execute to the breakpoint. After waiting several moments, the application stops at the statement.

If you look at the processor standard register watch window, you will see that the portb register value is blue. Click the Step icon and notice that when portb is changed, the color of the register changes to red to indicate that one or more bits within it were changed. If you look at the contents of the register, you will see that bit 1 (RB1) changed from 1 to 0.

In the application, this would correspond to the LED being turned on.

Looking at the stopwatch, I can see that 524.368 instruction cycles have executed to get to this point. I should also point out that if you are going to use a PIC16F627 with this sample interface, the internal 4 MHz timer can also be off by 5 percent or so. The total possible error could be as high as 10 percent (which is somewhat more than the ideal 500,000 cycles that should pass before the LED changes state), but since the flashing LED is for humans, I don't see the need to time it more precisely.

I have just gone through about 50 percent of what you need to know to simulate an application. For all the sample output device applications in this chapter, you have just learned the basic skills necessary for simulating the application and looking for potential problems.

Now that the simulated application code is working correctly, it is time build the circuit and test it out. I always assemble my application circuit before programming the part and trying it out. This makes sure that it is possible to build the circuit; sometimes parts interfere with one another in such a way that the application cannot be assembled, which means you must go back and select different pins for different functions.

After the part is programmed, place it into the circuit and apply power. If you have assembled the circuit correctly, the LED should flash at a rate of approximately one time per second. If the LED does not flash, then you should go over the circuit in detail, making sure that the voltage levels are correct at the different PICmicro MCU pins and the LED is oriented correctly.

PWM Power-Level Control

The PICmicro, like most other digital devices, does not handle analog voltages very well. This is especially true for situations where high current voltages are involved. The best way to handle analog voltages is to use a string of varying wide pulses to indicate the actual voltage level. This string of pulses is known as a *PWM analog signal* and can be used to pass analog data from a digital device, control DC devices, or even output an analog voltage.

A PWM signal is a repeating signal that is on for a set period of time that is proportional to the voltage that is being output. A PWM signal is shown in Figure 4-33. The *pulse width* is the time that the signal is on and the *duty cycle* is the percentage of the on time relative to the PWM signal's period.

To output a PWM signal, use the following code:

```
Period = PWMPeriod;          //  Initialize the Output
On = PWMOn;                  //  Parameters

while (1 == 1) {
```

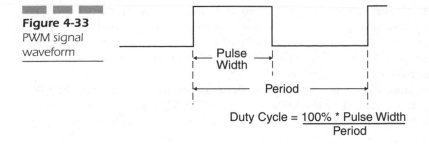

Figure 4-33
PWM signal
waveform

Pulse Width

Period

$$\text{Duty Cycle} = \frac{100\% * \text{Pulse Width}}{\text{Period}}$$

```
PWM = ON;                    //   Start the Pulse
for (i = 0; i < On; i++ );   //   Output ON for "On" Period of
                             //     Time
PWM = off;                   //   Turn off the Pulse
For ( ; i < PWMPeriod; i++ );//   Output off for the rest of the
                             //     PWM Period
} //  end while
```

This code can be implemented easily in the PICmicro MCU controlling the robot, but running it in the mainline of the program makes it impossible for any other mechalogic or electronic interface code to run. In this situation, it should be obvious that the biologic code cannot run either.

To avoid this problem, I could put code like the following into the TMR0 interrupt handler:

```
void interrupt tmr0_int(void)              //   TMR0 Interrupt Handler
{

    :

if (PWM == ON) {     //  If PWM is ON, Turn it off and Set Timer
    PWM = off;       //     Value
    TMR0 = PWMPeriod - PWMOn;
} else {             //  If PWM is off, Turn it ON and Set Timer
    PWM = ON;        //     Value
    TMR0 = PWMOn;
} //  end if

    :

} //  End Interrupt Handler
```

The sample LED dimming application shown in the following section is another method of implementing the PWM that uses the 1-millisecond interrupt. The previous method falls into a trap that you should avoid when peripheral interfaces are designed; the code changes a central resource that could be used by other peripheral interfaces. As shown in the next section, there is a way to implement a timer-interrupt-based PWM output that works with other peripheral interfaces easily.

When using a PWM for driving a motor, it is important to make sure that the frequency is outside the range of human hearing. For motors and other devices that may have an audible whine, the PWM signal should have a frequency of 20 KHz or more to ensure that the signal does not bother users (although it may cause problems with their dogs).

The problem with the higher frequencies is that the granularity of the PWM signal decreases. This is due to the inability of the PICmicro MCU (or whatever digital device is driving the PWM output) to change the output in relatively small time increments from on to off in relation to the size of the PWM signal's period.

In the sample code presented in this chapter, it is possible to change the TMR0 interrupt period to a value that enables you to run a reasonable 20 KHz PWM period. However, this causes problems with the timing of any other interfaces that are built into the PICmicro MCU and robot. With this in mind, you might think that you have to come up with a hardware circuit that can be controlled by the PICmicro MCU and output a 20 KHz PWM signal. This probably seems like a lot of work.

Actually, if you are in this situation, you should try using a PICmicro MCU (or other microcontroller) that has a built-in PWM capability to run your motors with a 20 KHz PWM frequency. Many midrange microcontrollers have this capability with a very modest increase in cost.

Another solution to the problem can be implemented in many robots—you can implement a PWM that is *below* the range of human hearing. Many toy manufacturers use a 30 to 60 Hz PWM instead of a PWM that is measured in thousands of hertz. The only stipulation to using this trick is that low-current motors (500 mA or less) can only be used to avoid problems with the motor drivers due to the strain experienced when the motors are turned on and off. The advantage of using a very low-frequency PWM signal is that you have the opportunity to use simple microcontrollers (like the PIC16F84 used in this example) without any kind of modification.

In the sample LED dimming application shown in the next section, 30 TMR0 interrupts are counted out to create a 32 Hz timebase. When you learn how to create a small robot base to test out different interfaces, you will see that this same timebase is used.

ledpwm—Dimming the Output of an LED

The second sample application expands upon how the TMR0 interrupt can be used to sequence the operation of a peripheral interface and investigate different methods of adding peripheral functions to the robot controller. Starting with the circuit presented in the flashing LED example application (ledflash), I will demonstrate code in this sample output device that can independently control up to three LEDs.

The basic circuit is similar to the one used by ledflash, but has two more LEDs wired to RB2 and RB3, respectively (hopefully you haven't disassembled the ledflash circuit yet). The circuit diagram is shown in Figure 4-34 and the recommended breadboard wiring is shown in Figure 4-35.

The parts required by this application are listed in Table 4-3.

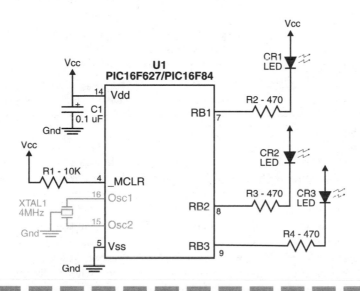

Figure 4-34 ledpwm application circuit. XTAL1 is only required if U1 is a PIC16F84.

Figure 4-35
lcdpwm
breakboard
circuit

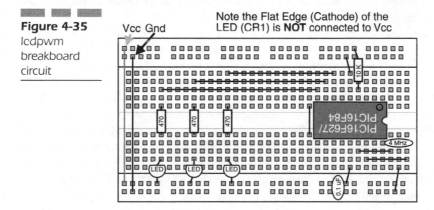

Table 4-3

Parts of the
ledflash
Application
Circuit

Label	Part Number	Comment
U1	PIC16F627	Can use a PIC16F84
CR1–CR3	Visible light LED	Uses any type
R1	10K, ¼ watt	Uses _MCLR pull-up
R2–R4	470 ohm, ¼ watt	Uses CR1 current-limiting resistors
C1	0.1 µF	Uses any type of U1 decoupling capacitor
XTAL1	4 MHz ceramic resonator with internal capacitors	Required for PIC16F84
Misc.		Requires a breadboard, wiring, and +5-volt power supply

The code presented here counts through 30 instances of the 1-millisecond interrupt to produce a 32 Hz PWM. This low-frequency PWM is probably very surprising, especially in light of the availability of the PWM signal generator built into the PIC16F627.

The reasons for using a 32 Hz software PWM instead of the built-in PWM of the PICmicro MCU center on flexibility. Later in this chapter, I will work through the different ways of adding additional PWMs to the circuit. Multiple and independent PWMs are an important feature of a controller when individual control of multiple motors is required. Individual control of robot motors is a useful feature to have as it makes sure that a differentially driven robot is tracking in a straight line and no wheels are skidding. This capability is not available as part of the PIC16F627 because there is only one PWM output. By not using the built-in PWM of the PICmicro MCU, I can use the PWM for different peripherals, including an IR collision detection system.

When the ledpwm application was first implemented, the PIC16F627 was connected to one LED on RB1 so the code could run on the same hardware as required by the ledflash code. In the following application code (which can be found in the robot/ledpwm directory), the PWM controller has been implemented in the 1-millisecond interrupt handler as a counter that repeatedly counts to 30 and keeps the output LED on while a duty cycle value is less than the counter.

The code to specify the PWM operation is located in the biologic code section of the application and waits for the PWM generator code, outputs the current duty cycle twice, and then decrements the duty cycle to zero (with PWM off 100 percent of the time). After turning off the LED, a 100 percent PWM duty cycle is specified and the process restarts:

```c
#include <pic.h>

//   Output a PWM Signal on a single LED (on RB1)
//
//   Use the 1 msec TMR0 interrupt to create a 32 Hz PWM signal
//
//   02.04.10 - Created Application
//   02.03.28 - Updated to allow PIC16F627/PIC16F84 PICmicro MCUs
//   02.01.23 - Originally created by myke predko
//
//
//   Hardware Notes:
//   PIC16F84/PIC16F627 running at 4 MHz
//   PIC16F627 uses internal 4 MHz oscillator
//   External _MCLR connection required
//   "LED" - RB1/LED Control Output
//


//   Configuration Fuses
#if defined (_16F84)
#warning PIC16F84 selected
        CONFIG(0x03FF1);        //   PIC16F84 Configuration Fuses:
                                //    - XT Oscillator
                                //    - 70 msecs Power Up Timer On
                                //    - Watchdog Timer Off
                                //    - Code Protection Off
#elif defined(_16F627)
#warning PIC16F627 with External oscillator selected
        CONFIG(0x03F61);        //   PIC116F627 Configuration Fuses:
                                //    - External Oscillator
                                //    - RA6/RA7 Digital I/O
                                //    - External Reset
                                //    - 70 msecs Power Up Timer On
                                //    - Watchdog Timer Off
                                //    - Code Protection Off
                                //    - BODEN Enabled
#else
#error Unsupported PICmicro MCU selected
#endif

//   Global Variables
volatile int  RTC = 0;          //   Real Time Clock Counter
char PWMCycle;                  //   Cycles are from 0 to 29
char PWMDuty;                   //   Start with full on
volatile int  PWMLoop;          //   PWM Loop Count

static bit trisLED @ (unsigned) &TRISB*8+1;    //   LED Physical Bits
static bit    LED @ (unsigned) &PORTB*8+1;
const int LEDon  = 0;           //   Declare values for LED ON/off
const int LEDoff = 1;


//   Interrupt Handler
void interrupt tmr0_int(void)   //   TMR0 Interrupt Handler
{

        if (T0IF) {
```

```
            TOIF = 0;              //  Reset Interrupt Flag

            RTC++;                 //  Increment the Clock

//  Put additional interface code for 1 msec interrupt here

            switch(PWMDuty) { //  Look for extremes first
                case 0:       //  All off
                     LED = LEDoff;
                    break;
                case 29:      //  Always On (100%)
                    LED = LEDon;
                    break;
                default:      //  Else, something in between
                              //   If the cycle count is less
                              //     than the current then LED on
                    if (PWMCycle <= PWMDuty)
                         LED = LEDon;
                    else
                         LED = LEDoff;
            }  //  endswitch

            if (++PWMCycle == 30) { //  Roll around cycle count?
                PWMCycle = 0;       //    Yes
                PWMLoop++;          //    Increment the Loop Counter
            }  //  endif

        }  //  endif

//  Put Different Interrupt Handlers here

}

void enableLED(int LEDstate)   //  Set "eLED" according to
{                              //    "LEDstate"

    LED = LEDoff;              //  Start with LED Off
    if (LEDstate) {
        PWMCycle = 0;          //  Cycles are from 0 to 29
        PWMDuty = 29;          //  Start with full on
        PWMLoop = 0;           //  PWM Loop Count starting at zero
        trisLED = 0;           //  Make LED Bit Output
    } else
        trisLED = 1;           //  Make LED Bit Input/Stop Display

}  //  End enableLED


//  Mainline
void main(void)                //  Template Mainline
{

    OPTION = 0x0D1;            //  Assign Prescaler to TMR0
                              //   Prescaler is /4
    TMR0 = 0;                  //  Reset Timer before starting
                              //  application
    T0IE = 1;                  //  Enable Timer Interrupts
    GIE = 1;                   //  Enable Interrupts
```

```
//  Put in Interface initialization code here

    enableLED(1);              //  Enable the LED Bit

    while (1 == 1) {           //  Loop forever

//  Put in Robot high level operation code here

            if (PWMLoop == 2) {      //  Change PWM Duty Cycle
                if (PWMDuty == 0)
                        PWMDuty = 29;
                else
                        PWMDuty--;
                PWMLoop = 0;         //  Reset Counter
            }  //  endif

    }  //  endwhile

}  //  End of Mainline
```

There should not be anything surprising about this code. When you compile and burn it into a PICmicro MCU, it should dim the LED over a period of 2 seconds. You may see a bit of flashing as the LED dims. This is due to the speed at which you can perceive changes in light. As the time the LED is off becomes longer in each PWM cycle, it becomes more discernable instead of being averaged by your brain with the time the LED is on, making the level dimmer.

I was surprised after I looked at different ways of adding additional PWM outputs to the code. I worked through 4 different ways of adding PWM controls to the 1-millisecond interrupt handler, which enables the independent control of more LEDs. The results were quite surprising.

To demonstrate that the PWM controls were working, I output different LED PWM changing rates. For 3 LEDs, the LED on RB1 dimmed over 2 seconds, the LED on RB2 dimmed over 5 seconds, and the LED on RB3 brightened over 7 seconds. By changing the dimming/brightening rates over the three LEDs, I could be sure visually that the code was working.

The first method of adding more PWM controls to the interrupt handler was to replicate the code for one LED for one or two more. This method is called *summation*. The two-LED application (also in the robot\ledpwm director) is called *ledpwm2*, and the three-LED application is called *ledpwm3*.

I don't like repeating code when it can be handled by a loop. To make this code more efficient, I put in a loop that accessed the different parameters for the three different PWM outputs as arrays. To avoid extra calculations, when I declared the array variables, I added one more element than needed; this way I address the PWM parameter variables as 1, 2, and 3, which corresponds directly with the RB1, RB2, and RB3 I/O pins of the PICmicro MCU.

The loop code created for handling the multiple PWM controllers in the interrupt handler is as follows:

```
for (i = 1; i < 4; i++) {       //  Repeat 3x for 3 LEDs
    switch(PWMDuty[i]) {        //  Look for extremes
                               //  first
        case 0:                //  All off
            writeLED(i, LEDoff);
            break;
        case 29:               //  Always On (100%)
            writeLED(i, LEDon);
            break;
        default:               //  Else, something in
                               //  between
                               //  If the cycle count is
                               //  less than the current
                               //  then LED on
            if (PWMCycle[i] <= PWMDuty[i])
                writeLED(i, LEDon);
            else
                writeLED(i, LEDoff);
    } // endswitch

    if (++PWMCycle[i] == 30) {  //  Roll around
                               //  cycle
                               //  count?
        PWMCycle[i] = 0;       //  Yes
        PWMLoop[i]++;          //  Increment Loop Counter
    } // endif
} // endfor
```

You can see that this code is a simple reproduction of the LED PWM control with arrayed parameters. However, you might be surprised by the apparent writeLED function that provides a write to the selected PORTB I/O pin. This is not a function; it is actually the defined macro:

```
#define writeLED(bit, value) ((PORTB) = (PORTB & ~(1 << (bit)) _
| (value * (1 << bit))
```

When invoked, this defined macro reads in PORTB, clears the output bit, and then sets it according to the value (which can only be 1 or 0). Note that by starting the array indexes at 1 (instead of the standard 0), the code will access both the array element and the corresponding PORTB pin that has the same bit number as the array element.

I have used this defined macro in my PC programming for years—the C programming language does not handle the modification of individual bits very well. The algorithm used in the defined macro enables you to write to a bit without affecting any of the others. This is why I call this method of adding PWM controllers the *algorithmic loop*.

When I ran the algorithmic loop, I was surprised at the number of cycles required to perform the PORTB update and I assumed that this was due to an inefficiency in how the writeLED-defined macro executed. To see if there was some way to improve the code's execution, I created ledpwm2b and ledpwm3b, which use the explicitly set bits of the first application and write to them based on the current bit using the following function:

```
void writeLED(int LEDbit, int value)      // Write to Specified LED
{

    switch(LEDbit) {
         case 1:
                LED1 = value;
                break;
         case 2:
                LED2 = value;
                break;
         case 3:
                LED3 = value;
                break;
    } // endswitch

} // End writeLED
```

I called this test the *switch loop*.

I came up with the last method of looking at the issue of making the loop as efficient as possible when I was proofreading this sample application. When I looked at the original defined macro, I wondered how efficiently PICC Lite would convert the value $* (1 << $ bit) expression, so I changed the defined macro to the following:

```
#define writeLED(bit, value) if (value == 0) PORTB &= ~(1 << bit);
_ else PORTB | = (1 << bit);
```

This defined macro tests the value before setting or clearing the bit. It eliminates the need for multiplying the value by $(1 << $ bit). I copied the three-LED algorithmic loop solution and called it ledpwm3c. To check the operation of the two-LED loop with this new defined macro, I simply changed the number of times the loop executed from three to two. As you will see in the following results, the new defined macro does not change my overall conclusions regarding this experiment.

Table 4-4 shows the end product from using the results from the compilation operation and simulating the length of the interrupt handler.

From a gross perspective, for the summation method of adding PWM controllers, the interrupt handler code size increased by 23 instructions for each LED and the interrupt cycles increased by 14 cycles. In order to change the loop with the bit algorithmically, the interrupt handler's size does not change (which is to be expected) and the number of interrupt cycles increases by 63 cycles for each additional LED. In order to change the loop with the bit by a switch statement, the interrupt handler's size does not change (again, which is to be expected) and the number of interrupt cycles increases by 66 for each additional LED. In the last case with the improved bit change algorithm and loop, the number of instruction cycles per LED increases by 63.

The improved bit change algorithm improved the number of instructions and the increase in interrupt handler looping marginally. For this reason, I will ignore it in my observations and conclusions, but leave you with the defined macro in case you ever require it for your own applications.

Table 4-4

ledpwm Implementation Comparison Table

Application	LEDs	Comments	Interrupt Size	Main Size	Total Code Size	Interrupt Cycles
ledpwm	1	Original	63	44	107	36
ledpwm2	2	2 summation	86	80	166	50
ledpwm3	3	3 summation	109	116	225	64
ledpwm2a	2	2 algorithm loop	179	140	319	151
ledpwm3a	3	3 algorithm loop	179	152	331	214
ledpwm2b	2	2 switch loop	116	210	326	157
ledpwm3b	3	3 switch loop	116	256	372	223
ledpwm2c*	2	2 mod. alg. loop	133	*		147
ledpwm3c	3	3 mod. alg. loop	133	102	235	210

*ledpwm2c is not created, but results are taken from the modified ledpwm3c.

Figure 4-36 shows the effect that the number of LEDs (or the number of PWM peripheral controllers built into the interrupt handler) had on the space required by the application. In Figure 4-36, notice that the summation (replicating the code) is the most efficient method in terms of interrupt handler space utilization for three LEDs. When more than three LED PWM controllers are required, the switch loop becomes more efficient. Only after more than six LED PWM controllers are required does the algorithmic loop become more efficient in terms of space rather than summation.

In practical terms, the summation method of adding LED PWM controllers is most efficient in an application because few robot applications require PWM control of more than three motors. As shown later in the chapter, this PWM algorithm is acceptable for controlling small electric motors and will be used as part of a sophisticated motor controller application.

From the number of instruction cycles added to the application for each PWM controller, the advantages of using the summation method to add PWM controllers are even more noticeable (see Figure 4-37). The higher slope of the other two methods indicates a significant imposition in terms of the number of instruction cycles available for the mechalogic and elelogic peripheral interfaces to execute in.

The results of the different methods of adding the multiple LED PWM controllers to the application came as a bit of a relief to me. As I indicated previously, if I have to perform multiple operations of a similar type, I prefer putting them in a loop and letting the loop execute. The reasons why I avoid putting in the

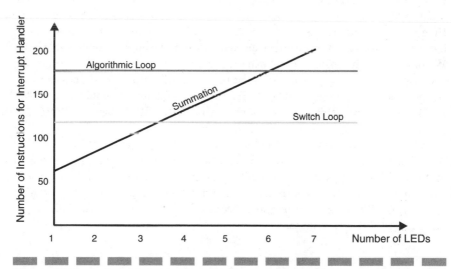

Figure 4-36 *ledpwm interrupt handler size according to the method used to add LEDs*

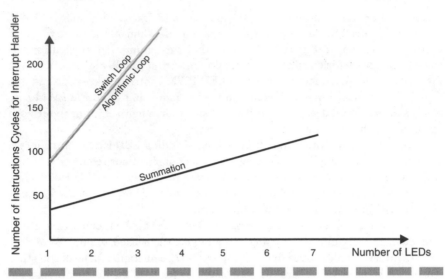

Figure 4-37 ledpwm interrupt handler duration according to the method used to add LEDs

same code over and over again are the same ones you probably heard when you first learned to program—the number of keystrokes (and thus the opportunity for errors) is reduced and the code is shorter and easier to read.

I was relieved by the results because I was worried that adding multiple peripheral interfaces to an interrupt handler would take up an unreasonable amount of room in the application. I was also somewhat concerned that the multiple peripheral interfaces would potentially take up more instruction cycles in between 1-millisecond timer interrupt requests than I would be comfortable with. When I started this experiment, I expected the results to be the complete opposite because I was dealing with a somewhat unreasonable situation—in a real robot, the peripheral interfaces would be different and could not be combined in a loop.

The results from this experiment show that for a PIC16F627 (or PIC16F84) executing C code compiled by PICC Lite, the most efficient method of adding multiple peripherals is to put in a unique peripheral interface that references I/O pins explicitly in the 1-millisecond interrupt handler. Note that in this conclusion statement, I have provided many qualifications. If you are developing robot software in a different microcontroller using a different compiler (or programming language) or not basing your peripheral control code on a 1-millisecond interrupt, you may get different results.

■■■ Piezo Electric Buzzers and Speakers for Audible Output

When discussing a microcontroller's processing capabilities with regards to audio, I tend to be quite disparaging. This is because of the lack of hardware multipliers in the low-end and midrange MCUs and the inability of most small microcontrollers to natively handle numbers larger than 8 bits or even data in a floating point format. Most microcontrollers have been optimized for responding to digital inputs and cannot implement the real-time processing routines needed for complex analog I/O.

Despite this, adding a simple speaker (which is capable of a little more than simple beeps and boops) using a circuit like the one shown in Figure 4-38 can help you understand the current state of the robot. Listening for simple tones to indicate what is happening is much easier than running alongside the robot to read an LCD.

The circuit in Figure 4-38 passes DC waveforms through the capacitor (which filters out the *kickback spikes*) to the speaker or piezo buzzer. When a tone is output, you will hear a reasonably good tone, but if you were to look at the actual signal on an oscilloscope, you would see the waveform shown in Figure 4-39, both from a microcontroller's CMOS I/O pin and the piezo buzzer.

The microcontroller pin, capacitor, and speaker actually make up a rather complex analog circuit. Note that the voltage output on the I/O pin is changed from a fairly straightforward digital waveform to dual spikes. This is due to the inductive effects of the piezo buffer (if a speaker was used, you would see a similar waveform). The important thing to note in Figure 4-39 is that the upward spikes appear at the correct period for the output signal.

Timing the output signal can be accomplished in a number of different ways. For the most reliable tone, you should look at using a built-in PWM circuit or a timer interrupt, (which takes place at twice the desired output frequency). The reason for interrupting at twice the output frequency is to make sure that a full wave is produced.

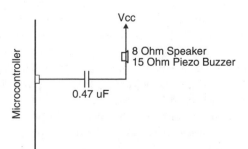

Figure 4-38
Circuit for driving an external speaker

Figure 4-39
PICmicro MCU
driving a
speaker output

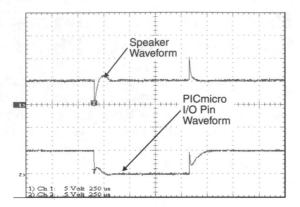

For the 1-millisecond interrupt handler that the robot interfaces are based on, the resulting output frequency is approximately 500 Hz and produces a slightly sharp B. This sound isn't terrible, but it is noticeably different from an A (440 Hz), which is typically used by electronic devices.

To produce different frequencies with the PIC16F84 sample circuits shown in this chapter, you must modify the interrupt handler to reload the TMR0 register with a new value, which causes the interrupt to take place after fewer counts than the 256 that it currently uses. Modifying the code to do this is PICmicro MCU specific and you must retime some of the other interfaces (such as the motor PWM and button debouncing).

Beeper Application

As shown in the previous section, adding audio output to your application is quite simple. With the code presented in the first two sample interfaces, you should have a pretty good idea of how to implement audio in a PIC16F627 using PICC Lite and put it in the 1-millisecond interrupt handler. In addition to showing how a piezo electric speaker works in a robot application in this sample interface, this section shows how delays can be implemented for the biologic code.

The application circuit is similar to the ledflash circuit; however, the LED and its current limiting resistor are replaced by a piezo electric speaker and capacitor, as shown in Figure 4-40. Wiring the application on a breadboard is also similar to the ledflash's wiring (see Figure 4-41). The parts required for the application are shown in Table 4-5.

Implementing an application where the speaker outputs a signal of approximately 500 Hz is based on the previous two sample interface applications and can be found in robot\beeper:

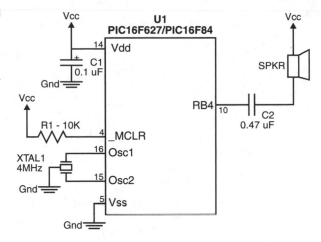

Figure 4-40
Beeper
application
circuit

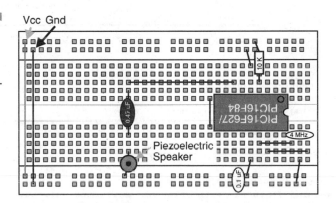

Figure 4-41
Beeper
breadboard
circuit

```
#include <pic.h>

//   "Beeper" - Beep a speaker for 1 Second On, followed by
//     One Second off
//
//   TMR0 1 msec interrupt used to produce ~500 Hz Signal
//
//   02.03.28 - Updated to allow PIC16F627/PIC16F84 PICmicro MCUs
//   02.01.23 - Originally created by myke predko
//
//
//   Hardware Notes:
//   PIC16F84/PIC16F627 running at 4 MHz
//   PIC16F627 uses internal 4 MHz oscillator
//   External _MCLR connection required
//   RB4 - Connected to a 0.47 µF cap and piezo electric speaker
//
```

Table 4-5

Parts for the
Beeper
Application
Circuit

Label	Part Number	Comment
U1	PIC16F627	Can use a PIC16F84
R1	10K, ¼ watt	Uses _MCLR pull-up
C1	0.1 µF	Uses any type of U1 decoupling capacitor
C2	0.47 µF	Uses any type
SPKR	15 ohm piezo electric transducer	Requires piezo electric speaker
XTAL1	4 MHz ceramic resonator with internal capacitors	Required for PIC16F84
Misc.		Requires a breadboard, wiring, and +5-volt power supply

```
//  Configuration Fuses
#if defined (_16F84)
#warning PIC16F84 selected
        CONFIG(0x03FF1);        //  PIC16F84 Configuration Fuses:
                                //   - XT Oscillator
                                //   - 70 msecs Power Up Timer On
                                //   - Watchdog Timer Off
                                //   - Code Protection Off
#elif defined(_16F627)
#warning PIC16F627 with External oscillator selected
        _CONFIG(0x03F61);       //  PIC116F627 Configuration Fuses:
                                //   - External Oscillator
                                //   - RA6/RA7 Digital I/O
                                //   - External Reset
                                //   - 70 msecs Power Up Timer On
                                //   - Watchdog Timer Off
                                //   - Code Protection Off
                                //   - BODEN Enabled
#else
#error Unsupported PICmicro MCU selected
#endif


//  Global Variables
volatile unsigned int RTC = 0;      //  Real Time Clock Counter
unsigned char BeeperFlag = 0;       //  Enable Beeper Flag

static bit trisBeeper @ (unsigned) &TRISB*8+4;  //  RB4 - Beeper
static bit     Beeper @ (unsigned) &PORTB*8+4;  //    Physical Bits
```

```
//  Interrupt Handler
void interrupt tmr0_int(void) //  TMR0 Interrupt Handler
{

     if (T0IF) {

          T0IF = 0;          //  Reset Interrupt Flag

          RTC++;             //  Increment the Clock

//  Put mechalogic/elelogic interface code for 1 msec interrupt
                             //  here

          if (BeeperFlag)    //  Toggle the Beeper I/O Pin
               Beeper ^= 1;//   If Flag != 0

     } //  endif

//  Put interrupt handlers for other mechalogic/elelogic interface
                             //  code

} //  End Interrupt Handler


void Dlay(unsigned int msecs) //  Delay a specified number of msecs
{

unsigned int DlayEnd;

     DlayEnd = RTC + msecs + 1;    //  Setup Delay so that it
                                   //  waits a
                                   //   MINIMUM number of "msecs"
     while (DlayEnd != RTC);

} //  End Dlay


void enableBeeper(void)       //  Enable the Beeper Output Bits
{

     trisBeeper = 0;          //  Set the Beeper I/O Pin to Output

} //  End enableBeeper


//  Mainline
void main(void)               //  Template Mainline
{

     TMR0 = 0;                //  Reset the Timer for Start
     OPTION = 0x0D1;          //  Assign Prescaler to TMR0
                              //   Prescaler is /4
     T0IE = 1;                //  Enable Timer Interrupts
     GIE = 1;                 //  Enable Interrupts

//  Put hardware interface initialization code here

     enableBeeper();          //  Enable the Beeper Output
```

```
        while (1 == 1) {           //  Loop forever

//   Put robot biologic code here

              BeeperFlag = 1;      //  Start Beeping
              Dlay(1000);          //   For 1 Second

              BeeperFlag = 0;      //  Stop Beeping
              Dlay(1000);          //   For 1 Second

        }  //  endwhile

}  //  End of Mainline
```

In each invocation of the 1-millisecond timer interrupt handler, the Beeper-Flag variable is tested, and if it is not equal to zero, then the beeper-assigned I/O pin is toggled. By toggling the I/O pin once every second, the interrupt handler drives out a square wave of 2 milliseconds in duration or at 500 Hz.

In the beeper.c application, you can see that I created a very simple 1-second delay function by loading the current value of RTC and adding 1,001 to it. The Dlay function waits for RTC to be incremented to this target value.

The Dlay function is simple and works very well; however, be careful when using it in robot applications. Using the Dlay function for biologic code delays may prevent some sensor inputs from being processed in a timely manner. When Dlay is executing, the biologic code cannot respond to any other inputs. Chapter 5, "Bringing Your Robot to Life," discusses some strategies that can minimize the problems with Dlay when it is used in a robot application.

▰▰▰ LCDs

LCDs can be valuable for debugging your robot application. They can be used for status data or displaying sensor input data. Unfortunately, many of them are difficult to read across the room so you might find yourself running after a robot to read the LCD to understand exactly what is happening. One way to avoid this predicament is to use a *Vacuum Florescent Display* (VCD). A number of different ones are available that have the same 14-pin interface as the LCD displays described in this section and are programmed exactly the same way.

The most common type of LCD controller is the Hitatchi 44780, which provides a relatively simple interface between a processor and LCD. This type of LCD controller is often reluctantly used by inexperienced designers and programmers because they think it is difficult to find good documentation on the interface. Initializing the interface is also seen as a problem and that the displays are expensive.

I have worked with Hitatchi-44780-based LCDs for years and I don't agree with any of these perceptions. LCDs can be added quite easily to an application and use as few as three digital output pins for control. As for cost, LCDs can often be pulled out of old devices or found in surplus stores for less than a dollar.

The most common connector used for the 44780-based LCDs is 14 pins in a row, with pin centers 0.100 inch apart. The pins are wired as shown in Table 4-6.

As you can probably guess from this description, the interface is a parallel bus, which enables the simple and fast reading/writing of data to and from the LCD. Figure 4-42 shows the input waveform into the LCD.

The waveform shown in Figure 4-42 writes an ASCII byte out to the LCD's screen. The ASCII code that is displayed is 8 bits long and is sent to the LCD either 4 or 8 bits at a time. If 4-bit mode is used, 2 nybbles of data (send high 4 bits first and then low 4 bits with an E clock pulse to latch in each nybble) are sent to make up a full 8-bit transfer. The E clock is used to initiate the data transfer within the LCD.

Sending parallel data as either 4 or 8 bits represents the two primary modes of operation. Although there are secondary considerations and modes, deciding how to send the data to the LCD is the most critical decision to be made for an LCD interface application. Eight-bit mode is best used when speed is required in

Table 4-6

Wiring for the Pins in 44780-Based LCDs

Pins	Description
1	Ground
2	Vcc
3	Contrast voltage
4	R/S _Instruction/Register Select
5	R/W _Read/Write LCD Registers
6	E clock
7–14	Data I/O pins (pin 7—D0 and pin 14—D7)

Figure 4-42

LCD data write waveform

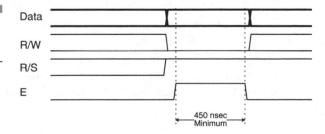

an application and at least 10 I/O pins are available. Four-bit mode requires a minimum of 6 bits. To wire a microcontroller to an LCD in 4-bit mode, only the top 4 bits (DB4-7) are written to.

The R/S bit is used to select whether data or an instruction is being transferred between the microcontroller and LCD. If the bit is set, then the byte at the current LCD cursor position can be read or written. When the bit is reset, an instruction is sent to the LCD or the execution status of the last instruction is read back (whether or not it has completed).

The different instructions (which I sometimes call "commands") available for use with the 44780 are shown in Table 4-7.

Reading data back is best used in applications that require data to be moved back and forth on the LCD (such as in applications that scroll data between lines). The Busy Flag can be polled to determine when the last instruction that has been sent has completed processing.

In most of my applications, the R/W line is tied to the ground because there is no requirement to read data back. This simplifies the application because when data is read back, the microcontroller I/O pins have to be alternated between input and output modes. When transferring data to the LCD with the R/W line tied to the ground, always wait the maximum amount of time required for each instruction to complete (which is 4.1 milliseconds for clearing the display or moving the cursor/display to the home position and 160 µsecs for all other commands). In addition to simplifying the application software, it also frees up a microcontroller pin for other uses. Different LCDs execute instructions at different rates. To avoid problems later on (such as if the LCD is changed to a slower unit), I recommend using the maximum delays given previously.

In terms of options, I have never seen a 5×10 LCD display. This means that the F bit in the Set Interface Instruction should always be reset (equal to 0).

Before you can send commands or data to the LCD module, the module must be initialized. For 8-bit mode, this is done using the following series of operations:

1. Wait more than 15 milliseconds after power is applied.

2. Write 0x030 to the LCD and wait 5 milliseconds for the instruction to complete.

3. Write 0x030 to the LCD and wait 160 µsecs for the instruction to complete.

4. Write 0x030 again to the LCD and wait 160 µsecs (or poll the busy flag).

5. Set the operating characteristics of the LCD.

   a. Write Set Interface Length.

   b. Write 0x008 to turn off the display.

   c. Write 0x001 to clear the display.

   d. Write Set Cursor Move Direction to set the cursor behavior bits.

   e. Write Enable Display/Cursor and enable the display and optional cursor.

Table 4-7

Instructions for 44780

R/S	R/W	D7	D6	D5	D4	D3	D2	D1	D0	Instruction	Bit Description
4	5	14	13	12	11	10	9	8	7	Pins	
0	0	0	0	0	0	0	0	0	1	Clear Display	
0	0	0	0	0	0	0	0	1	*	Return Cursor and LCD to Home Position	
0	0	0	0	0	0	0	1	ID	S	Set Cursor Move Direction	ID—Increment the cursor after each byte written to display is set. S—Shift the display when each byte is written to display.
0	0	0	0	0	0	1	D	C	B	Enable Display/Cursor	D—Turn display on (1)/off (0). C—Turn cursor on (1)/off (0). B—Turn cursor blink on (1)/off (0).
0	0	0	0	0	1	SC	RL	*	*	Move Cursor/Shift Display	SC—Display shift on (1)/off (0). RL—Direction of shift right (1)/left (0).
0	0	0	0	1	DL	N	F	*	*	Set Interface Length	DL—Set data interface length 8(1)/4(0). N—Number of display lines 1(0)/2(1). F—Character font 5—10(1)/5—7(0).
0	0	0	1	A	A	A	A	A	A	Move Cursor into *Character Generator RAM* (CGRAM)	A—Address.
0	0	1	A	A	A	A	A	A	A	Move Cursor to Display	A—Address.
0	1	BF	*	*	*	*	*	*	*	Poll the Busy Flag	BF—This bit is set while the LCD is processing.
1	0	D	D	D	D	D	D	D	D	Write a Character on the Display at the Current Cursor Position	D—Data.
1	1	D	D	D	D	D	D	D	D	Read the Character on the Display at the Current Cursor Position	D—Data.

*Not used/ignored. This bit can be either 1 or 0.

When describing how the LCD should be initialized in 4-bit mode, I will specify writing to the LCD in terms of nybbles. This is because initially only single nybbles are sent (not two, which make up a byte and full instruction). As I mentioned previously, when a byte is sent, the high nybble is sent before the low nybble and the E pin is toggled each time a nybble is sent to the LCD. To initialize in 4-bit mode, perform the following steps:

1. Wait more than 15 milliseconds after power is applied.
2. Write 0x03 to the LCD and wait 5 milliseconds for the instruction to complete.
3. Write 0x03 to the LCD and wait 160 μsecs for the instruction to complete.
4. Write 0x03 again to LCD and wait 160 μsecs (or poll the busy flag).
5. Set the operating characteristics of the LCD.
   a. Write 0x02 to the LCD to enable 4-bit mode.
      (All of the following instruction/data writes require two nybble writes).
   b. Write Set Interface Length.
   c. Write 0x00/0x08 (0x008) to turn off the display.
   d. Write 0x00/0x01 (0x001) to clear the display.
   e. Write Set Cursor Move Direction to set the cursor behavior bits.
   f. Write Enable Display/Cursor and enable the display and optional cursor.

Once the initialization is complete, the LCD can be written to with data or instructions as required. Each character to display is written like the control bytes, except that the R/S line is set. During initialization, by setting the S/C bit during the Move Cursor/Shift Display command, after each character is sent to the LCD, the cursor built into the LCD will increment to the next position (either right or left). Normally, the S/C bit is set (equal to 1) along with the R/L bit in the Move Cursor/Shift Display command for characters to be written from left to right (as with a teletype video display).

One area of confusion is how to move to different locations on the display and therefore how to move to different lines on an LCD display. Table 4-8 shows how different LCD displays that use a single 44780 can be set up with the addresses for specific character locations. The LCDs listed are the most popular arrangements available and the layout is given as a number of columns by a number of lines.

The ninth character is the position of the ninth character on the first line.

Most LCD displays have a 44780 and support chip to control the operation of the LCD. The 44780 is responsible for the external interface and provides sufficient control lines for 16 characters on the LCD. The support chip enhances the I/O of the 44780 to support up to 128 characters on an LCD. From Table 4-8, it should be noted that the first two entries (8×1 and 16×1) only have the 44780, not the support chip. This is why the ninth character in the 16×1 does not appear at address 8 and shows up at the address that is common for a two-line LCD.

Table 4-8

Most Popular LCD Arrangements

LCD Layout	Top-Left Character	Ninth Character	Second Line	Third Line	Fourth Line	Comments
8×1	0	N/A	N/A	N/A	N/A	Single 44780/no support chip
16×1	0	0x040	N/A	N/A	N/A	Single 44780/no support chip
16×1	0	8	N/A	N/A	N/A	44780 with support chip (this is quite rare)
8×2	0	N/A	0x040	N/A	N/A	Single 44780/no support chip
10×2	0	8	0x040	N/A	N/A	44780 with support chip
16×2	0	8	0x040	N/A	N/A	44780 with support chip
20×2	0	8	0x040	N/A	N/A	44780 with support chip
24×2	0	8	0x040	N/A	N/A	44780 with support chip
30×2	0	8	0x040	N/A	N/A	44780 with support chip
32×2	0	8	0x040	N/A	N/A	44780 with support chip
40×2	0	8	0x040	N/A	N/A	44780 with support chip
16×4	0	8	0x040	0x020	0x060	44780 with support chip
20×4	0	8	0x040	0x020	0x060	44780 with support chip

Cursors for the 44780 can be turned on as a simple underscore at any time using the Enable Display/Cursor LCD instruction and setting the C bit. I don't recommend using the B (block mode) bit because it causes a flashing full character square to be displayed and really isn't very attractive.

The LCD can be thought of as a teletype display because in normal operation, after a character has been sent to the LCD, the internal cursor is moved one character to the right. Note that when the cursor reaches the end of a line, the display does not scroll. The Clear Display and Return Cursor and LCD to Home Position instructions are used to reset the cursor's position to the top-right character on the display.

To move the cursor, use the Move Cursor to Display instruction. For this instruction, bit 7 of the instruction byte is set and the remaining 7 bits are used as the address of the character on the LCD that the cursor should move to. These 7 bits provide 128 addresses, which matches the maximum number of LCD character addresses available. Table 4-8 should be used to determine the address of a character offset on a particular line of an LCD display.

The character set available in the 44780 is basically ASCII. I say basically because some characters do not follow the ASCII convention fully (probably the most significant difference is 0x05B or \ is not available). The ASCII control characters (0x008 to 0x01F) do not respond as control characters and may display unexpected (Japanese) characters.

Eight programmable characters are available, which use codes 0x000 to 0x007. They are programmed by pointing the LCD's cursor to the CGRAM area at eight times the character address. The next 8 bytes written to the RAM are each lines of the programmable character, starting at the top. The user-defined character line information is saved in the LCD's CGRAM area. These 64 bytes of memory are accessed using the Move Cursor into CGRAM instruction in a similar manner to that of moving the cursor to a specific address in the memory; however, there is one important difference.

The difference is that each character starts at eight times its character value. This means for the user-defined character 0, the data starts at address 0 of the CGRAM, character 1 starts at address 8, character 2 starts at address 0x010 (16), and so on. To get a specific line within the user-defined character, its offset from the top (the top line has an offset of 0) is added to the starting address. In most applications, characters are written to all at one time with character 0 first. In this case, the instruction 0x040 is written to the LCD followed by all of the user-defined characters.

The last aspect of the LCD to discuss is how to specify a contrast voltage to the display. I typically use a potentiometer wired as a voltage divider, as shown in Figure 4-43. This provides an easily variable voltage between the ground and Vcc, which is used to specify the contrast (or darkness) of the characters on the LCD screen. You may find that different LCDs work differently with lower and higher voltages, providing darker characters in some voltages.

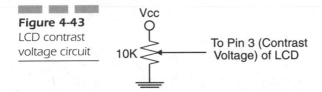

Figure 4-43
LCD contrast
voltage circuit

There are a variety of different ways to wire up an LCD. Previously, I noted that the 44780 can interface with 4 or 8 bits. To simplify the demands in microcontrollers, a shift register is often used to reduce the number of I/O pins to three. This can be further reduced by using the circuit shown in Figure 4-44, which combines the incoming serial data line with the previously loaded contents of the shift register to produce the E strobe after the shift register has been loaded.

This circuit ANDs (using the 1K resistor and IN914 diode) the output of the sixth D-flip-flop of the shift register (I normally use a 74LS174) and the data bit from the device writing to the LCD to form the E strobe. This circuit is ideal for use in robots as it minimizes the pins that the microcontroller is expected to drive.

I normally use a 74LS174 wired as a shift register instead of a serial-in/parallel-out shift register. This circuit should work without any problems with a dedicated serial-in/parallel-out shift register chip, but the timings/clock polarities may be different. When the 74LS174 is used, note that the data is latched on the rising (from logic low to high) edge of the clock signal. The data and clock waveforms that output to the LCD from the 74LS174 are shown in Figure 4-45.

Before data can be written to it, the shift register is cleared by loading every latch with zeros. Next, a 1 (to provide the E gate) is written followed by the R/S bit and the 4 data bits. Once the data is loaded in correctly, the data line is pulsed to strobe the E bit. The biggest difference between the 3- and 2-wire interface is that the shift register has to be cleared before it can be loaded and the 2-wire operation requires more than twice the number of clock cycles to load 4 bits into the LCD.

It is important to watch for the E strobe's timing to make sure that it is within specification (that is, greater than 450 nanoseconds). The shift register loads can be interrupted without affecting the actual write. This circuit does not work with open-drain-only outputs (something that catches up many people).

One note about the LCD's E strobe is that in some documentation it is specified as high-level active, whereas in others it is specified as falling-edge active. It seems to be falling-edge active, which is why the two-wire LCD interface presented in the following section works even if the line ends up being high at the end of data being shifted in. If the falling edge is used (like in the two-wire interface), then make sure that the E line is output on 0. There is at least a 450-nanosecond delay with no lines changing state.

Figure 4-44
Two-wire LCD
interface circuit

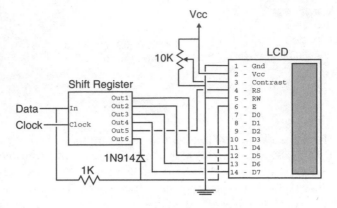

Figure 4-45
Two-wire LCD
interface circuit
waveform

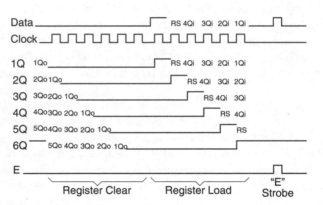

LCD—Sample Two-Wire LCD Interface

The two-wire interface presented in the previous section will be used to demonstrate the operation of an LCD. This interface was selected because it requires a limited number of I/O pins. I'm sure this interface seems like it is quite (and perhaps needlessly) complex, but I would disagree. With a bit of planning, this interface can be wired quite easily into a robot controlled by a PIC16F627. The same comments apply to the software. You might implement an LCD interface as if it was the only peripheral interface in the system, but the interface can be made to operate exactly the same way as other peripheral interfaces and not affect other peripherals or the biologic code.

The circuit that I used for my prototypes is shown in Figure 4-46 and uses a 74LS174 as a shift register. I tend to use 74LS174s and 74LS374s (as well as 74LS373s, 74LS573s, and 74LS574s) for as many applications as I can, including,

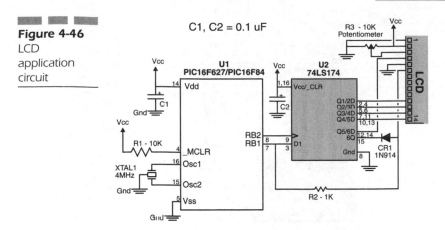

Figure 4-46
LCD
application
circuit

as in this case, shift registers. This cuts down on the number of chips that I have to stock and enables me to be as comfortable with the chips as possible. I'm pointing this out because there are some TTL shift registers that could be used in place of the 74LS174 if you are more comfortable with using them.

The two-wire LCD interface fits on a small breadboard, as shown in Figure 4-47. I have a few of notes for wiring this application on a small breadboard. The first is that I have put the AND circuit (the resistor and diode) on an unused section of breadboard, along with the contrast potentiometer. Secondly, to eliminate the need to run wires over the LCD, I moved the LCD to the outside row of holes where the LCD's Vcc and Gnd connections are shared with the contrast potentiometer. The breadboard's second Vcc and Gnd rail connections are hidden beneath the LCD in Figure 4-47. The circuit is a bit tight, but everything fits on it without too many wires being obscured. Table 4-9 lists the bill of materials for the circuit.

To test out this circuit, I created some write software that would be appropriate to use in an application where the LCD display is the only microcontroller interface. When the LCD is initialized or data is displayed upon it, the biologic code is responsible for the data I/O operations. So far in this chapter, I have emphasized the importance of designing your interface software so that it can be easily integrated with other interfaces in the robot. I deviated from this rule to demonstrate how the application could be written and test out my circuitry to make sure there were no mistakes.

The initial application code can be found in robot\lcd and is called lcd.c:

```
#include <pic.h>

//   02.01.27 - LCD: Display "Hello World" on an LCD
//
//   This is a two wire interface using a 74LS174 as a shift
//     register and the code called from the mainline (and timed
//     from TMR0 interrupt)
```

Figure 4-47
LCD
breadboard
circuit

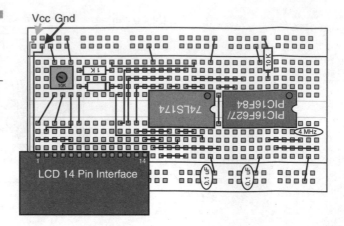

Vcc Gnd

LCD 14 Pin Interface

Table 4-9

Materials for
the LCD
Application
Circuit

Label	Part Number	Comment
U1	PIC16F627	Can use PIC16F84
U2	74LS174	
CR1	Silicon diode	Uses any type, but 1N914 is used for prototypes
R1	10K, ¼ watt	Uses _MCLR pull-up
R2	1K, ¼ watt	
R3	10K, 1 turn pot	Requires LCD contrast
C1 and C2	0.1 µF	Uses any type of decoupling capacitor
LCD		Hitachi-44780-controlled LCD display with a 14-pin interface
XTAL1	4 MHz ceramic resonator with internal capacitors	
Misc.		Requires a breadboard, wiring, and +5-volt power supply

```
//
//
//   Hardware Notes:
//   PIC16F84/PIC16F627 Running at 4 MHz
//   PIC16F84 implementation requires a 4 MHz Ceramic Resonator
//   RB1 - CLock Signal to the LCD
//   RB2 - Data Signal to the LCD
//
```

```c
// Global Variables
int RTC = 0;                            // Real Time Clock Counter

int Dlay;                               // Initialization Delay Value
static volatile bit Clock      @ (unsigned)&PORTB*8+1;
static volatile bit ClockTRIS @ (unsigned)&TRISB*8+1;
static volatile bit Data       @ (unsigned)&PORTB*8+2;
static volatile bit DataTRIS  @ (unsigned)&TRISB*8+2;

char Message[13] = "Hello World!";  // Message to Display


// Configuration Fuses
#if defined (_16F84)
#warning PIC16F84 selected
        CONFIG(0x03FF1);                // PIC16F84 Configuration
                                        // Fuses:
                                        //  - XT Oscillator
                                        //  - 70 msecs Power Up Timer
                                        //  - Watchdog Timer Off
                                        //  - Code Protection Off
#elif defined(_16F627)
#warning PIC16F627 with external oscillator selected
        _CONFIG(0x03F61);               // PIC116F627 Configuration
                                        // Fuses:
                                        //  - External Oscillator
                                        //  - RA6/RA7 Digital I/O
                                        //  - External Reset
                                        //  - 70 msecs Power Up Timer
                                        //  - Watchdog Timer Off
                                        //  - Code Protection Off
                                        //  - BODEN Enabled
#else
#error Unsupported PICmicro MCU selected
#endif


// Interrupt Handler
void interrupt tmr0_int(void)       // TMR0 Interrupt Handler
{

     if (T0IF) {

             T0IF = 0;                  // Reset Interrupt Flag

             RTC++;                     // Increment the Clock

// Put additional interface code for 1 msec interrupt here

     } // endif

// Put Different Interrupt Handlers here

} // End Interrupt Handler
```

```
// Subroutines
LCDNybble(char Nybble, char RS) {    // Send Nybble to LCD

unsigned int i;

        Data = 0;                      // Clear the '174
        for (i = 0; i < 6; i++) {      // Repeat for six bits
             Clock = 1; Clock = 0;     // Write the "0"s into the '174
        }  // endfor

        Data = 1;                      // Output the "AND" Value
        Clock = 1; Clock = 0;

        Data = RS;                     // Output the RS Bit Value
        Clock = 1; Clock = 0;

        for (i = 0; i < 4; i++) {      // Output the Nybble
            if ((Nybble & 0x008) != 0)
                  Data = 1;            // Output the High Order Bit
            else
                  Data = 0;
            Clock = 1; Clock = 0;      // Strobe the Clock
            Nybble = Nybble << 1;      // Shift up Nybble for Next
                                       // Byte
        }  // endfor

        Data = 1; Data = 0;            // Toggle the "E" Clock Bit

}  // End LCDNybble

LCDByte(char Byte, char RS) {          // Send Byteto LCD
int LBDlay;

        LCDNybble((Byte >> 4) & 0x00F, RS);    // Send High Nybble
        LCDNybble(Byte & 0x00F, RS);  // Send Low Nybble

        if ((Byte < 4) && (RS == 0))  // Type of Instruction?
              LBDlay = RTC + 6;        // Wait at Least 5 msecs
        else                           // Minimum 160 µsec Dlay
              LBDlay = RTC + 2;
        while(LBDlay != RTC);          // Command to Complete Wait

}  // End LCDByte



// Mainline
void main(void)                                // Template Mainline
{
int i;

        OPTION = 0x0D1;                // Assign Prescaler to TMR0
                                       //  Prescaler is /4
        TMR0 = 0;                      // Reset the Timer for Start
        T0IE = 1;                      // Enable Timer Interrupts
        GIE = 1;                       // Enable Interrupts

// Put in Interface initialization code here
```

```
    Clock = 0; Data = 0;              // Low I/O Bits
    ClockTRIS = 0; DataTRIS = 0;

    Dlay = RTC + 20;                 // Step 1 - Wait > 15 msecs
    while (Dlay != RTC);             //   for LCD Power up

    LCDNybble(0x003, 0);             // Step 2 - Send Init Char
    Dlay = RTC + 6;                  // Wait > 5 msecs for
    while (Dlay != RTC);             //   LCD to Accept Command

    LCDNybble(0x003, 0);             // Step 3 - Send Init Char
    Dlay = RTC + 1;                  // Wait > 160 µsecs for
    while (Dlay != RTC);             //   LCD to Accept Command

    LCDNybble(0x003, 0);             // Step 4 - Send Init Char
    Dlay = RTC + 1;                  // Wait > 160 µsecs for
    while (Dlay != RTC);             //   LCD to Accept Command

    LCDNybble(0x002, 0);             // Step 5 - Set Operating
    Dlay = RTC + 1;                  //   Interface Size (4 Bits)
    while (Dlay != RTC);

    LCDByte(0x028, 0);               // Step 6 - Set Operating

    LCDByte(0x008, 0);               // Step 7 - Display Off

    LCDByte(0x001, 0);               // Step 8 - Clear Display

    LCDByte(0x006, 0);               // Step 9 - Shift

    LCDByte(0x00E, 0);               // Step 10 - Display On

    for (i = 0; i < 12; i++)         // Print out "Hello World!"
        LCDByte(Message[i], 1);

    while (1 == 1) {                 // Loop forever

//  Put in Robot high level operation code here

    }  //  endwhile

}  //  End of Mainline
```

Not many aspects of this application code should surprise you. I retained the template's 1-millisecond interrupt with the RTC variable increment to have a simple timer for putting in delays and sequencing the operations of the LCD. This code is quite simple and reliable. I would not want to put it into a robot's controller because it uses the section of the application devoted to the robot's bio-logic code.

To convert this application to execute as an elelogic interface during the 1-millisecond interrupt handler, I converted the initialization and string output code into a state machine that executes every time the interrupt handler is invoked. In some cases, it simplifies the operation of the delays, whereas other times it simplifies how the interface code executes.

The updated application is known as lcd2.c, which can be found in the same directory as lcd.c (robot\lcd):

```c
#include <pic.h>

//   02.04.17 - LCD2: Handle LCD Functions as part of the robot
//    Paradigm
//   02.01.27 - LCD: Display "Hello World" on an LCD
//
//   This is a two wire interface using a 74LS174 as a shift
//    register and the code called from the mainline (and timed
//    from TMR0 interrupt)
//
//
//   Hardware Notes:
//   PIC16F84 Running at 4 MHz
//   RB1 - CLock Signal to the LCD
//   RB2 - Data Signal to the LCD
//

//   Global Variables
int RTC = 0;                    //   Real Time Clock Counter

volatile char LCDDlay = 20;     //   Initialization Delay Value
volatile char LCDState = 1;     //   Current LCD State

static volatile bit Clock     @ (unsigned)&PORTB*8+1;
static volatile bit ClockTRIS @ (unsigned)&TRISB*8+1;
static volatile bit Data      @ (unsigned)&PORTB*8+2;
static volatile bit DataTRIS  @ (unsigned)&TRISB*8+2;

char * MessageOut;              //   Message to be Sent Out
volatile char MessageOuti = 0;         //   Index to current Message
Byte
char Message[13] = "Hello World!";  //  Message to Display
char Message2[11] = "\376\3002nd Line";


//   Configuration Fuses
#if defined (_16F84)
#warning PIC16F84 selected
        CONFIG(0x03FF1);        //   PIC16F84 Configuration Fuses:
                                //   - XT Oscillator
                                //   - 70 msecs Power Up Timer On
                                //   - Watchdog Timer Off
                                //   - Code Protection Off
#elif defined(_16F627)
#warning PIC16F627 with external oscillator selected
        CONFIG(0x03F61);        //   PIC116F627 Configuration Fuses:
                                //   - External Oscillator
                                //   - RA6/RA7 Digital I/O
                                //   - External Reset
                                //   - 70 msecs Power Up Timer On
                                //   - Watchdog Timer Off
                                //   - Code Protection Off
                                //   - BODEN Enabled
#else
#error Unsupported PICmicro MCU selected
```

```
#endif


//  Subroutines
LCDNybble(char Nybble, char RS) {    //  Send Nybble to LCD

unsigned int i;

        Data = 0;               // Clear the '174
        for (i = 0; i < 6; i++) {      // Repeat for six bits
            Clock = 1; Clock = 0;    // Write the "0"s into the '174
        }  //  endfor

        Data = 1;                   // Output the "AND" Value
        Clock = 1; Clock = 0;

        Data = RS;               // Output the RS Bit Value
        Clock = 1; Clock = 0;

        for (i = 0; i < 4; i++) {       // Output the Nybble
            if ((Nybble & 0x008) != 0)
                Data = 1;    // Output the High Order Bit
            else
                Data = 0;
            Clock = 1; Clock = 0;    // Strobe the Clock
            Nybble = Nybble << 1;    // Shift up Nybble for Next
                                     // Byte
        }  //  endfor

        Data = 1; Data = 0;       // Toggle the "E" Clock Bit

}  //  End LCDNybble

LCDByte(char Byte, char RS) { //  Send Byteto LCD

        LCDNybble((Byte >> 4) & 0x00F, RS); //  Send High Nybble
        LCDNybble(Byte & 0x00F, RS);         //  Send Low Nybble

}  //  End LCDByte


LCDInit()                     //  Initialize the LCD I/O Pins
{

        Clock = 0; Data = 0;     //  Low I/O Bits
        ClockTRIS = 0; DataTRIS = 0;

}  //  End LCDInit


LCDOut(char * const LCDString)     //  Output the Data String
{

        while (LCDState);      //  Wait for LCD Available

        MessageOut = LCDString; //  Load up the string
        LCDState = 100;         //  Start Sending the String

}  //  End LCDOut
```

```
//   Interrupt Handler
void interrupt tmr0_int(void) //  TMR0 Interrupt Handler
{

char temp;

     if (T0IF) {

               T0IF = 0;           //  Reset Interrupt Flag

               RTC++;              //  Increment the Clock

//  Put additional interface code for 1 msec interrupt here

//   LCD State Machine

               switch(LCDState) {       //  Process State Machine
                    case 1:       //  Start LCD Initialization Process
                         if (--LCDDlay == 0)
                                LCDState++;
                         break;       //  Wait 20 msecs for LCD to
                                      //  reset
                    case 2:
                         LCDNybble(0x003, 0);      //  Step 2 - Send
                                                   //  Init
                         LCDDlay = 5;
                         LCDState++;
                    case 3:       //  Wait for Command to Complete
                         if (--LCDDlay == 0)
                                LCDState++;
                         break;
                    case 4:
                         LCDNybble(0x003, 0);      //  Step 3 - Send
                                                   //  Init
                         LCDState++;
                         break;
                    case 5:
                         LCDNybble(0x003, 0);      //  Step 4 - Send
                                                   //  Init
                         LCDState++;
                         break;
                    case 6:
                         LCDNybble(0x002, 0);      //  Step 5 - Set
                         LCDState++; //   Operating Interface Size
                         break;      //   (4 Bits)
                    case 7:
                         LCDByte(0x028, 0);        //  Step 6
                         LCDState++;
                         break;
                    case 8:
                         LCDByte(0x008, 0);        //  Step 7 - Disp.
                                                   //  Off
                         LCDState++;
                         break;
                    case 9:
                         LCDByte(0x001, 0);        //  Step 8 - Clear
                         LCDState++;
```

```
                        LCDDlay = 5;        //  Wait 5 msecs to
                                            //  Complete
                break;
           case 10:
                if (--LCDDlay == 0)
                     LCDState++;
                break;
           case 11:
                LCDByte(0x006, 0);       //  Step 9 - Shift
                LCDState++;
                break;
           case 12:
                LCDByte(0x00E, 0);       //  Step 10 - Disp.
                                         //  On
                LCDState = 0;     //  Ready to Run
                break;

           case 100:          //  Output Character in
                              //  MessageOut
                switch (temp = MessageOut[MessageOuti++]) {
                     case '\0':  //  At the End of the
                                 //  Message
                          LCDState = 0;
                          MessageOuti = 0;  //  Reset
                                            //  Index
                          break;
                     case '\f':  //  Clear Display
                          LCDByte(0x001, 0);
                          LCDState++;
                          LCDDlay = 5;
                          break;
                     case 254:   //  Command Byte follows
                          if ((temp = _
          MessageOut[MessageOuti++]) == 0)
                                    LCDState = 0;
                          else {
                               if (temp < 4) {
                                    LCDState++;
                                    LCDDlay = 5;
                               } //  endif
                               LCDByte(temp, 0);
                          } //  endif
                          break;
                     default:    //  All other characters
                          LCDByte(temp, 1);
                } //  endswitch
                break;
           case 101:            //  Message Delay
                if (--LCDDlay == 0)
                     LCDState--;
                break;
      } //  endswitch

   } //  endif

//  Put Different Interrupt Handlers here

} //  End Interrupt Handler
```

```
//  Mainline
void main(void)                         //  Template Mainline
{

    OPTION = 0x0D1;                     //  Assign Prescaler to TMR0
                                        //   Prescaler is /4
    TMR0 = 0;                           //  Reset the Timer for Start
    T0IE = 1;                           //  Enable Timer Interrupts
    GIE = 1;                            //  Enable Interrupts

    LCDInit();                          //  Initialize the LCD Port

//  Put in Interface initialization code here

    LCDOut(Message);                    //  Pass String to Output

    LCDOut(Message2);                   //  2nd Line Message

    while (1 == 1) {                    //  Loop forever

//  Put in Robot high level operation code here

    } //  endwhile

} //  End of Mainline
```

If you compare lcd2.c with lcd.c, you will see that the LCDNybble and LCDByte functions are used between the two with basically no modification. The only modification to the functions is the elimination of the computed delay in LCDByte.

Another addition to the application is the LCDOut function, which copies the pointer to the message to be displayed when the display state machine has finished displaying the current string. It is possible that by invoking this function, the biologic code might hang up so you may want to add the following function:

```
int LCDPoll()                          //  Indicate when the LCD is ready to
{                                       //   display another message

    if (LCDState)
            return 0;                  //  LCD Active, Can't send a new
string
        else
            return 1;                  //  LCD NOT Active.  Can send new
string

} //  End LCDPoll
```

This function polls the LCD state and returns true (not zero) when the LCD state machine is not initializing the LCD display or displaying a string.

Notice that some enhancements were made to the LCD interface. The first one is that the string passed to LCDOut prints all the characters in the string until it encounters a Null (\0) string. If you put a \f (form feed) character in the string, the LCD will be cleared.

LCD commands can be sent to the LCD display by prefacing them with the character decimal 254 (hex 0x0FE) followed by the character of the command. Character decimal 254 is a convention used by many commercial serial LCD interfaces to indicate that an LCD command should be passed instead of a character. In this application, if the 254 character is encountered, then the next character is sent to the LCD as an instruction. Normally, every character encountered in the string will be sent as a character.

The easiest way of adding the LCD commands with the PICC Lite compiler is to use the C ASCII character specification (a backslash followed by three octal digits). For example, to clear the LCD display (send instruction 0x001 to the LCD), the string \376\001 would be sent to LCDOut.

Table 4-10 lists some of the commonly used LCD commands in the 254 character format.

To use these commands to clear the LCD display and write the following two-line message to the LCD with the cursor turned off afterward,

```
Robot Moving
Forward
```

use the following string:

```
"\fRobot Moving\376\300Forward\376\014\0"
```

One last point needs to be made about putting interface code in the elelogic/mechalogic format. You may be under the impression that this format is less efficient than writing interface code normally without regard to other interfaces or code running in the system. If you build both lcd.c and lcd2.c, you will be surprised to discover that lcd2.c requires two fewer instructions than lcd.c, even

Table 4-10

LCD Commands in the 254 Character Format

Command	Character 254 Command String
Clear display	\376\001
Return cursor to home	\376\002
Turn off display	\376\010
Turn on display, no cursor	\376\014
Turn on display, cursor active	\376\016
Move cursor to address 0x040	\376\300

though lcd2.c provides more functions (handling \f and the character 254 command strings) than lcd.c.

I make this point because you may want to follow the guidelines I have used for all of your microcontroller interfacing (not just for robots). When you develop your interfaces using a whole-system approach, you will end up with multiple interfaces that can coexist without major concerns and will probably be smaller than the sum of interfaces that are designed for unique applications.

If you are going through these applications and building them to see how the sample interfaces work, you should leave this one alone as it is used for most of the remaining applications in this chapter. To save space and avoid confusing the code presented for the other applications, the LCD function code is deleted in the other applications.

Sensors

I always think of the television show "Star Trek" when I hear the word sensors because they represented nebulous devices that could detect substances, energy signatures, and, of course, life forms. These sensors generally fit the description of Alfred Hitchcock's MacGuffins—a plot device to move the story forward. They weren't central to the plot, but they were requested to pass information regarding the surrounding environment to characters and did not make conclusions about what's out there or decide on an appropriate reaction. When designing robots, the sensors that are built into the robots follow this model; they provide information to biologic code and the sensor code does not initiate actions (the role of biologic code).

Robot sensors can provide information from many different sources, but the actual mechanism isn't important (and is hidden from the biologic code). The following sections show some examples of sensors with code. There are many different ways of implementing these functions and other books in this series delve into the topic of sensors in much more detail. The purpose here is to introduce you to different types of sensors, explain how software is written to access the encompassed hardware, and discover how sensors interface to the biologic code of the robot application.

Although I discuss many different devices, still many other sensor types are not addressed in this book, including the following:

- *Global Positioning System* (GPS)
- Compass
- Tilt
- Video

Despite the many different types of robot sensors, each type of sensor should be implemented as shown in this chapter, following the model of the three different code spectrums and the Star Trek model.

It is a good idea to put LEDs, buzzers, or other output devices on the robot to indicate the sensor states. These output devices can either be connected directly to the sensors or switched by the input devices themselves. By providing these feedback mechanisms to the robot, you can gain a better idea of why the robot responds in a particular way instead of trying to decode what is happening from what you believe the inputs actually are.

Just one word of caution—remember that you can affect the operation of the robot if the robot senses you in some way or if your body affects the environment in which the robot is running (that is, if you block the light or cast a shadow). Ideally, your robot's sensors would only try to understand what is around it and ignore any closeby life forms.

Whiskers for Physical Object Detection

Virtually all robots that I have seen have some form of object detection. The purpose of the object detection is to help the robot navigate in its environment and prevent the robot from becoming stalled by some object. Personally, I find whiskers to be very awkward to implement and difficult to make robust. For these reasons, I try to avoid using them unless there is no way around them. This does not mean that I do not have some form of object detection in my robots; it just means that I do not use whiskers.

Object detection using whiskers or physical object sensors is implemented similarly to and serve the same purpose on robots as they do on cats. A long, light, flexible rod is mounted to a sensor (a microswitch) that is secured to the robot's body, as shown in Figure 4-48. The purpose of the whisker is to detect objects in front of the robot in time for the robot to respond and avoid hitting the object. Ideally, the object is detected with enough time for the robot to respond either by stopping or turning away from the object.

A majority of developers place whiskers at the front of the robot, but in many applications, this does not provide enough protection for the full operation of the robot. Whiskers can be added to the rear and corners of the robot to indicate that a collision is possible when it is moving in reverse or turning. The side-sensing whiskers are always used for maze-solving robots, which keep one side's whisker in contact with a wall of the maze.

The best whisker/bump sensor implementation is a wire connected to a microswitch (as shown in Figure 4-48). When implemented, the fulcrum provides

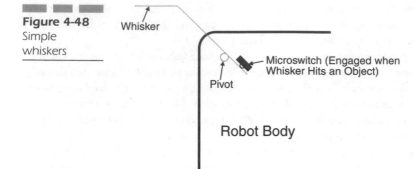

Figure 4-48
Simple
whiskers

amplified force to microswitch contact from the whisker/bump sensor. This reduces the force applied to the whisker for a positive contact and lessens switch bouncing.

When a switch opens or closes, the contacts within it bounce (as shown in Figure 4-49), and open and close the connection very quickly. Even if you use a microswitch to minimize the opportunities for the whiskers to return intermittent contacts during motion, you will still have to filter out bounces. Depending on the time between the bounces, the robot's controller may have issued a response command after noting the first bounce contact and invalidly responds when the second bounce is received.

The rule of thumb for standard switches is that the contacts should not change for at least 20 milliseconds before the signal can be considered debounced. This is a reasonable delay for robots with microswitch-based whiskers. The standard method that I use for reading and debouncing buttons in a microcontroller system is as follows:

```
while (1 == 1) {            //  Loop Forever

    :                       //  Perform tasks outside of button
                            //  press

    if (Button == Press) {  //  Wait for Button Press to stay in
                            //  the
        Debounce = 0;       //    same state for 20 msecs
        while ((Debounce != 20msecs) && (Button == Press))
            for (Debounce = 0; (Button == Press) && _
                (I < 20msecs ); Debounce++);

//  Put in Code that responds to the button being pressed

        Debounce = 0;       //  Wait for Button Released for 20
                            //  msecs
        while ((Debounce != 20msecs) && (Button == Release))
            for (Debounce = 0; (Button == Release) && _
```

Figure 4-49
Oscilloscope
picture of a
switch bounce

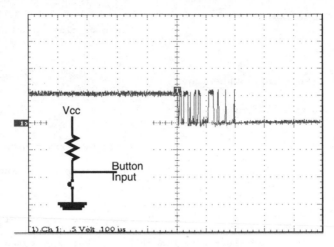

```
                        (I < 20msecs ); Debounce++);

        } // endif

    :                           //  Tasks outside of button press

} // endwhile
```

Depending on how you have implemented your robot, you may find that you have intermittent contacts on your whiskers as the robot runs across the floor. When using a microswitch with an internal spring, keeping the contacts open is the best way to avoid these intermittent contacts as the robot is running. If you do not use a microswitch (that is, a grounded whisker comes in contact with a pulled-up I/O pin contact), then you should make sure that your debounce routine ignores any intermittent bounces. If it doesn't and you use the code previously outlined, you will find yourself in a situation where the robot controller responds to the first bounce and waits for the contact to be debounced.

To avoid this problem, use flags that are set for a collision only when 20 milliseconds have gone by without a change in the whisker's state. Taking advantage of the 1-millisecond interrupt handler that the interface code is based on, the mainline code is as follows:

```
while (1 == 1) {                // Loop Forever

    :                           // Perform tasks outside of button
                                // press

    if (ButtonPressFlag) { //  Wait for Debunced Button Press

//  Put in Code that responds to the button being pressed

        } // endif
```

```
      :                          //  Tasks outside of button press

}  //   endwhile
```

with the interrupt handler code:

```
void tmr0_int interrupt(void)
{

      if (T0IF) {              //  1 msec Interrupt handler

          T0IF = 0;            //  Reset Timer 0 Interrupt Request

          RTC++;

      :                        //  Other TMR0 Intt Handler Functions

          if (Button == Press) {  //  Button Pressed
              if (ButtonReleaseCounter != 0) {  //  First Time
                  ButtonPressCounter = 0;
                  ButtonReleaseFlag = 0;
              } else if (++ButtonPressCounter >= 20) {
                  ButtonPressFlag = 1;    //  Debounced
                  ButtonPressCounter = 19;
              } else;
          } else {                 //  Button Released
              if (ButtonPressCounter != 0) {  //  First Time
                  ButtonReleaseCounter = 0;
                  ButtonPressFlag = 0;
              } else if (++ButtonReleaseCounter >= 20) {
                  ButtonReleaseFlag = 1;    //  Debounced
                  ButtonReleaseCounter = 19;
              } else;
          }  //   endif
      }  // endif

}  //   End Interrupt Handler
```

In this interrupt-based code, if the whisker has an intermittent contact, then the ButtonRelease flag is cleared and the ButtonPressCounter starts incrementing until the whisker is released. When the whisker is released, the ButtonReleaseCounter is reset and counts up to 20 for the ButtonRelease flag to set again. The advantage of this method is that the positive setting of the debounced state of the whisker that the biologic code can poll and respond to when there is a debounced whisker condition.

The whiskers you put on a robot should *always* be metal and grounded. Running a robot across the floor can cause a static charge buildup. This is especially important if the robot is used to follow a maze (and is constantly rubbing against something). The whiskers are an excellent way of discharging the robot when it comes in contact with another object. If plastic whiskers are used, then the static charge may not be dissipating, leading to possible damage to the robot when you pick it up or it comes into contact with another object.

Along with the simple wire and microswitch whiskers presented in this section, the following sensors provide the same functions to the robot:

- A metal ring around the robot could be an easily deformable grounded ring that changes the logic level pulled-up sensor when the robot collides with an object. I have seen this method of providing mechanical collision detection implemented in quite a few robots. This type of sensor is good for sensing the direction of the collision (which is useful in sumo matches), although care must be taken to make sure that it doesn't register a collision on robot stops or starts. Depending on how the ring is mounted, inertia may cause it to come in contact with the contacts.

- Another type of contact sensor could be a rubber tube run around a robot, which changes its internal pressure on contact. This can be difficult to implement a sensor on, even though it acts like a bumper for the robot and is very insensitive to robot motion, inertia, and vibration.

- Motor stall detectors can be used to detect when the robot has collided with an object and cannot continue moving in the required direction. The problems with this method are the increased current required by the motors when they are stalled and the possibility that the robot was damaged by the collision. Stalling the motors can cause problems with the motors and motor drivers, potentially burning out and definitely resulting in a lower battery life for the robot.

Whiskers are difficult to implement and work with because they require a lot of experimentation in order to properly place the switch and the pivot as well as determine the correct length and shape of wire. Along with this, I find that if a wire whisker is used, it often becomes deformed when the robot runs under something, hits something at full speed, or one of my kids picks up the robot by the wire.

When given a choice, I prefer a noncontact method of object detection using either the IR object detectors described in the following section or ultrasonic ranging modules. This preference is somewhat atypical because many people consider using whiskers as the most reliable way of detecting objects around the robot (the noncontact methods may ignore objects that do not reflect the transmitted signal back to the robot).

Whisker Debounce

When I first created the 1-millisecond TMR0 interrupt as a basis for robot elelogic and mechalogic interfaces, I was thinking in part of the need for debouncing whisker inputs, which act like buttons commonly used in electronic circuits. Having code that executes once every millisecond enables the application to poll input pins connected to whiskers without being a large drain of processor resources. This was an important requirement because I wanted to make sure that I could add as many whiskers to the robot as possible.

The task of continuously polling a whisker pin and incrementing/decrementing a counter to 20 to debounce the input is accomplished by the following lines of code:

```
if (!Pin)          //  If pin pull low
     if (ButtonPress < 20)
          ButtonPress++;
     else;         //  Button press debounced
else              //  Pin floating high
     if (ButtonPress != 0)
          ButtonPress--;
     else;         //  Button release debounced
```

The first line polls the whisker input pin and if it is pressed (pulled low), the ButtonPress counter is incremented to 20. If the counter is already 20, then the counter is left alone. When the whisker is released (the pin returns to a high state), the counter is decremented to 0.

The biologic code that processes the data can be as simple as the following:

```
if (ButtonPress == 0) {

//  Code to execute when the button is released

} else if (ButtonPress == 20) {

//  Code to execute when the button is pressed

} else {

//  Code to execute when the button state is being debounced

}  //  endif
```

This code and the previous snippet can be implemented as many times as required for each whisker.

To demonstrate the operation of this whisker debounce code, I created the whisker application circuit, which is a simple modification of the LCD circuit (see Figure 4-50) with the addition of a button, LED, and resistor. Although the LED and current-limiting resistor are not really needed for this application, they can be extremely useful for your robot applications and enable you to see what kind of input the robot's processor is receiving.

The LCD display is used to display the current status of the button press. When the button is released and RB0 is being held at a high voltage level, the LCD is blank. When the button state is being debounced, the LCD displays debouncing. PRESSED is displayed when the button is down and debounced. The ButtonPressed counter is continuously polled in the biologic code and its state according to these three conditions is displayed on the LCD. When you are using

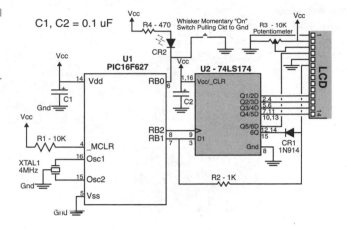

Figure 4-50
Whisker
application
circuit

the application, you should see the debouncing message display briefly when you press and release the button.

To wire the circuit, you can use the LCD base, as shown in Figure 4-51. When selecting parts, it is recommended that you find as poor a switch as possible. This application demonstrates the operation of the debounce capabilities of the code as much as possible. In a robot, I recommend using a microswitch with a positive click, which minimizes contact bouncing during collisions. By using a switch that does not have a positive click, you should maximize the bouncing that the microcontroller has to eliminate. Table 4-11 shows the bill of materials for this application.

The code I created to demonstrate the operation of the whisker debounce is whisker.c, which can be found in robot\whisker:

```
#include <pic.h>

//   02.04.19 - WHISKER: Debounce Pulled Up Switch Input.
//
//   This is a two wire interface using a 74LS174 as a shift
//    register and the code called from the mainline (and timed
//    from TMR0 interrupt)
//
//
//   Hardware Notes:
//   PIC16F827 Running at 4 MHz with External Oscillator
//   RB0 - Pulled Up Switch Input (NOT Using Internal PORTB Pull
//   Ups)
//   RB1 - CLock Signal to the LCD
//   RB2 - Data Signal to the LCD
//

//   Global Variables
int RTC = 0;                        //   Real Time Clock Counter
```

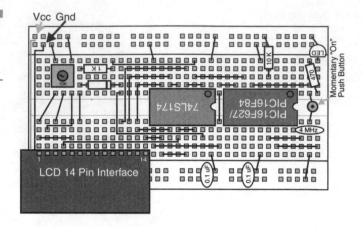

Figure 4-51
Whisker
breadboard
circuit

Table 4-11

Materials for
the Whisker
Breadboard
Circuit
Application

Label	Part Number	Comment
U1	PIC16F627	Can use a PIC16F84
U2	74LS174	
CR1	Silicon diode	Uses any type, but 1N914 is used for prototypes
CR2	LED	Uses any type
R1	10K, ¼ watt	Uses _MCLR pull-up
R2	1K, ¼ watt	
R3	10K, 1 turn pot	Requires LCD contrast
R4	470 ohm, ¼ watt	
C1 and C2	0.1 µF	Uses any type of decoupling capacitor
Switch	Momentary on	See text
LCD		Has Hitachi-44780-controlled LCD display with a 14-pin interface
XTAL1	4 MHz ceramic resonator with internal capacitors	
Misc.		Requires a breadboard, wiring, and +5-volt power supply

```
volatile char LCDDlay = 20;      //  Initialization Delay Value
volatile char LCDState = 1;      //  Current LCD State

static volatile bit Clock     @ (unsigned)&PORTB*8+1;
static volatile bit ClockTRIS @ (unsigned)&TRISB*8+1;
static volatile bit Data      @ (unsigned)&PORTB*8+2;
static volatile bit DataTRIS  @ (unsigned)&TRISB*8+2;

char * MessageOut;               //  Message to be Sent Out
volatile char MessageOuti = 0;      //  Index to current Message
Byte
char Message1[2] = "\f";         //  Clear Display
char Message2[12] = "\fdebouncing"; //  Display "debouncing"
char Message3[9] = "\fPRESSED";     //  Display "PRESSED"

volatile char ButtonPress = 0;      //  Count of the Button Presses


//  Configuration Fuses
#if defined (_16F84)
#warning PIC16F84 selected
    __CONFIG(0x03FF1);           //  PIC16F84 Configuration Fuses:
                                 //   - XT Oscillator
                                 //   - 70 msecs Power Up Timer On
                                 //   - Watchdog Timer Off
                                 //   - Code Protection Off
#elif defined(_16F627)
#warning PIC16F627 with external XT oscillator selected
    __CONFIG(0x03F61);           //  PIC116F627 Configuration Fuses:
                                 //   - External "XT" Oscillator
                                 //   - RA6/RA7 Digital I/O
                                 //   - External Reset
                                 //   - 70 msecs Power Up Timer On
                                 //   - Watchdog Timer Off
                                 //   - Code Protection Off
                                 //   - BODEN Enabled
#else
#error Unsupported PICmicro MCU selected
#endif


//  Subroutines
LCDNybble(char Nybble, char RS) {   //  Send Nybble to LCD
:

LCDByte(char Byte, char RS) { //  Send Byteto LCD
:


LCDInit()                        //  Initialize the LCD I/O Pins
:


LCDOut(char * const LCDString)      //  Output the Data String
:


//  Interrupt Handler
```

```
void interrupt tmr0_int(void) //  TMR0 Interrupt Handler
{

char temp;

     if (TOIF) {

           TOIF = 0;           //  Reset Interrupt Flag

           RTC++;              //  Increment the Clock

//  Put additional interface code for 1 msec interrupt here

//  Increment Counter if Button Pressed

           if (!RB0)           //  Button Pressed
               if (ButtonPress < 20)
                           //  Button Press being debounced
                   ButtonPress++;
           else;         //  Button Pressed for 20 msecs
       else                   //  Button Released
           if (ButtonPress != 0)
                           //  Button Release being debounced
               ButtonPress--;
           else;         //  Button released for 20 msecs

//  LCD State Machine
:

     }  //  endif

//  Put Different Interrupt Handlers here

}  //  End Interrupt Handler


//  Mainline
void main(void)                     //  Template Mainline
{

     OPTION = 0x0D1;               //  Assign Prescaler to TMR0
                                   //   Prescaler is /4
     TMR0 = 0;                     //  Reset the Timer for Start
     TOIE = 1;                     //  Enable Timer Interrupts
     GIE = 1;                      //  Enable Interrupts

     LCDInit();                    //  Initialize the LCD Port

//  Put in Interface initialization code here

     while (1 == 1) {              //  Loop forever

//  Put in Robot high level operation code here

           switch(ButtonPress) {   //  Process Button Press
               case 0:             //  Released - Fully Debounced
                   LCDOut(Message1);
                   break;
               case 20:            //  Pressed - Fully Debounced
```

```
                              LCDOut (Message3);
                              break;
                      default:
                              //  Between 0 and 20 - Button being
    debounced
                              LCDOut (Message2);
                              break;
               }  //  endswitch

       }  //  endwhile

 }  //  End of Mainline
```

Before burning the PICmicro MCU with this application code, I simulated it to make sure it worked correctly.

To simulate the operation of the button, I added the Asynchronous Stimulus dialog box to the whisker MPLAB IDE project. To do this, after setting up the project, click Debug | Asynchronous Stimulus . . . , as shown in Figure 4-52.

After clicking on this pull-down menu, the Asynchronous Stimulus dialog box appears on the MPLAB IDE desktop (see Figure 4-53). Right-click on the first button and select the operation type (toggle) followed by RB0, as shown in Figure 4-54.

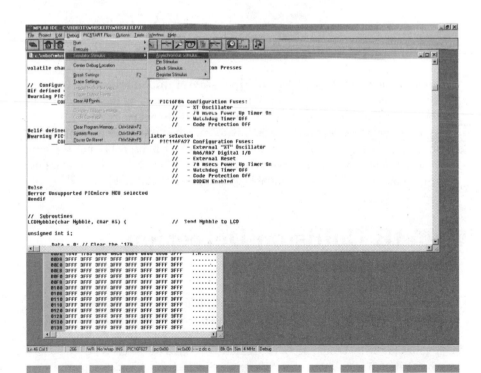

Figure 4-52 Enabling Asynchronous Stimulus

Figure 4-53

Asynchronous
Stimulus dialog
box

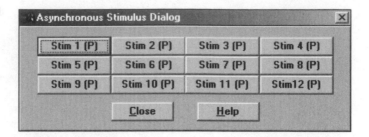

Figure 4-54

Selecting the
Asynchronous
Stimulus
function

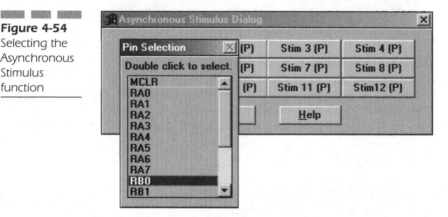

When you left-click this button, the input state of RB0 flips between 0 and 1. This operation simulates the action of the whisker pushing against the contacts and button presses can happen at any time to simulate bouncing.

Asynchronous Stimulus is a nice feature to use when you are simulating asynchronous inputs like the one used in this example. It is probably not the best method of testing applications that have a timed interface. For this type of input, other methods of stimulus are available within the MPLAB IDE.

IR Collision Detection Sensors

I'm sure you are familiar with IR remote controls for your TV or other electronic devices in your home. When I bought my first one 20 years ago, I ended up trying it in different circumstances to see how well it would work. At first, I tried seeing how far away I could be for the remote control to work. Then I tried bouncing the signal off a wall and other materials like pictures, windows, plants, and so on.

I admit doing things like this may seem strange, but it gave me the background to try the IR detector as an object/collision detector. My theory was to reduce the energy passed through the IR LED to the point where the reflected energy to the IR detector would only be strong enough at a set distance. Figure 4-55 shows the IR LED sending out a series of pulses that are reflected from an object and returned to the IR detector.

Note that in order to implement this type of object detection, an opaque barrier has been placed between the IR LED and IR detector. I'm pointing this out because there are many materials that you might want to use that are actually transparent to IR light. A piece of sheet metal acting as a shield between the IR LED and detector is probably the best opaque barrier that you can use. Other materials that work appropriately include black rubber electrical tape and black heat shrink tubing.

This is a noncontact method of detecting objects. Using this method along with ultrasonic ranging has several advantages over physical contact sensors (such as whiskers or bump sensors). The robot doesn't actually hit anything before an object is detected, which means there is less opportunity for damage to the robot, no bending is required to keep whiskers in the proper shape, and there is no whisker button bouncing to deal with. The noncontact methods of object detection are important if an *odometry sensor* is used; if objects are detected before a collision, there is no chance for the wheels to spin when colliding with an object.

There are some disadvantages to using IR or ultrasonic object detection. Most notably, the implementation of these methods requires substantial hardware and

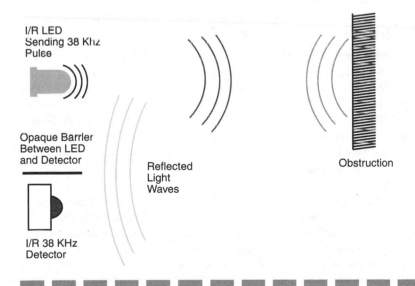

I/R LED
Sending 38 KHz
Pulse

Opaque Barrier
Between LED
and Detector

I/R 38 KHz
Detector

Reflected
Light
Waves

Obstruction

Figure 4-55   IR object detection

software investments. These investments are not all financial. You can buy an IR LED and detector for less than $2. However, understanding and applying the theory behind this method places restrictions on the microcontroller that you use as well as the software. The IR object detection method presented in this section has a very wide field of view, which may be an advantage or a problem, depending on the application.

The reason why this method can detect objects at a set distance from an object is due to the property of light sources, where the power of the light decreases as a square with distance. If you look at a light from 1 foot away, you will find that it is 4 times brighter than if you look at the light from 2 feet away. The light from 1 foot will be 16 times as bright as the light seen from 4 feet away.

Using this property, the light from the IR LED is attenuated to a point where the brightness of the reflected light is sufficient at a set distance to trigger the IR detector. Figure 4-55 shows the light's strength diminishing as it travels through space by minimizing the current passing through the IR LED which reduces its output.

Three methods are used to reduce the LED's output:

- Use a potentiometer with an LED as its current-limiting resistor. This lowers the signal output strength by increasing the resistance, which decreases the current through the LED. Once you have established the resistance that gives detection at the appropriate distance for your robot, the potentiometer can be replaced with a fixed resistance.

- By driving the LED at a frequency other than 38 KHz, the sensitivity of the IR detector to the IR light is decreased.

- The last method of having attenuation in the IR LED's output is to use a PWM duty cycle that is less than 50 percent. This is generally easy to implement because of built-in microcontroller hardware that provides a PWM output working at varying duty cycles.

As you undoubtedly know, IR light is everywhere. The IR detector filters out the background IR light and lets through any IR signals that are present. It does this with a built-in *band-pass filter*, which only passes signals that are modulated at 38 KHz.

For the LED source, you should be working with a 38 KHz modulated IR signal. A reasonably accurate source can be produced using the PIC16F627's PWM output. When I was looking at the different components that can be built on a robot, I considered this one to be very important, which is why I did not use the built-in PWM generator for the motor PWM source.

If most of the IR LED's energy is directed to the front of the robot, the IR detector indicates the presence of objects farther away from the side. This property can be used to combine a single IR LED and detector on each side of the robot to detect objects in front of the robot and to the sides.

The IR detector that you use requires anywhere from two to six or seven cycles of the modulated IR signal to respond. This is shown in the picture of an oscilloscope in Figure 4-56. To enable an IR remote control to be used with the IR

Figure 4-56
IR
diode/detector
operation

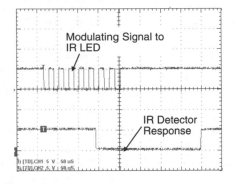

Modulating Signal to
IR LED

IR Detector
Response

object detector circuitry shown in this example, I often send out eight modulated pulses once every millisecond. Before sending additional pulses for collision detection, the IR detectors are polled to see if a remote control is sending data to the robot.

irdetect—IR LED and Detector-Based Object Sensor

Although the concept of this project is one of the simplest projects presented in this book, getting this code to work properly was quite difficult. When I finally got the application working, I was surprised at how simple the problem was and happy at how well it let me add other interfaces (such as the remote control) to an application that used the IR LED and detector for object detection.

For this application, I decided to stay with the LCD display for the simple reason that after showing how the circuitry is used for a collision detection system and an IR TV remote control receiver, I wanted to show how the two interfaces could be implemented together. This has been a central theme of this book, and although at first glance combining the functions of the IR collision detection system and IR remote control seems quite difficult (even though they use the same hardware), actually implementing the two functions together is quite easy.

To generate the 38 KHz square wave needed by the IR detector, use the PIC16F627's built-in PWM generator. When I was originally planning on implementing this function, the PICC Lite compiler would not work with a PICmicro MCU that had a built-in PWM generator. My original solution was to use a PIC16F84 running an assembly language application that output a 38 KHz square wave.

With the PIC16F627 running at 4 MHz, driving out a 38 KHz square wave is as simple as the following:

```
CCPR1L = 13;                    //  50% Duty Cycle
CCP1CON = 0b000001111;          //  Turn on PWM Mode

PR2 = 25;                       //  26 µsec Cycle for 38 KHz
TMR2 = 0;                       //  Reset TMR2
T2CON = 0b000000100;            //  TMR2 is On, Scalers 1:1

TRISB3 = 0;                     //  Put RB3 (PWM Output) into
                                //   Output Mode
```

Figure 4-57 shows the circuit that was used for testing out the IR collision detection. I connected an IR LED with a 10K potentiometer on RB3 (which is the PWM output pin). I could have changed the PWM's duty cycle, but I went with the potentiometer because it is much easier to change when the application is running.

For the IR detector, a standard 38 KHz IR TV remote control receiver was used. To help minimize the interference that the IR detector gets from noise in the system, use the filter shown in Figure 4-57 along with the pull-up on the output pin (most devices have open collector outputs).

The circuitry continues to fit on the small breadboard, as shown in Figure 4-58. Some of the wires have been relocated to the LCD to make space for the IR LED and its potentiometer, but the connections are the same as they were for the previous applications. Table 4-12 shows the bill of materials for the circuit.

The Waitrony IR LED and IR detector are for reference only. Many 38 KHz IR detectors and IR LEDs that you can use for this project are available (from Radio Shack and Digi-Key) so the phrase "or equivalent" should not be that frightening. I can get the Waitrony parts easily and inexpensively in Toronto which leads me to believe that these parts are fairly easy to find in other medium and large cities.

When the IR LED is used in an actual robot (or even for this application), you will probably want to bend the IR LED so that it is facing the same direction as

Figure 4-57
irdetect
application
circuit

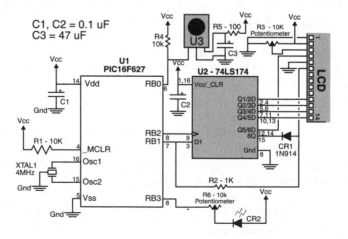

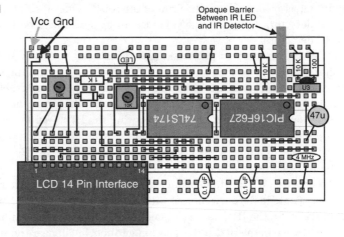

Figure 4-58
irdetect
breadboard
circuit

Table 4-12

Material for
the irdetect
Breadboard
Circuit

Label	Part Number	Comment
U1	PIC16F627	Can use a PIC16F84
U2	74LS174	
U3	3-pin IR detector	Requires Waitrony PIC-1018SCL or equivalent
CR1	Silicon diode	Uses any type, but 1N914 is used for prototypes
CR2	IR LED	Requires Waitrony IE-0530HP or equivalent
R1 and R4	10K, 1/4 watt	Uses _MCLR pull-up
R2	1K, 1/4 watt	
R3 and R6	10K, 1 turn pot	Requires LCD contrast
R5	100 ohm, 1/4 watt	
C1 and C2	0.1 μF	Uses any type of decoupling capacitor
C3	47 μF, 16 volt	Requires an electrolytic capacitor
LCD		Has Hitachi-44780-controlled LCD display with a 14-pin interface
XTAL1	4 MHz ceramic resonator with internal capacitors	
Misc.		Requires a breadboard, wiring, and +5-volt power supply

the IR detector (most likely the front of the robot). One IR LED can be used with two IR detectors, but remember to place an opaque shield between the IR LED and IR detectors.

Once this circuit was put together, I discovered that it wouldn't work even though there was nothing in it that I thought was challenging or difficult (the other application seemed to work the first time without any problems). I spent a number of frustrating weeks trying to figure out what went wrong. I discovered the solution by going back and implementing the IR object detection software in exactly the same way as a working version that I had created a year before. When I just ran the signal for a few cycles, the problems with the IR signal (not receiving both transmitted and reflected) went away.

It seems that if a constant signal is passed to the IR detector, it starts filtering it out as it adapts to the environment with it running constantly. By stopping the signal periodically, the IR detector does not adapt to the signal and filter it out. This isn't mentioned in the IR detector's documentation, but appears once I added the following code to the 1 msec interrupt handler, the detection circuit worked perfectly.

```
TRISB3 = 0;             //  Enable PWM Output

while (TMR0 < 64);      //  For a Limited Number of
                        //  Cycles

if (!RB0)               //  If Low, then Collision
        Collision = 1;
else
        Collision = 0;

TRISB3 = 1;
```

This code drives out the 38 KHz PWM signal for a TMR0 count of 64, which translates to 256 μsecs. During this time, eight or nine full cycles of the PWM are output. From my previous experiments with different IR detectors, I had a pretty good idea that the IR detector would synch up to this signal and output an indication that the signal was received. Once the signal was turned off, the IR detector would stop driving out its signal a few microseconds later.

To solve the problem, I compared the PIC16F627 PWM code with the application code I had written a year before for another robot. Electrically, the working implementation was identical to this one and I could even bring up the robot with the receiver used in this example and show that it was working. In regards to software, I believed that running the PWM constantly wouldn't be a problem, but almost as an act of desperation, I tried out the application with the eight or nine blocks of PWM pulses.

This is explained to illustrate the importance of trying *everything* when you are debugging an application and taking nothing for granted or assuming because something makes sense, it can't be wrong.

The final irdetect.c code is listed in the following section and can be found in the robot\irdetect directory. This code outputs the 38 KHz pulses. If there is a reflected signal, it displays COLLISION; if there is no reflected signal, it displays a blank LCD. To make sure the circuit detects objects at the desired distance, the potentiometer (R6) must be calibrated.

Before calibrating R6, make sure there is an opaque barrier (to IR light) between the IR LED and IR detector. Figure 4-58 shows where I put a piece of black antistatic foam. You may find that it is difficult to find a material that is truly IR opaque. IR light passes through most paper and cloth tape. I have found that the best barriers are built from materials like tinfoil and black electrical tape.

To calibrate the potentiometer, place a piece of white paper (like a business card) 4 or 5 inches away from and facing the IR receiver and IR detector, and adjust the potentiometer value until the IR LCD message flashes "COLLISION". Now, when an object is at the same distance away from the IR LED and IR detector, it detects the object and updates the LCD display.

The irdetect.c application is as follows:

```
#include <pic.h>

//  02.04.17 - IRDETECT: Sense Objects in front of IR LED/IR
//  Detector
//
//  This is a two wire interface using a 74LS174 as a shift
//    register and the code called from the mainline (and timed
//    from TMR0 interrupt)
//
//
//  Hardware Notes:
//  PIC16F827 Running at 4 MHz with External Oscillator
//  RB0 - IR Detector Input
//  RB1 - CLock Signal to the LCD
//  RB2 - Data Signal to the LCD
//  RB3 - PWM Output
//

//  Global Variables
int RTC = 0;                        //  Real Time Clock Counter

volatile char LCDDlay = 20;         //  Initialization Delay Value
volatile char LCDState = 1;         //  Current LCD State

static volatile bit Clock     @ (unsigned)&PORTB*8+1;
static volatile bit ClockTRIS @ (unsigned)&TRISB*8+1;
static volatile bit Data      @ (unsigned)&PORTB*8+2;
static volatile bit DataTRIS  @ (unsigned)&TRISB*8+2;

char * MessageOut;                  //  Message to be Sent Out
volatile char MessageOuti = 0;      //  Index to current Message
                                    //  Byte
char Message1[2] = "\f";            //  Clear Display
char Message2[11] = "\fCOLLISION";  //  Display "Collision"

volatile char Collision = 0;        //  Current Collision
volatile char OldCollision = 0;
```

```
//  Configuration Fuses
#if defined(_16F627)
#warning PIC16F627 with external XT oscillator selected
      __CONFIG(0x03F61);              //  PIC116F627 Configuration
                                      //  Fuses:
                                      //  - External "XT" Oscillator
                                      //  - RA6/RA7 Digital I/O
                                      //  - External Reset
                                      //  - 70 msecs Power Up Timer
                                      //  - Watchdog Timer Off
                                      //  - Code Protection Off
                                      //  - BODEN Enabled
#else
#error Unsupported PICmicro MCU selected
#endif


//  Subroutines
LCDNybble(char Nybble, char RS) {    //  Send Nybble to LCD
:

LCDByte(char Byte, char RS) {        //  Send Byteto LCD
:


LCDInit()                            //  Initialize the LCD I/O Pins
:


LCDOut(char * const LCDString)       //  Output the Data String
:


//  Interrupt Handler
void interrupt tmr0_int(void)        //  TMR0 Interrupt Handler
{

char temp;

     if (T0IF) {

          T0IF = 0;                  //  Reset Interrupt Flag

          RTC++;                     //  Increment the Clock

//  Put additional interface code for 1 msec interrupt here

//  Output a Series of Pulses for Collision Detection

          TRISB3 = 0;                //  Enable PWM Output

          while (TMR0 < 64);         //  For a Limited Number of
                                     //  Cycles

          if (!RB0)                  //  If Low, then Collision
               Collision = 1;
          else
```

```
                        Collision = 0;

                    TRISB3 = 1;

//  LCD State Machine
:

        }  //  endif

//  Put Different Interrupt Handlers here

}  //  End Interrupt Handler


//  Mainline
void main(void)                      //  Template Mainline
{

        OPTION = 0x0D1;              //  Assign Prescaler to TMR0
                                     //   Prescaler is /4
        TMR0 = 0;                    //  Reset the Timer for Start
        T0IE = 1;                    //  Enable Timer Interrupts
        GIE = 1;                     //  Enable Interrupts

        LCDInit();                   //  Initialize the LCD Port

//  Put in Interface initialization code here

        CCPR1L = 13;                 //  50% Duty Cycle
        CCP1CON = 0b000001111;       //  Turn on PWM Mode

        PR2 = 26;                    //  26 µsec Cycle for 38 KHz
        TMR2 = 0;                    //  Reset TMR2
        T2CON = 0b000000100;         //  TMR2 is On, Scalers 1:1

        while (1 == 1) {             //  Loop forever

//  Put in Robot high level operation code here

                if (Collision != OldCollision) {
                        OldCollision = Collision;
                                     //  Save Current Collision
                                     //  State
                        if (Collision)  //  Put in Appropriate Message
                            LCDOut(Message2);
                        else
                            LCDOut(Message1);
                        while (LCDState);  //  Wait for LCD Available
                }  //  endif

        }  //  endwhile

}  //  End of Mainline
```

When you are finished with this application, please do not dismantle the circuit because it is required for the next two applications.

IR Remote Controls

I have found that most robots I have worked on have required a method of wireless control providing basic functions that enable you to explicitly control the operation of the robot and start specific applications. The best solution that I have found so far is to use a universal TV remote control. These remote controls rely on an IR signal (which is similar to the one used for detecting objects presented earlier) to send a multibit data packet. When I use a TV remote control for controlling a robot, I use the matrixed 1 through 9 buttons for motion and the volume control for adjusting the speed of the robot. Any other buttons that are available to you (power and 0 come to mind) can be used for direct commands of the robot.

The most popular remote control standard used by electronics designers for remote controls is the Sony TV standard. You can buy a universal remote control from your local Radio Shack for less than $10 with this capability and you are ready to go. Data is sent to an IR receiver (or detector) in Manchester-encoded format, with the length of each of the three different pulses representing a different data type or bit in the packet. Figure 4-59 shows each type of bit and its important parameters.

A data packet consists of 13 pulses. The first, the long data start, indicates that there is a packet coming followed by 12 data bits. You can see the different pulse types by looking at the data coming in using an oscilloscope (see Figure 4-60).

Assuming that the first bit (after the data start) is bit 0, the buttons shown in Table 4-13 send the following data packets out.

Each packet is repeated once every 50 milliseconds or so. There is no autorepeat built into the remote control; instead, the autorepeat function must be built into the receiving circuit. This is actually not a problem for robots, where the multiple data packet output is useful for controlling motion as long as a key is held down. I have created a number of robot functions where the motors are driven for 200 milliseconds after receiving each IR data packet; this keeps the motors running to the next data packet while the button is held down. The 200 milliseconds give a margin to keep the motors running even if it is late from the remote control or if the next packet is not received properly and is thrown out.

Figure 4-59
Sony IR data
packet pulse
timing

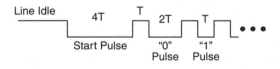

T = 0.60 msecs
4T = 2.40 msecs
2T = 1.20 msecs

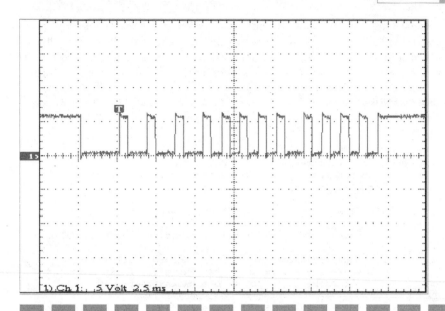

Figure 4-60   Picture of Sony TV IR remote control

Table 4-13

Buttons that
Send Data
Packets Out

Button	Data Packet
0	0b0011011101111
1	0b0111111101111
2	0b0011111101111
3	0b0101111101111
4	0b0001111101111
5	0b0110111101111
6	0b0010111101111
7	0b0100111101111
8	0b0000111101111
9	0b0111011101111
Vol up	0b0101101101111
Vol down	0b0001101101111
Chan up	0b0111101101111
Chan down	0b0011101101111
Prev chan	0b0001000101111
Mute	0b0110101101111
Power	0b0010101101111

Despite the theory of how IR remote controls work, many people have problems trying to figure out how to process the incoming data. The most obvious method of receiving data is to continuously poll the incoming data and record the lengths of each of the pulses using code that looks something like the following:

```
void main(void)              //  Read and Output Data from an IR
Receiver
{

unsigned int i;
unsigned int DataPacketStart; //  Pulse Start Time
unsigned int DataPacketEnd;   //  Pulse End Time
unsigned int DataPacket;      //  Actual Data Packet

    while (1 == 1) {          //  Loop forever

        while (IRData == High); //  Wait for Incoming Pulse

        DataPacketStart = TMR0; //  Record Start Time
        while (IRData == low);  //  Wait for Pulse End
        DataPacketEnd = TMR0 - DataPacketStart;
                                //  Get Packet Time
        if ((DataPacketEnd > 2.2msecs) && _
            (DataPacketEnd < 2.6 msecs)) {
                            //  Wait for Next 12 Bits
            DataPacket = i = 0;       //  Initialize Data
                                      //  Variable
            DataPacketEnd = 1.8msecs;

            while (( i < 12) && (DataPacketEnd != 0)) {

                while (IRData == High); // Start Pulse Wait
                DataPacketStart = TMR0; //  Record Start
                                        //  Time
                while (IRData == low);  //  Wait for Pulse
                                        //  End
                DataPacketEnd = TMR0 - DataPacketStart;
                        //  Time Packet
                if ((DataPacketEnd > 0.45msec) && _
                    (DataPacketEnd < 0.75msec)
                    DataPacket = (DataPacket << 1) + 1;
                else if ((DataPacketEnd > 0.95msec) && _
                    (DataPacketEnd < 1.35msec) {
                    DataPacket = (DataPacket << 1) + 1;
                else            //  Invalid
                DataPacketEnd = 0;       //  Stop Operation
                i++;           //  Point to the next bit
            } //  endwhile

            if (DataPacketEnd != 0)      //  Output Data
                                         //  Packet
                printf("Data Packet = %i\n", DataPacket);

        } //  endif
    } //  endwhile

} //  End IR Remote Control Read Code
```

This method, despite looking ungainly, actually works very well as a method of demonstrating how IR TV remote controls work. However, the problem with this method is that it cannot be used in a robot because it continuously polls the IR receiver line.

Instead of continually polling the line, pass the IR detector's output to an interrupt request pin on the controller that is used with the robot. Within the interrupt request handler, you can perform the pulse width checks and data-packet build quite efficiently without impeding the operation of any other functions within the robot. The next section demonstrates how this can be done.

▬▬ Sony Remote Control Receiver Application

As indicated previously, many people do not use IR television remote controls for their robots because they are afraid of the difficulty of writing the code. You may be concerned about the complexity involved with adding a remote control interface to your robot that follows the requirements that have been set out in this section.

I am happy to say that these fears are misplaced—adding a remote control interface to a robot using the template code presented in this chapter is actually quite easy and works very well with other applications in the robot. In my other books (such as *Programming and Customizing PICmicro® Microcontrollers*), I have implemented a remote control receiver in several different ways using different microcontrollers. Once you have solved a problem one way, other methods of solving the problem become obvious and quite easy to implement.

To successfully implement a remote control receiver, you must have a clear understanding of your goal of using the remote control, the appearance of the incoming signal, and the built-in features of the microcontroller that you are using that will make this application easier. Once you have this information, several solutions to the problem become obvious.

My definition of what I would like the remote control receiver to do is as follows:

Return valid 12-bit Sony TV control codes to the biologic code.

You should be aware of a couple of points. The first centers on the word *valid* in this definition. In the previous section, I outlined the expected timings from the Sony TV remote control. If any of the data timings are out of specification (that is, either longer or shorter than what 0 or 1 can be), I will discard the entire packet. The second requirement is that I am looking for 12 bits—I don't care about the start pulse. This part of the definition helped me come to a solution.

I won't go over the data or its timings. They were defined in the previous section along with the picture of the oscilloscope, which showed you what you can expect.

Finally, I have to look at the different built-in features of the PICmicro MCU that I am using to see if there is anything that I can take advantage of. Going over the PICmicro MCU's interrupt sources, I can see that pin 6 (RB0/INT) can be used as either a digital I/O pin or as an interrupt pin, which requests an interrupt when the state of the pin input goes high or low.

Using this information, I decided to find out if I could use the RB0/INT pin to interrupt the operation of the PICmicro MCU every time the IR detector transitions from low to high. Figure 4-61 shows that every time the interrupt line transitions from low to high (rising edge) in the data packet coming from the remote control, the time from the previous transition is calculated. Based on this information and assuming that the time that the data line is high is constant, you can determine whether each bit was 0 or 1.

To calculate the time between the interrupt transitions, record the current TMR0 value as the base. When the next low-to-high transition comes in, subtract the current time from the new time to determine the actual delay between the bits.

The advantage of this method is that you can completely ignore the start pulse except for its usefulness as a point from which the length of the next transition (bit 0) is taken. In the code in the previous section, the start pulse was timed to make sure that it was valid; in this case, its timing is ignored completely. The resulting code is much simpler.

To check for valid data, if any of the times between the transitions are invalid or if more than 9 milliseconds have passed since the last transition, I will stop the data receive operation. The calculation of whether 9 milliseconds have passed is quite simple and there are opportunities for invalid errors to be flagged, but the data is repeated by the remote control every 50 milliseconds. This means that if I miss one packet, chances are I will pick up the next one. Going back to the previous section, this is why I assume that an operation will require 200 milliseconds before timing out because no more pulses are sent.

To demonstrate the operation of the code that I have come up with and to avoid the need to rewire the circuit from an application that is virtually identical to this one, the same circuit and wiring as the IR object detection system (irdetect) was used.

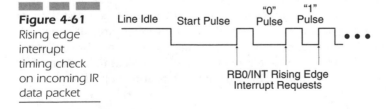

Figure 4-61
Rising edge
interrupt
timing check
on incoming IR
data packet

Line Idle   Start Pulse   "0" Pulse   "1" Pulse

RB0/INT Rising Edge
Interrupt Requests

To display the incoming data, the LCD code developed earlier was used. The source code for the application can be found in the robot/remote directory:

```c
#include <pic.h>

// 02.04.17 - Updated for New Compiler Format
// 02.02.15 - Remote: Display Sony IR Commands on an LCD
//
// Take Input from Sony IR Remote Control and Display on LCD.
//
// To display data, two wire interface using a 74LS174 as a shift
//   register is used and processed by the biologic code into a
//   string displayed by the LCD.
//
// Data Stream:
//
//              Leader/Header   High "1"          "0"
// -------+                 +----+   +----+               +--------
//        |                 |    |   |    |               |
//        +-----------------+     +-----+ +------------+
//
//        |    2.4 msecs    |540u|660us|540u|  1.2 msecs  |
//
//                          |1200 msecs|   1.76 msecs     |
//
//                          |   300    |       440        |
//
//
// Hardware Notes:
// PIC16F84/PIC16F627 Running at 4 MHz with an external ceramic
// resonator
// RB0 - Sony IR Input, Use Falling Interrupt
// RB1 - CLock Signal to the LCD
// RB2 - Data Signal to the LCD
//

// Global Variables
int RTC = 0;                    // Real Time Clock Counter

volatile char LCDDlay = 20;     // Initialization Delay Value
volatile char LCDState = 1;     // Current LCD State

static volatile bit Clock     @ (unsigned)&PORTB*8+1;
static volatile bit ClockTRIS @ (unsigned)&TRISB*8+1;
static volatile bit Data      @ (unsigned)&PORTB*8+2;
static volatile bit DataTRIS  @ (unsigned)&TRISB*8+2;

char * MessageOut;                          // Message to be Sent
                                            // Out
volatile char MessageOuti = 0;      // Index to current Message
                                    // Byte

unsigned int  DataIn;           // Data Input
unsigned char DataInCount = 0;      // Number of Characters to Read
                                    // In
unsigned char DataReady = 0;    // Data to be Read in
int  SaveRTC;                   // Saved RTC/TMR0 for Interrupt Pins
```

```
int   CurrentRTC;
char Message[20] = "\f-> 0b0xxxxxxxxxxxxx";        //  Bit Message
                                                   //  Header



//  Configuration Fuses
#if defined (_16F84)
#warning PIC16F84 selected
       __CONFIG(0x03FF1);       //  PIC16F84 Configuration Fuses:
                                //   - XT Oscillator
                                //   - 70 msecs Power Up Timer On
                                //   - Watchdog Timer Off
                                //   - Code Protection Off
#elif defined(_16F627)
#warning PIC16F627 with external XT oscillator selected
       __CONFIG(0x03F61);       //  PIC116F627 Configuration Fuses:
                                //   - External "XT" Oscillator
                                //   - RA6/RA7 Digital I/O
                                //   - External Reset
                                //   - 70 msecs Power Up Timer On
                                //   - Watchdog Timer Off
                                //   - Code Protection Off
                                //   - BODEN Enabled
#else
#error Unsupported PICmicro MCU selected
#endif



//  Subroutines
LCDNybble(char Nybble, char RS) {    //  Send Nybble to LCD
:


LCDByte(char Byte, char RS) { //  Send Byteto LCD
:


LCDInit()                       //  Initialize the LCD I/O Pins
:


LCDOut(char * const LCDString)       //  Output the Data String
:



//  Interrupt Handler
void interrupt tmr0_int(void) //  TMR0 Interrupt Handler
{

char temp;

     if (T0IF) {

            T0IF = 0;          //  Reset Interrupt Flag

            RTC++;             //  Increment the Clock

//  Put additional interface code for 1 msec interrupt here

//  Check for Packet which was lost and stop waiting unless packet
```

```
is active

                CurrentRTC = (RTC & 0x0FF) - ((SaveRTC >> 8) & 0x0FF);
                if (CurrentRTC < 0)       //  Calculate Time Since Save
                    CurrentRTC = 0 - CurrentRTC;
                if ((DataInCount != 0) && (CurrentRTC > 9))
                    DataInCount = 0;  //  Reset and Wait for Next
Char

//   LCD State Machine
:


        }  //  endif

//   Put Different Interrupt Handlers here

        if (INTF) {                      //  RB0/INT Pin Interrupt
            if (DataReady)               //  Waiting to Process Previous
                ;                        //  Ignore
            else if (DataInCount == 0) {  //  New Message Coming In
                DataInCount = 12;  //  12 Bits to Come in
                SaveRTC = ((RTC & 0x0FF) << 8) + TMR0;
                                    //  Save the Current Time
            } else {                     //  Bit to Process
                CurrentRTC = ((RTC & 0x0FF) << 8) + TMR0;
                                    //  Get the Current Time
                if ((SaveRTC - CurrentRTC - SaveRTC) < 0)
                    SaveRTC = 0 - SaveRTC;
                if ((SaveRTC > 250) && (SaveRTC < 350)) {
                    DataIn = (DataIn << 1) + 1;     //  "1"
                                    //  Received
                    if (--DataInCount == 0)
                        DataReady = 1;
                } else if ((SaveRTC > 390) && (SaveRTC < 490)) {
                    DataIn = DataIn << 1;          //  "0"
                                    //  Received
                    if (--DataInCount == 0)
                        DataReady = 1;
                } else                 //  Invalid - Delete
                    DataInCount = 0;
                SaveRTC = CurrentRTC;   //  Save the Execution
                                        //  Time
            }  //  endif
            INTF = 0;                   //  Reset Interrupt
        }  //  endif

}   //  End Interrupt Handler



//   Mainline
void main(void)                    //  Template Mainline
{
int i;

     OPTION = 0x0D1;                    //  Assign Prescaler to TMR0
                                        //     Prescaler is /4
     TMR0 = 0;                          //  Reset the Timer for Start
     T0IE = 1;                          //  Enable Timer Interrupts
```

```
        GIE = 1;                        //  Enable Interrupts

//  Put in Interface initialization code here

        LCDInit();                      //  Initialize the LCD Port

        INTEDG = 1;                     //  Interrupt on Rising Edge of
                                        //    RB0/IR TX'r
        INTE = 1;                       //  Enable RB0/INT Pin
                                        //  Interrupts

        while (1 == 1) {                //  Loop forever

//  Put in Robot high level operation code here

            if (DataReady) {
                while (LCDState != 0);  //  Wait for LCD
                                        //  Available
                for (i = 0; i < 12; i++) {
                    if (DataIn & (1 << 11))
                        Message[i + 7] = '1';
                    else
                        Message[i + 7] = '0';
                    DataIn = DataIn << 1;
                } //  endfor
                LCDOut(Message);  //  Pass String to Output
                DataReady = 0;    //  Reset the Data Flag
            } //  endif
        } //  endwhile

} //  End of Mainline
```

Although this code is apparently very complex, it is actually reasonably simple. When an IR data packet is coming in, the IR receiver interrupt (INTF code) executes on each low-to-high transition on the RB0/INT pin. If it is the first low-to-high transition, the code assumes that this was the start pulse, loads the counter with 12, and records the current time. For each subsequent interrupt, the time between interrupts is calculated and the bit value is recorded until all 12 bits have been read.

After all 12 bits have been read, the DataReady flag is set, indicating to the biologic code that a valid remote control packet has been received. The remote control receive code then ignores all incoming data until the biologic code clears it. This feature of the code was put in to eliminate the possibility that subsequent packet receives would affect any received data packets that biologic code has not yet acted upon.

As part of using the LCD, after sending the data for the current command packet, wait for its information to be displayed on the LCD before reloading the Message with the new command packet. This delay was used to prevent the current command packet from being overwritten by a new packet's data while it was being displayed. This was done largely by force of habit to make sure that data being output would not be changed during the output operation. In this application, I do not believe that there is any possibility of the invalid data being displayed, but in other applications, where the Message that is displayed can be different, this is a good practice to follow.

These possibilities may seem remote, but you should think of the case where the Message is being updated with DataIn's contents and another packet comes in. The new packet starts shifting in as the old one is being read, resulting in the old packet not being reported properly and the new packet not being stored properly because both code functions are shifting the data without consulting each other.

▓▓▓ Combining irdetect and remote

With the IR LED and detector-based object detector and remote control applications working, I thought it would be a simple project to combine the two applications into one that performed both interface functions. My original thought was to modify the 1-millisecond interrupt handler code to first check to see if the RB0/INT pin (which I was using for the IR detector) was active (low), and if it was, then read the incoming IR packet as I did in remote. If the RB0/INT pin was high upon entry into the 1-millisecond interrupt handler, then I would drive out the 38 KHz IR signal to detect objects close by the robot. The concept seems quite simple, but in actuality, it turned into several hours of frustrating work trying to get the application to work reliably. The solution to the problem was a result of listing the assumptions again and working through them until I properly understood what I would expect in the system.

Along with combining the IR object detection and remote control receive interrupt code, I also modified the messages that were printed to the LCD with commands to place the strings at specific locations within the LCD rather than just clearing the LCD as I did in the previous applications. This is actually a very powerful feature of the LCD control code. By using the 254 commands to place strings at different locations on the LCD, you can display multiple inputs easily and quickly without affecting other data on the LCD.

I made actually made two mistakes in my original set of assumptions. The first was that I forgot that my design for the remote control receiver was actually quite elegant. I totally ignored the length of the start pulse and instead started looking at the bits after the start pulse had finished by using its rising edge as a basis for the timing of the first bit. In my initial assumptions, I correctly noted that the length of the start pulse was more than 1-millisecond, and I could expect to poll it during the start of the 1-millisecond interrupt handler and then start looking at the transitions. The problem with this assumption was that I didn't take into account that I would still have to use its rising edge as the start timing of the first bit.

When I originally created the code, when I polled RB0/INT to be low, I set the INTE bit of INTCON, which enables the RB0/INT pin interrupts, and forgot that I would have 12 bits following it. When INTE was enabled, my bit counter was loaded with 12, which was incorrect as I would have 13 rising edges to deal with. Once I changed the bit counter to 13 and ignored the timing for the first transition in the RB0/INT pin interrupt request source handler, the application started working much better.

Although I was receiving correct information on the LCD, I noticed that the collision LCD message would come on occasionally when I was pressing the remote control command button. This problem was due to the rare cases when the start pulse came in while the IR LED was active. At the end of this interval, the start pulse would still be active and the code would behave as if there was an object before it.

The solution to this problem was to change the object detector from reporting a collision when an object was detected after a single 1-millisecond interrupt poll to 3 milliseconds. The start pulse is 2.4 milliseconds in length so if the IR detector is polled for 3 milliseconds, a remote control packet should never come in; instead, a legitimate object should be in front of the IR LED and detector. I reported a collision on the basis of one poll reporting an object because of the reliability that I have found with the IR object detection system. In a mechanical system, I would never have been so cavalier, but with the IR object detection, I have never seen any situations where an invalid contact caused any bouncing.

With the check changed from 1 millisecond to 3 milliseconds, the application worked almost perfectly. When an object was in front of the IR LED and detector and I was sending a remote control packet, the incorrect packet would occasionally be received. The error was about once every 10 times and was always a 0 incorrectly read for a 1.

I theorized that the problem was caused because the LCD elelogic code in the interrupt handler executed while the rising edge was being received and caused enough of a delay for the bit to be incorrectly read. By turning off either the collision message or the packet LCD messages, the problem went away completely.

A potential solution to this problem is to disable remote control receives when data is being sent to the LCD. This should not affect the operation of the robot because the remote control resends data every 50 milliseconds. If one packet is ignored, then the next one coming in will be received and processed correctly.

The final application code (combine.c, which is found in robot\combine) uses the same hardware that was used for irdetect:

```
#include <pic.h>

//   02.04.19 - Created from "irdetect" and "remote"
//   02.02.15 - Remote: Display Sony IR Commands on an LCD
//
//   A 38 KHz IR Signal to be used as an object detection system
//
//   Take Input from Sony IR Remote Control and Display on LCD.
//
```

```
//   To display data, two wire interface using a 74LS174 as a shift
//    register is used and processed by the biologic code into a
//    string displayed by the LCD.
//
//   Data Stream:
//
//               Leader/Header  High  "1"          "0"
//   -------+                   +----+   +----+          +--------
//          |                   |    |   |    |          |
//          +-------------------+    +-----+  +------------+
//
//          |    2.4 msecs      |540u|660us|540u|  1.2 msecs |
//
//                              |1200 msecs|    1.76 msecs   |
//
//                              |    300    |        440     |
//
//
//   Hardware Notes:
//   PIC16F627 Running at 4 MHz with an external ceramic resonator
//   RB0 - Sony IR Input, Use Falling Interrupt
//   RB1 - CLock Signal to the LCD
//   RB2 - Data Signal to the LCD
//

//   Global Variables
int RTC = 0;                    //   Real Time Clock Counter

volatile char LCDDlay = 20;     //   Initialization Delay Value
volatile char LCDState = 1;     //   Current LCD State

static volatile bit Clock    @ (unsigned)&PORTB*8+1;
static volatile bit ClockTRIS @ (unsigned)&TRISB*8+1;
static volatile bit Data     @ (unsigned)&PORTB*8+2;
static volatile bit DataTRIS @ (unsigned)&TRISB*8+2;

char * MessageOut;              //   Message to be Sent Out
volatile char MessageOuti = 0;      //   Index to current Message
                                    //   Byte

unsigned int  DataIn;           //   Data Input
unsigned char DataInCount = 0;      //   Number of Characters to
                                    //   Read In
unsigned char DataReady = 0;    //   Data to be Read in
int  SaveRTC;                   //   Saved RTC/TMR0 for Interrupt Pins
int  CurrentRTC;
char Message[22] = "\376\002-> 0b0xxxxxxxxxxxxx!";
                                //   Incoming Packet Display
char Message1[12] = "\376\300          ";    //   Clear Display
char Message2[12] = "\376\300COLLISION";     //   Display
                                             //   "Collision"

volatile char Collision = 0;  //   Current Collision
volatile char OldCollision = 0;


//   Configuration Fuses
#if defined(_16F627)
```

```
#warning PIC16F627 with external XT oscillator selected
    __CONFIG(0x03F61);        //  PIC116F627 Configuration Fuses:
                              //   - External "XT" Oscillator
                              //   - RA6/RA7 Digital I/O
                              //   - External Reset
                              //   - 70 msecs Power Up Timer On
                              //   - Watchdog Timer Off
                              //   - Code Protection Off
                              //   - BODEN Enabled
#else
#error Unsupported PICmicro MCU selected
#endif


//  Subroutines
LCDNybble(char Nybble, char RS) {   //  Send Nybble to LCD
:

LCDByte(char Byte, char RS) { //  Send Byteto LCD
:

LCDInit()                     //  Initialize the LCD I/O Pins
:

LCDOut(char * const LCDString)     //  Output the Data String
:


//  Interrupt Handler
void interrupt tmr0_int(void) //  TMR0 Interrupt Handler
{

char temp;

    if (T0IF) {

          T0IF = 0;           //  Reset Interrupt Flag

          RTC++;              //  Increment the Clock

//  Put additional interface code for 1 msec interrupt here

//  Check for Packet lost and stop waiting unless packet is active

          if (DataInCount != 0) {   //  Check/Update Time for New
                              //   IR Packet Coming In

              if ((CurrentRTC = _
          (RTC & 0xFF) - ((SaveRTC >> 8) & 0xFF)) < 0)
                  CurrentRTC = 0 - CurrentRTC;
                              //  Calculate Time Since Save
              if (CurrentRTC > 9) {   //  Delay too Long?
                  DataInCount = 0;
                              //  Reset and Wait for the Next
                              //  Packet
                  INTE = 0;
                              //  Disable Incoming Data Interrupts
```

```
                   } // endif

          } else if ((!RB0) && (!DataReady)) {
                              // Data Packet Coming in?

               DataInCount = 13;   // 12 Bits to Come in/13
                                   // Edges
               SaveRTC = ((RTC & 0x0FF) << 8) + TMR0;
                              // Save the Current Time
               INTF = 0;    // Enable Pin Change Interrupt
               INTE = 1;

          } else {          // Do Collision Detection

               TRISB3 = 0; // Enable PWM Output

               while (TMR0 < 64);      // For Limited Number
                                       // Cycles

               if (!RB0)    // If Low, then Collision
                     Collision++;
                              // Must Have 3 Collisions in a Row
               else
                     Collision = 0;

               TRISB3 = 1;

          } // endif

// LCD State Machine
:

     } // endif

// Put Different Interrupt Handlers here

     if (INTF) {                      // RB0/INT Pin Interrupt
          CurrentRTC = ((RTC & 0x0FF) << 8) + TMR0;
                              // Get the Current Time
          if ((SaveRTC = CurrentRTC - SaveRTC) < 0)
               SaveRTC = 0 - SaveRTC;
          if ((SaveRTC > 250) && (SaveRTC < 350)) {
               DataIn = (DataIn << 1) + 1;   // "1" Received
               if (--DataInCount == 0) {
                    DataReady = 1;
                    INTE = 0;    // Finished, Disable
                                 // Interrupts
               } // endif
          } else if ((SaveRTC > 390) && (SaveRTC < 490)) {
               DataIn = DataIn << 1;   // "0" Received
               if (--DataInCount == 0) {
                    DataReady = 1;
                    INTE = 0;    // Finished, Disable
                                 // Interrupts
               } // endif
          } else if (--DataInCount != 12) {   // Invalid -
                                              // Delete
               DataInCount = 0;
```

```
                    INTE = 0;    // Finished, Disable Interrupts
              } // endif
              SaveRTC = CurrentRTC;  // Save the Execution Time
              INTF = 0;              // Reset Interrupt
        } // endif

} // End Interrupt Handler



//  Mainline
void main(void)                    // Template Mainline
{
int i;

      OPTION = 0x0D1;              // Assign Prescaler to TMR0
                                   //  Prescaler is /4
      TMR0 = 0;                    // Reset the Timer for Start
      T0IE = 1;                    // Enable Timer Interrupts
      GIE = 1;                     // Enable Interrupts

//  Put in Interface initialization code here

      LCDInit();                   // Initialize the LCD Port

      INTEDG = 1;                  // Interrupt on Rising Edge of
                                   //  RB0/INT IR TX'r
      CCPR1L = 13;                 // 50% Duty Cycle
      CCP1CON = 0b000001111;       // Turn on PWM Mode

      PR2 = 26;                    // 26 µsec Cycle for 38 KHz
      TMR2 = 0;                    // Reset TMR2
      T2CON = 0b000000100;         // TMR2 is On, Scalers 1:1

      while (1 == 1) {             // Loop forever

//  Put in Robot high level operation code here

            if (DataReady) {
                  while (LCDState != 0); // Wait for LCD
                                         // Available
                  for (i = 0; i < 12; i++) {
                        if (DataIn & (1 << 11))
                              Message[i + 8] = '1';
                        else
                              Message[i + 8] = '0';
                        DataIn = DataIn << 1;
                  } // endfor
                  LCDOut(Message);  // Pass String to Output
                  DataReady = 0;    // Reset the Data Flag
            } // endif

            if ((Collision == 0) && (OldCollision == 3)) {
                  OldCollision = Collision;
                                    // Save Current Collision
                                    // State
                  LCDOut(Message1);
            } // endif
            if ((Collision == 3) && (OldCollision == 0)) {
```

```
                 OldCollision = Collision;
                                 // Save Current Collision
                                 // State
                 LCDOut(Message2);
         } // endif

     } // endwhile

 } // End of Mainline
```

◼ Ultrasonic Distance Measurement

Ultrasonic distance measuring (or *range finding*) hardware, a third method of detecting objects around a robot, has different characteristics from whiskers and IR light reflection. I like to see ultrasonic object detection and distance measuring as similar to the methods that bats use to hunt insects—a fairly narrow beam of sound waves is directed in a specific direction. The time of flight of the sound waves gives quite an accurate distance measurement to an object (to 1 inch of accuracy or less). Ultrasonic object detection complents IR object detection; objects which are invisible to IR (ie black plastic) will reflect ultrasonics very well. Many objects which cannot be detected by ultrasonics (such as fine cloth) can be easily detected using the IR techniques presented in this chapter.

The ultrasonic distance measurement equipment described here is different from the sonar used by ships and submarines. When a ship's sonar is used to detect a submarine, sound waves are usually broadcast omnidirectionally (which means in every possible direction rather than in a narrow beam) and a directional microphone is used to find the angle of the submarine to the ship. Distance is measured by moving the ship and triangulating on the position of the reflected sound waves rather than measuring the time of flight; the speed of sound in water changes according to its temperature and salinity (as any reader of Tom Clancy's thrillers would know).

A minor drawback to the ultrasonic range finders is their narrow field of view. For the Polaroid 6500, you can expect the nearest object within ±5 degrees of the direction the transducer points will be found in—anything closer but outside this range of angles is ignored. This is a minor drawback because the sensor can be mounted on an RC servo with the servo moved a few degrees at a time and the robot's surroundings can be scanned in, giving the robot a picture of its surroundings.

A much more significant drawback to the ultrasonic range finders is the power required for them to operate. The Polaroid 6500 draws 1 amp of current when it generates the output pulses. This draw of current can be mitigated by the use of large capacitors, but it is still a significant drain on the robot's power resources.

To measure the distance from an object, an initial pulse is sent to the IR range finder module and the time required for the echo to return is measured. Assuming that sound travels at 1,087.4 feet/second (331.4 meters/second), it takes 153.3 μsecs to travel 1 inch. If you work with metric units, the speed of sound is 331.4 meters/second and the time required for sound to travel 1 cm is 30.2 μsecs. One practical aspect of the Polaroid 6500 (and some other ultrasonic range finders) is its masking of echoes for 18 inches—it is assumed that the body of the unit the range finder is mounted on will return echoes when a pulse is sent. By masking 18 inches (or ignoring returns for 2.76 milliseconds), internal reflections are ignored.

The Polaroid 6500 is a popular device for measuring distances within robots. It was first introduced in the 1970s and is quite expensive ($50 or more per unit). However, the Polaroid 6500 is very reliable and robust. If you look on the Internet, you will find simple designs for ultrasonic range finders, but using the 6500 eliminates the hassle of building them and trying to get them to work.

The Polaroid 6500 consists of two parts: a small PCB board and a round black (with gold interior) transducer. The PCB is connected to the transducer by a cable approximately 15 inches in length. The PCB also has a 9-pin ribbon cable with a rather unique connector soldered to it, which connects to the hardware that provides power and control signals to the unit. When I am working with the 6500, I desolder this cable and put in individual 22-gauge wires to enable me to control the unit easily. Table 4-14 shows the 9-pin connector pinout.

In normal operation, BLNK and BINH are held low, and the INIT pulse is used to initiate an operation with the system polling ECHO to see how far away

Table 4-14

The 9-pin
Connector
Pinout

Pin	Label	I/O	Comments
1	Gnd		
2	BLNK	I	When driven high, any reflected signal is blanked out.
3	N/C		
4	INIT	I	When driven high, the pulse is driven out of the 6500.
5	N/C		
6	OSC	O	This has 49.4 KHz oscillator output.
7	ECHO	O	Open collector is output from 6500. You can pull-up with a 4.7K resistor.
8	BINH	I	When asserted, the masking will not be active.
9	Vcc		This takes up +5 volts and a 1,000 μF capacitor is connected to it.

an object is. Figure 4-62 shows the output of the INIT pulse followed by the ECHO pulse response. Note that the INIT pulse is driven low after the ECHO was received.

In the datasheet, Polaroid recommends that you connect a 1,000 μF capacitor across the power supply to minimize the power transients throughout the system when the 6500's transducer is operating.

When working with any ultrasonic range finder, always be careful of the amount of energy that is available within the circuit. I like to tell the story about my first time using a Polaroid 6500—I picked up the transducer by its ring, where it is connected to the PCB. After the first measurement took place, I was surprised to see the transducer hanging by its wiring and my hand and arm nowhere to be seen. A few seconds later the pain came—a lot of energy is used to send the ultrasonic pulses.

I was fortunate that I was not holding onto the transducer with both hands when this was happening—if I had been holding onto the transducer with both hands, I'm sure I would have had a pretty good shock through my heart. As a precaution when working with the Polaroid 6500, make sure that you only use one hand (keep the other in your pocket).

The maximum detection range for the 6500 is 35 feet. The time of flight for 35 feet is 64.4 milliseconds so you should put in a timeout at 66 or more milliseconds. If the ECHO pulse has not been received by this time, chances are there are no objects within range of the ultrasonic range finder.

ultra—Example Interface

When I presented the IR TV remote control receiver, I included a complex application that measured the interval between two RB0INT interrupt requests. This

Figure 4-62
INIT/ECHO
signals to/from
the 6500

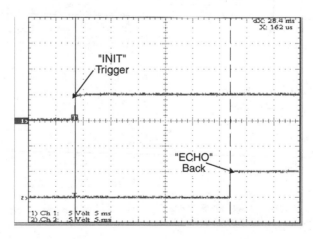

code was used as the basis for this application in which an ultrasonic pulse, initiated by a command pin from the PICmicro, is timed from leaving a Polaroid 6500 to when it is received. After the pulse has been received, the distance is converted into feet and inches and displayed on an LCD.

Adding the Polaroid 6500 to the LCD circuit is quite simple, as shown in Figure 4-63. There are just a couple of things to note. The first is that the INIT and ECHO pins are pulled up to Vcc using 4.7K resistors and the second is that the 1,000 µF capacitor is used to filter power to the Polaroid 6500 ultrasonic range finder.

The outputs of the 6500 are open collector and pull-up resistors are specified for all inputs into the circuit. The large capacitor minimizes the disruption to the PICmicro MCU and the other circuitry in the application when the 6500 sends an ultrasonic burst to measure the distance to the next object. The 6500 outputs a lot of energy when it is firing and the 1,000 µF capacitor must be placed as close as possible to the 6500's power wiring to help source the current needed rather than drawing it from the power supply or other components in the circuit.

It is very easy to wire the Polaroid 6500 ultrasonic range finder into a circuit, as shown in Figure 4-64. You will see that I have removed the custom wire and connected and added labeled wires to the 9-pin D-shaped hole layout. This modification to the 6500 makes it very easy to add the range finder to a circuit or wire it into a breadboard. Table 4-15 shows the bill of materials for this application.

It is very important to use a high current-sourcing-capable power supply for this application. As I have said, the 6500 draws a lot of current when it fires and

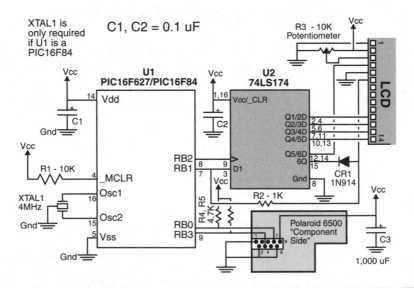

Figure 4-63 Ultra application circuit

Figure 4-64
Ultra
breadboard
circuit

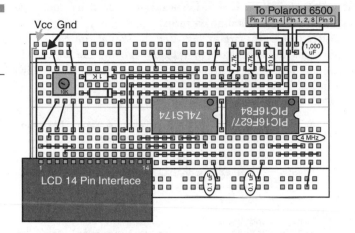

Table 4-15

Materials for
Ultra
Breadboard
Circuit

Label	Part Number	Comment
U1	PIC16F627	Can use a PIC16F84
U2	74LS174	
CR1	Silicon diode	Uses any type, but 1N914 is used for prototypes
R1	10K, ¼ watt	Uses _MCLR pull-up
R2	1K, ¼ watt	
R3	10K, 1 turn pot	Requires LCD contrast
R4 and R5	4.7K, ¼ watt	
C1 and C2	0.1 µF	Uses any type of decoupling capacitor
C3	1,000 µF, 16 volt	Requires an electrolytic capacitor
LCD		Hitachi-44780-controlled LCD display with a 14-pin interface
XTAL1	4 MHz ceramic resonator with internal capacitors	Required for PIC16F84
Polaroid 6500	Polaroid 6500 ultrasonic range finder	
Misc.		Requires a breadboard, wiring, and +5-volt power supply

you will find that the application does not work with a simple power supply made out of AA or 9-volt alkaline batteries.

Before I wired the circuit, I created the application code ultra.c (which is found in the robot\ultra directory) and tested it using the MPLAB IDE simulator and a stimulus file. The source code for this application is as follows:

```
#include <pic.h>

//   02.04.19 - ULTRA: Get the distance from the Polaroid 6500
//     Transducer and Display it on a LCD
//
//   This is a two wire interface using a 74LS174 as a shift
//     register and the code called from the mainline (and timed
//     from TMR0 interrupt)
//
//
//   Hardware Notes:
//   PIC16F827 Running at 4 MHz with External Oscillator
//   RB0 - Pulled Up "ECHO" from Polaroid 6500
//   RB1 - CLock Signal to the LCD
//   RB2 - Data Signal to the LCD
//   RB3 - "INIT" Line to start the distance measurement
//

//   Global Variables
unsigned int RTC = 0;              //  Real Time Clock Counter

volatile char LCDDlay = 20;    //  Initialization Delay Value
volatile char LCDState = 1;    //  Current LCD State

static volatile bit Clock     @ (unsigned)&PORTB*8+1;
static volatile bit ClockTRIS @ (unsigned)&TRISB*8+1;
static volatile bit Data      @ (unsigned)&PORTB*8+2;
static volatile bit DataTRIS  @ (unsigned)&TRISB*8+2;

char * MessageOut;               //  Message to be Sent Out
volatile char MessageOuti = 0;       //  Index to current Message
                                     //  Byte
char Message[9] = "\fxx\' xx\"";  //  Distance Measurement
char Message2[9] = "\fInvalid";   //  "Invalid" Distance
                                  //  Measurement

unsigned int  CheckDlay = 50; //  Wait 500 msecs between Range
                              //  Findings
unsigned int  PulseStartRTC;  //  RTC at Pulse Send
unsigned int  PulseEndRTC;    //  RTC at Pulse End
unsigned char PulseEndTMR0;   //  TMR0 Value for Pulse End
unsigned char PulseState = 0; //   0 - Waiting to Check Distance
                              //   1 - Distance Measurement Pending
                              //   2 - Have Distance
                              //   3 - Invalid Distance
unsigned long PulseTime;      //  Number of msecs for the Pulse
                              //  Time
unsigned long PulseInches;    //  Number of Inches measured
unsigned long PulseFeet;      //  Number of Feet measured
```

```
// Configuration Fuses
#if defined (_16F84)
#warning PIC16F84 selected
      __CONFIG(0x03FF1);          // PIC16F84 Configuration Fuses:
                                  //   - XT Oscillator
                                  //   - 70 msecs Power Up Timer On
                                  //   - Watchdog Timer Off
                                  //   - Code Protection Off
#elif defined(_16F627)
#warning PIC16F627 with external XT oscillator selected
      __CONFIG(0x03F61);          // PIC116F627 Configuration Fuses:
                                  //   - External "XT" Oscillator
                                  //   - RA6/RA7 Digital I/O
                                  //   - External Reset
                                  //   - 70 msecs Power Up Timer On
                                  //   - Watchdog Timer Off
                                  //   - Code Protection Off
                                  //   - BODEN Enabled
#else
#error Unsupported PICmicro MCU selected
#endif


// Subroutines
LCDNybble(char Nybble, char RS) {  // Send Nybble to LCD
:


LCDByte(char Byte, char RS) { // Send Byteto LCD
:


LCDInit()                          // Initialize the LCD I/O Pins
:



LCDOut(char * const LCDString)       // Output the Data String
:



// Interrupt Handler
void interrupt tmr0_int(void) // TMR0 Interrupt Handler
{

char temp;

      if (T0IF) {

            T0IF = 0;           // Reset Interrupt Flag

            RTC++;              // Increment the Clock

// Put additional interface code for 1 msec interrupt here

// Ultra sonic operations

            if (--CheckDlay == 0) { // Start the Pulse Output
                  PulseStartRTC = RTC;
                  PulseState = 1; // Doing Distance Measurement
```

```
                    RB3 = 1;          //  Start Pulse
                    INTF = 0;         //  Make sure no Pending
                                      //  Interrupts
                    INTE = 1;         //  Enable RB0/INTF
                    CheckDlay = 500;  //  Wait another 1/2 Second
                    if (RTC > 64900)
                    //  Make sure no issue with Crossing RTC Roll
                    //  Over
                            CheckDlay += 200;
             }  //  endif

             if ((PulseState == 1) && ((PulseStartRTC + 60) < RTC))
{
                    RB3 = 0;      //  Invalid delay, Turn off
                    INTF = 0;     //    6500 and requests
                    INTE = 0;
                    PulseState = 3;   //  Invalid
             }  //  endif

//  LCD State Machine
:

      }  //  endif

//  Put Different Interrupt Handlers here

      if (INTF) {
             PulseEndTMR0 = TMR0;     //  Recored the Pulse End
             PulseEndRTC = RTC;
             RB3 = 0;                 //  Reset Measurement Request
             PulseState = 2;          //  Have the Pulse
             INTF = 0;                //  Turn Off Interrupts
             INTE = 0;
      }  //  endif

}  //  End Interrupt Handler


//  Mainline
void main(void)                       //  Template Mainline
{

unsigned int temp;                    //  Temporary Storage Value

      OPTION = 0x0D1;                 //  Assign Prescaler to TMR0
                                      //   Prescaler is /4
      TMR0 = 0;                       //  Reset the Timer for Start
      T0IE = 1;                       //  Enable Timer Interrupts
      GIE = 1;                        //  Enable Interrupts

      LCDInit();                      //  Initialize the LCD Port

//  Put in Interface initialization code here

      RB3 = 0;                        //  Enable RB3 for Output
      TRISB3 = 0;                     //  Keep Low Initially

      while (1 == 1) {                //  Loop forever
```

```
// Put in Robot high level operation code here

                switch(PulseState) {      // Process according to State
                   case 2:
                        // Have Distance, Print Measurement
                        PulseTime = (((long) PulseEndRTC * 256) + _
                  (long) PulseEndTMR0) * 40;
                        PulseTime -= ((long) PulseStartRTC * _
                  (256 * 40));
                        PulseInches = PulseTime / 1533;
                        PulseFeet = PulseInches / 12;
                        PulseInches = PulseInches % 12;
                        if ((temp = PulseFeet / 10) == 0)
                             Message[1] = ' ';
                                       // Distance Not 10s of feet
                        else
                             Message[1] = temp + '0';
                        Message[2] = (PulseFeet % 10) + '0';
                        if ((PulseInches / 10) == 0)
                             Message[5] = ' ';
                        else
                             Message[5] = temp + '0';
                        Message[6] = (PulseInches % 10) + '0';
                        LCDOut(Message);
                        PulseState = 0;     // Wait for the Next
                                            // Pulse
                        break;
                   case 3:                // Invalid for some reason
                        LCDOut(Message2);
                        PulseState = 0;     // Wait for the Next
                                            // Pulse
                        break;
                } // endswitch

          } // endwhile

    } // End of Mainline
```

This code delays for 50 milliseconds (approximately) after power up and measures the distance of the nearest object to the range finder. If the object is too close (less than 18 inches from the transducer) or too far (more than 35 feet from the transducer), an invalid message is displayed on the LCD. If the object is within this range, the distance between the transducer and the object will be measured.

The measurement process consists of multiplying the number of microseconds it took for the ultrasonic signal to be sent from the transducer and back again by 10 and then dividing it by 1,533 to find the number of inches between the 2. Multiplying the number of microseconds by 10 and dividing by 1,533 is the same as dividing the number of microseconds by 153.3, but without the need to use floating point values. When the distance in inches is calculated, the number of feet is calculated. This, along with the remaining inches, is displayed on the LCD. In the code following the "case 2:" statement in the biologic code, fairly basic mathematical operations were used to perform this conversion and create the characters that are displayed.

Before burning a PICmicro MCU with this code, I tested it out using the included file ultra.sti, which can be found in robot\ultra:

```
!   Ultra-Sonic Response
!
!   Assume Initial Pulse at 51,200 µsecs
!   For 12', 7", Time required is 23,148.3 µsecs
!   Assume 50 Clock Intervals (51,200 cycles) before First
!    Send so actual Pulse is at:
!    74,348
!
!   Myke Predko
!   02.04.19
!
Step              RB0   RB4   RB5   RB6   RB7
1                  0     0     0     0     0    !  Start with Line
Input Low
15000              0     0     0     0     1
30000              0     0     0     1     1
45000              0     0     1     1     1
60000              0     1     1     1     1
74348              1     1     1     1     1    !  Pulse Going High
75000              0     1     1     1     0    !  Pulse Going Low
Again
!
```

This is the stimulus file, which provides inputs to the application during simulation. It behaves like a Polaroid 6500 with an object at a target distance of 12 feet, 7 inches away, assuming that the first signal is sent 51,200 µsecs from power up.

When the stimulus file is active, it monitors the current cycle step count of the simulating application. When there is a match between it and a step count in the stimulus file, it drives the specified values to the simulated PICmicro MCU. In the ultra.sti file, when the stimulus file is enabled and when the step count displayed by the Stopwatch dialog box in the MPLAB IDE is 30,000, the stimulus file drives the simulated RB6 and RB7 pins high and holds the simulated RB0, RB4, and RB5 low.

The previous stimulus file includes input values for pins RB4, RB5, RB6, and RB7. This probably does not make any sense to you because these pins are not used in the application code.

I included them to help me understand if the stimulus file was working. I often find that during debugging I forget to enable the stimulus file and simulate the application, looking for a response from the program, but not finding any. By placing values on unused pins, I can check periodically to see how (and if) the stimulus file is working and if the inputs are progressing over time.

It is a personal choice whether or not to use this tool. I have found that it has saved me from wasting a lot of time on many occasions when I am so focused on the code that I don't carry out all the correct steps in simulating it. It is purely optional and I'm sure many people will not see the value in specifying values on unused pins to check if the stimulus file is working.

To see how the stimulus file works, click Debug | Simulator Stimulus | Pin Stimulus | Enable, and select ultra.c when the file selection box is displayed.

After clicking OK, the previous stimulus file is enabled. Next, add the P16F627.WAT file along with a watch file that monitors the 32-bit values of PulseTime, PulseInches, and PulseFeet along with the stopwatch.

Set a breakpoint at the first line of the biologic code switch's case 2 (PulseTime = (((long) PulseEndRTC * 256) + (long) PulseEndTMR0) * 40; statement), reset the simulated PICmicro MCU, and click the Start Executing icon.

When the application stops, you should single step the application code to the point where the calculated message is displayed (LCDOut("Message");) and click Window | File Registers. The window that comes up displays the contents of the RAM (file) registers. You should be able to find the output message quite easily when the registers are displayed in ASCII format. To display the registers' contents in ASCII format, click the top-left square of the window and click ASCII Display.

Now, single step through the statements that calculate the distance and convert them into a format that can be displayed as characters in the Message; you should notice two things.

The first thing is that each arithmetic statement, even though it seems to be quite simple, takes many mouse clicks to step through. When I stepped through this code, I found that almost each statement took several milliseconds (resulting in multiple invocations of the 1-millisecond interrupt handler) to execute.

This should be a vivid example of why I emphatically state that all complex calculations should take place in the biologic code and not in the interrupt handler. If this code was located in the 1-millisecond interrupt handler, you would discover that several 1-millisecond interrupt events would be missed, which could result in the robot application running incorrectly.

The other thing that you should notice when you look at the Message in the File Register window is that the answer is wrong. Instead of displaying the string

    12'   7"

it displays

    12'   1"

This error can be very disconcerting and hard to understand.

It is a result of being lazy when I created the stimulus file and it can be used to demonstrate that the code is actually working correctly.

If you reset the simulated PICmicro MCU, set a breakpoint in the 1-millisecond interrupt handler at the statement "RB3 = 1;", and run the application, the data is actually driven at step 52,037, not 51,200, as expected.

These extra 837 instruction steps cause a delay between the time the pulse was assumed to be output and when it actually was. Knowing that sound requires 153.3 μsecs to travel 1 inch, these 837 cycles represent 5.46 inches in the measurement.

If you subtract 5.46 from 7 and round down to the nearest decimal, you get 1, which is exactly what was stored in the Message variable.

If you were to go back to the stimulus file and add 837 to the calculated number of cycles for 12 feet, 7 inches (75,185 instead of 74,348) and simulated the application again, the result is 12 feet, 7 inches, as expected.

The simulation method shown here is a useful way of testing your application in the MPLAB IDE simulator when you are expecting a response after a set period of time. A stimulus file could have been created to test the whisker debounce or IR TV remote control code that was presented earlier in the chapter.

Although sometimes it seems tedious to develop a stimulus file, it can be a life-saver when your software is behaving erratically and you have no idea what the problem is.

I must caution you again to be very careful when the 6500 is operating. There is a lot of current that passes through the unit and I suggest that you only handle the operating unit with one hand and keep the other in your pocket. Along with this, if the unit falls, let it hit the ground—don't try to catch it. This eliminates any chance that current that passes through your body will pass through your heart.

Light-Level Sensors

You are probably not aware of it, but simple vision capabilities are often built into robots. When I say vision, I am not talking about video (as we and most fictional robots use); I'm referring to simple light monitoring around the perimeter of the robot. The earliest robots were built with just two photocells arranged in such a way that the robot would move toward the brightest light in its environment. This type of robot is known as a *photovore* (which literally translates into "light eater") or *moth* because it simulates an insect's attraction to the brightest location in its environment.

When implementing light sensors in a robot, they are normally given different fields of view (see Figure 4-65). In this diagram, you can see that by offsetting the two light sensors by 45 degrees from the front of the robot, the different light levels around the robot can be sampled. This gives the robot a differential field from which it can make decisions regarding what it should do next.

It should be obvious in Figure 4-65 that with only two light sensors arranged, not everything to the rear sides and direct rear can be seen. Some robots place sensors all around their bodies to avoid problems like this, whereas others turn and scan their environment before moving forward. Neither solution is 100 percent optimal for all cases.

Light sensing in robots is accomplished by using *cadmium sulfide* (CDS) cells, photodiodes, or phototransistors. A CDS cell's resistance depends on the amount of light that is hitting its surface and it normally decreases as it is exposed to more light. CDS cells are often known as *light-dependent resistors* (LDRs). A pho-

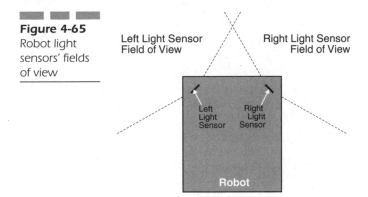

Figure 4-65
Robot light
sensors' fields
of view

todiode produces current that is proportional to the amount of light striking its surface. I have to point out that the current produced is infinitesimal and will have to be amplified in some way before it can be used in a robot. The current passing through the collector/emitter of a phototransistor is proportional to the amount of light striking it. There are quite a few different types of phototransistors. Some have Darlington outputs to amplify the current-passing capability of the device. Other phototransistors have base bias pins that enable current to be injected into the base of the transistor, providing control of the threshold level where the transistor turns on and off.

For all three of these different devices, the response to changes in light is quite slow (usually measured in tens or hundreds of milliseconds). This is an important point to consider when implementing simple light-measuring vision in a robot. You may want to keep the speed of the robot down to a minimum to better continuously scan your environment.

The CDS cell is the most popular type of light sensor used in most robot applications because it produces the most quantitative change for the least amount of light. Three different methods are used to convert the change in resistance of a CDS cell into a quantity that is useful for a robot controller.

The first method of monitoring the value of a CDS cell is to use an *analog-to-digital converter* (ADC). To convert the resistance of the CDS cell into a useable voltage for the ADC, a precision resistor is used along with the CDS cell (as shown in Figure 4-66) to create a voltage divider. The output of the voltage divider is passed to an ADC, where it is converted to a numeric value. In the circuit shown in Figure 4-66, as more light falls on the CDS cell, its resistance and the voltage read by the ADC decrease. The precision resistor should be of the same value as the largest resistance of the CDS cell.

If you would like to have the ADC value increase when more light strikes the CDS cell, you should switch the positions of the CDS cell and precision resistor in the circuit.

Another way of measuring the resistance of the CDS cell is to make it part of an RC network, as shown in Figure 4-67. In this circuit, the controller drives a

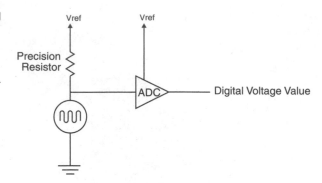

Figure 4-66
CDS resistance
check using an
ADC

high-voltage value into the circuit in order to charge the capacitor. The pin is then converted to an input pin and the charge in the capacitor is allowed to flow through the CDS cell to the ground. The time at which the voltage drops to the pin threshold voltage where the value read in changes from 1 to 0 depends on the resistance of the CDS cell.

The time for the voltage to drop from Vcc to the PICmicro's I/O pin threshold voltage can be approximated using the following formula:

```
Time = 2.2 * RC
```

This time is a very rough estimate and should only be used to calculate the degree of the time delay. The actual time delay varies from the calculated value due to variances in the expected to actual CDS cell resistance, which is the capacitance of C and the characteristics of the I/O pin used.

This method can be difficult to implement in applications that are timer interrupt based—especially if there are no more interrupt input pins available.

The third method is to make a voltage divider out of the two CDS cells and execute a comparison operation to see which one has the lowest resistance. Figure 4-68 shows two CDS cells used as a voltage divider with the output being passed to the positive input of a comparator, which has one-half of the reference voltage passed to its negative input. As long as the top CDS cell's resistance is equal to or less than the lower CDS cell's resistance a 1 is output. If the top CDS cell's resistance is greater than the lower CDS cell's, then a 0 is output from the comparator.

This last method is a very simple way of implementing a light-sensing operation and, as such, it can be difficult to work with. In the other methods of converting the CDS cell's value to a digital value, there was some attempt to calculate the actual value. In this method, it is binary with one CDS cell receiving more light than the other and the response to the data cannot be very proportional (measured).

It is important to remember that all methods of converting light into a digital value require a quiet power supply. Noisy power supplies can affect the operation of the ADC operation by changing charge voltages, comparison thresholds, and reference voltages. In many of my applications, I shut down the motors while I am checking the resistance of the CDS cells.

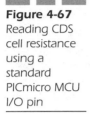

Figure 4-67
Reading CDS
cell resistance
using a
standard
PICmicro MCU
I/O pin

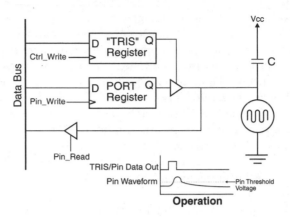

Figure 4-68
Two-CDS-cell
differential
resistance
check using a
voltage
comparator

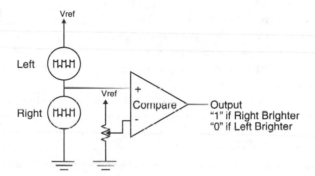

When you are experimenting with light sensors, you may have to set up some pretty unusual tests to check light-sensing robots. Having a robot hone in on a flashlight lying on the floor of a darkened room is not that unusual, nor is draping a room in black cloth to prevent the robot from being distracted by light objects in the room.

If you want to create a robot that is very tolerant of different room organizations, I suggest that you consider adding a small video camera to your robot and parse the resulting video signal for the brightest (or darkest) spot in the room.

light—Practical Sensor Examples

In the previous section, I presented three different ways to implement a light sensor for use on a robot. Each method has advantages and disadvantages. This section looks at some circuits that implement these light sensors and discusses how the different methods are best used in a robot application.

The most basic form of light sensor is the differential light sensor that consists of two CDS cells that are wired as a voltage divider. In this case, when one CDS cell is exposed to more light, the voltage drop across decreases because its resistance has decreased. The circuit is used in light-sensing robots because it can be easily compared to one-half of Vcc, and the CDS cell that has the most light striking it tips the voltage balance toward the other CDS cell.

This circuit can be implemented easily in a PIC16F627-based robot that has a simple analog voltage comparator, as shown in Figure 4-69. Along with the comparators, use the internally generated reference to create the one-half Vcc for the voltage divider comparator.

I have also noted pin RA2 in Figure 4-69 because it can be used to output the analog reference voltage produced within the PICmicro MCU. This is driven out to make sure that it is correctly used in the application.

It is very easy to wire this circuit into the LED circuit, as shown in Figure 4-70. I did not go through any great pains to make sure that the Gnd and Vcc used by the two CDS cells are common to the microcontroller. For absolute accuracy, I should have made sure the voltage and ground were as close as I could make them so the one-half Vcc value is common for both the microcontroller and CDS cell voltage divider. As discussed in the following section, this can be a major concern and you may want to use the optional trimmer potentiometer shown in Figure 4-69.

The long wires attached to the CDS cells are not well illustrated in Figure 4-70. The cells look like they are pressed flush against the breadboard, but when you build your own application, you might want to leave the leads as long as possible to make sure that they can bend away from each other. The action of the comparator can be illustrated by pointing a flashlight's beam at one CDS cell and then the other.

Table 4-16 lists the bill of materials for the circuit.

The application code was named light1.c, which can be found in the robot\light directory. The actual source code consists of the following:

```
#include <pic.h>

//   02.04.24 - LIGHT: Look at two CDS Cells in Series and indicate
//    brighter one
//
//   This is a two wire interface using a 74LS174 as a shift
//    register and the code called from the mainline (and timed
//    from TMR0 interrupt)
//
//
//   Hardware Notes:
//   PIC16F827 Running at 4 MHz with External Oscillator
//   RA0 - Comparator2, Connected to a 2 CDS Cell Voltage Divider
//    with "Left" on Top
//   RA1 - Comparator1, Tied to Vss
```

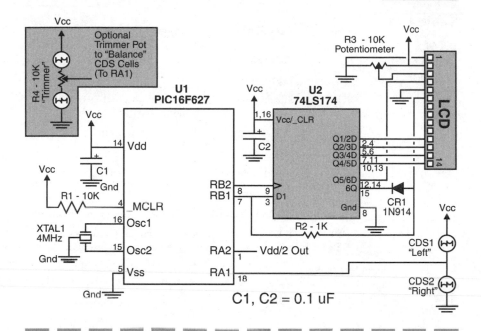

Vcc

Figure 4-69 light1 application circuit

Figure 4-70
light1
breadboard
circuit

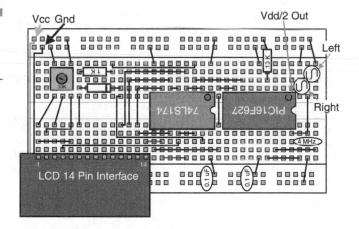

```
//   RA2 - 1/2 Vdd from Vref module, output to check Vref operation
//   RB1 - CLock Signal to the LCD
//   RB2 - Data Signal to the LCD
//

//   Global Variables
int RTC = 0;                    //   Real Time Clock Counter
```

Table 4-16

Bill of
Materials for
both light1
an light2
example
interfaces

Label	Part Number	Comment
U1	PIC16F627	
U2	74LS174	
CR1	Silicon diode	Uses any type, but 1N914 is used for prototypes
R1	10K, ¼ watt	Uses _MCLR pull-up
R2	1K, ¼ watt	
R3	10K, 1 turn pot	Requires LCD contrast
R4	10K, 1 turn pot	Optional, see text
R5 and R6	470 ohm, ¼ watt	Used for ADC circuit (light2.c)
CDS1 and CDS2	10K CDS cell	Is 10K when dark and resistance drops when exposed to light
C1 and C2	0.1 µF	Uses any type of decoupling capacitor
C3 and C4	0.01 µF	Used for any type of ADC circuit (light2.c)
LCD		Hitachi-44780-controlled LCD display with a 14-pin interface
XTAL1	4 MHz ceramic resonator with internal capacitors	
Misc.		Requires a breadboard, wiring, and +5-volt power supply

```
volatile char LCDDlay = 20;    // Initialization Delay Value
volatile char LCDState = 1;    // Current LCD State

static volatile bit Clock      @ (unsigned)&PORTB*8+1;
static volatile bit ClockTRIS  @ (unsigned)&TRISB*8+1;
static volatile bit Data       @ (unsigned)&PORTB*8+2;
static volatile bit DataTRIS   @ (unsigned)&TRISB*8+2;

char * MessageOut;                  // Message to be Sent Out
volatile char MessageOuti = 0;         // Index to current Message
                                       // Byte
char Message1[3] = "\f*";      // Clear Display, Put "*" on Left
                               // Clear Display, Put "*" on Right
char Message2[18] = "\f            *";
//                        1 2345678901234567


//  Configuration Fuses
#if defined(_16F627)
```

```
#warning PIC16F627 with external XT oscillator selected
        __CONFIG(0x03F61);        // PIC116F627 Configuration Fuses:
                                  //   - External "XT" Oscillator
                                  //   - RA6/RA7 Digital I/O
                                  //   - External Reset
                                  //   - 70 msecs Power Up Timer On
                                  //   - Watchdog Timer Off
                                  //   - Code Protection Off
                                  //   - BODEN Enabled
#else
#error Unsupported PICmicro MCU selected
#endif


//  Subroutines
LCDNybble(char Nybble, char RS) {    //  Send Nybble to LCD
:

LCDByte(char Byte, char RS) { //  Send Byteto LCD
:


LCDInit()                         //  Initialize the LCD I/O Pins
:


LCDOut(char * const LCDString)        //  Output the Data String
:


int ADCPoll()
{

        return C2OUT;

}  //  End ADCPoll


//  Interrupt Handler
void interrupt tmr0_int(void) //  TMR0 Interrupt Handler
{

char temp;

        if (T0IF) {

                T0IF = 0;          //  Reset Interrupt Flag

                RTC++;             //  Increment the Clock

//  Put additional interface code for 1 msec interrupt here

//  LCD State Machine
:

        }  //  endif

//  Put Different Interrupt Handlers here
```

```
}   //   End Interrupt Handler


//   Mainline
void main(void)                      //   Template Mainline
{

       OPTION = 0x0D1;               //   Assign Prescaler to TMR0
                                     //    Prescaler is /4
       TMR0 = 0;                     //   Reset the Timer for Start
       T0IE = 1;                     //   Enable Timer Interrupts
       GIE = 1;                      //   Enable Interrupts

//   Put in Interface initialization code here

       LCDInit();                    //   Initialize the LCD Port

       CMCON = 0x002;                //   Enable the Comparator
                                     //   Module
                                     //    C1/C2 Normal, not inverted
                                     //    CIS = 0
                                     //    Mode 2, compare against
                                     //    Vdd
       TRISA = 0x003;                //   Just bits RA0 and RA1 are
                                     //   I/Ps
       VRCON = 0x0EC;                //   Enable Vref module
                                     //    Vref output on RA2
                                     //    High Vref Range
                                     //    Ladder value of 12

       while (1 == 1) {              //   Loop forever

//   Put in Robot high level operation code here

              if (!ADCPoll())        //   Bit 7 of CMCOM
                   LCDOut(Message1); //   Light is left
              else
                   LCDOut(Message2); //   Light is right

       }  //   endwhile

}  //   End of Mainline
```

After the LCD is initialized, the code polls the comparator and the * character is placed on the display depending on whether the left or right side is exposed to the light. The PICmicro MCU's comparator works continuously, so there is no point in sequencing the operations; instead, the current value can just be polled.

Note that I did not enable the cursor in this application. The reason for disabling it was to eliminate the flickering along the bottom of the LCD's line when it is updating the display. Realistically, the cursor is only required for applications that require user input. If your application is just writing to the LCD, then if you dispense with the cursor, the displayed characters will be sharper and there won't be an annoying flashing on the bottom line of the LCD.

The ability to poll a comparator to get the current operating status is the circuit's biggest advantage. Comparison operations are essentially instantaneous and there is no requirement for complex code to sequence the operation or to perform an ADC.

The greatest disadvantage of this method of sensing light is the binary operation of the CDS cell voltage divider with the comparator. The difference between the two CDS cells cannot be determined, so if the robot is in a dark room, it behaves exactly the same way as it would in an evenly lighted room. If you had a slight difference in the resistance of two CDS cells, the robot responds in exactly the same way as if a spotlight was directed at one CDS cell while the other was wrapped in black electrical tape.

Another disadvantage of binary action of this circuit is its inability to let a robot run in a straight line toward a light source when both CDS cells have the same resistance (and are exposed to an equal amount of light). In an actual light-sensing robot application, this application turns from left to right as it approaches the source. This can be a problem with some applications because as the robot approaches the light source, if the robot turned too far in its oscillations, it can lose the brightest light source in its environment and lock onto another source that may not be the correct final destination.

Lastly, depending on the CDS cells that you use, you may find that they respond differently to different light levels. Figure 4-69 shows an alternative input circuit with a potentiometer between the two CDS cells. This potentiometer can be used as a trimmer to compensate for any differences in the returned values for the CDS cells and enable them to produce essentially the same values. With the trimmer, you should be able to tune the circuit in such a way that the display seems to be showing the splat (*) characters at both positions when an equal amount of light shines on the two CDS cells.

Reading the resistance of the CDS cells and then comparing the results is a much more effective method of sensing light around the robot. In the previous section, I presented two different ways of implementing an ADC for this task. The remainder of this section discusses the method of using the CDS cells as part of a timed-response RC network. The response of the RC network is dependent on the resistance of the CDS cell. The less light that is passed to the CDS cell, the higher its resistance and the longer it takes for the circuit to respond.

Figure 4-71 shows the schematic diagram used for this application and Figure 4-72 shows the wiring used to implement the circuit. Note that to wedge everything in on a small breadboard, I had to pass the 470 ohm resistors from RB4 to RB6 and RB5 to RB7. This is not good circuit assembly practice, but it does enable you to build the circuit on the small breadboard for testing.

The parts required for the application are essentially the same as the parts for the comparator version of the application; however, two additional capacitors and two resistors are included. For this reason, I listed all the required parts in Table 4-16.

This chapter does not include a sample application using an external ADC. The reasons for this are based on the assumption that the elelogic interface used for the ADC is similar to that used by other devices (such as the LCD), and with

Figure 4-71
Light 2
application
circuit

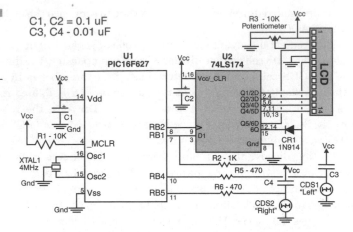

Figure 4-72
Light 2
breadboard
circuit

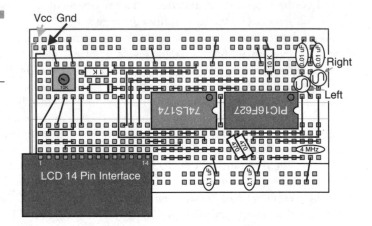

the LCD in place in a short breadboard, there was not enough space to add the ADC chip. The latter reason isn't quite true; with a bit of work, I could have put in an 8-pin ADC chip, but this would have made the circuitry much more complex and difficult to follow.

The code for this application (light2.c) can be found in the robot\source directory:

```
#include <pic.h>

//   02.04.24 - LIGHT2: Implement a dual RC Network ADC on the
//   PIC16F84/PIC16F627.  10K CDS Cell with a 0.01 µF Capacitor
//
//   This is a two wire interface using a 74LS174 as a shift
//   register and the code called from the mainline (and timed
//   from TMR0 interrupt)
//
```

```
//
//   Hardware Notes:
//   PIC16F827 Running at 4 MHz with External Oscillator
//   RB1 - CLock Signal to the LCD
//   RB2 - Data Signal to the LCD
//   RB4 - Left ADC
//   RB5 - Right ADC
//

//   Global Variables
int RTC = 0;                    //  Real Time Clock Counter

volatile char LCDDlay = 20;     //  Initialization Delay Value
volatile char LCDState = 1;     //  Current LCD State

static volatile bit Clock     @ (unsigned)&PORTB*8+1;
static volatile bit ClockTRIS @ (unsigned)&TRISB*8+1;
static volatile bit Data      @ (unsigned)&PORTB*8+2;
static volatile bit DataTRIS  @ (unsigned)&TRISB*8+2;

char * MessageOut;              //  Message to be Sent Out
volatile char MessageOuti = 0;      //  Index to current Message
                                    //  Byte
char MessageL[19] = "\376\200              *";
//                   1    2   34567890123456789
char MessageR[19] = "\376\300              *";

volatile char ADCState = 0;     //  Current ADC State Machine State
volatile char ADCDlay = 1;      //  Delay 10 ms between each ADC
//   Operation
volatile char LeftADC = 0;      //  Last Left ADC Value
volatile char RightADC = 0;     //  Last Right ADC Value


//   Configuration Fuses
#if defined(_16F627)
//   #warning PIC16F627 with external XT oscillator selected
      __CONFIG(0x03F61);        //  PIC116F627 Configuration Fuses:
                                //    - External "XT" Oscillator
                                //    - RA6/RA7 Digital I/O
                                //    - External Reset
                                //    - 70 msecs Power Up Timer On
                                //    - Watchdog Timer Off
                                //    - Code Protection Off
                                //    - BODEN Enabled
#else
#error Unsupported PICmicro MCU selected
#endif


//   Subroutines
LCDNybble(char Nybble, char RS) {   //  Send Nybble to LCD
    :


LCDByte(char Byte, char RS) { //  Send Byteto LCD
    :


LCDInit()                       //  Initialize the LCD I/O Pins
    :
```

```
LCDOut(char * const LCDString)      // Output the Data String
{

     while (LCDState);        // Wait for LCD Available

     MessageOut = LCDString; // Load up the string
     LCDState = 100;         // Start Sending the String

} // End LCDOut


// Interrupt Handler
void interrupt tmr0_int(void) // TMR0 Interrupt Handler
{

char temp;

     if (T0IF) {

          T0IF = 0;           // Reset Interrupt Flag

          RTC++;              // Increment the Clock

// Put additional interface code for 1 msec interrupt here

// LCD State Machine
:

// ADC State Machine Read Here

          if (!LCDState) {        // Only Execute if LCD NOT
                                  // Active
               switch (ADCState) {
                    case 0:     // ADC Start Delay
                         if (--ADCDlay == 0)
                              ADCState++;
                         break;
                    case 1:
                    // Start Charging Left and Right (for 1
                    // msec)
                         TRISB4 = 0;
                         TRISB5 = 0;
                         RB4 = 1;
                         RB5 = 1;
                         ADCState++; // Jump to next state
                         break;
                    case 2:
                         TRISB4 = 1; // Turn off Left Pin
                         // Driver
                         temp = PORTB;    // Clear pending
                         RBIF = 0;   //  Interrupts on Port
                         RBIE = 1;   //  Change
                         ADCState++;
                         break;
                    case 3:             // Roll Over, No
                                        // Interrupt
                         LeftADC = 0x0FF;
                         temp = PORTB; // Clear pending
```

```
                              RBIF = 0;    //   Interrupts on Port
                              RBIE = 0;    //   Change
                              ADCState++;
                              break;
                     case 4:
                              TRISB5 = 1; //  Change TRISB to Input
                              temp = PORTB;  //  Clear pending
                              RBIF = 0;    //   Interrupts on Port
                              RBIE = 1;    //   Change
                              ADCState++;
                              break;
                     case 5:              //   Roll Over, No
                                          //   Interrupt
                              RightADC = 0x0FF;
                              temp = PORTB;  //  Clear pending
                              RBIF = 0;    //   Interrupts on Port
                              RBIE = 0;    //   Change
                              ADCDlay = 10;
                              ADCState = 0;
                              break;
                  } //  endswitch
            } //  endif


      } //  endif

//  Put Different Interrupt Handlers here

      if (RBIF) {               //   Interrupt on Pin Change
            switch(ADCState) {
                  case 3:    //  Left ADC Active
                        LeftADC = TMR0;   //  Get the ADC Count
                        ADCState++;
                        break;
                  case 5:    //  Right ADC Active
                        RightADC = TMR0;  //  Get the ADC Count
                        ADCDlay = 10;
                        ADCState = 0;
                        break;
            } //  endswitch
            temp = PORTB;     //  Clear pending Interrupts on
            RBIF = 0;         //    Port Change
            RBIE = 0;
      } //  endif

} //  End Interrupt Handler


//  Mainline
void main(void)               //  Template Mainline
{

unsigned int i, j;
unsigned int tempLeft, tempRight;

      OPTION = 0x0D1;         //  Assign Prescaler to TMR0
                             //    Prescaler is /4
      TMR0 = 0;              //  Reset the Timer for Start
      T0IE = 1;             //  Enable Timer Interrupts
      GIE = 1;              //  Enable Interrupts
```

```
//  Put in Interface initialization code here

    LCDInit();                  //  Initialize the LCD Port

    while (1 == 1) {            //  Loop forever

//  Put in Robot high level operation code here

            while (ADCState != 2);  //  Wait for ADC Operation to
                                    //  Start
            while (ADCState);  //  Wait for ADC Operation to
                               //  Complete

            tempLeft = LeftADC;  tempRight = RightADC;
                               //  Save ADC Values

            j = (tempLeft / 8) + 1;  //  Calculate number of Left
                                     //  Splats
            for (i = 2; i < 18; i++ )
                if ((i - 2) <= j) //  Print a Splat
                    MessageL[i] = '*';
                else              //  Print a Blank
                    MessageL[i] = ' ';

            j = (tempRight / 8) + 1;     //  Calculate number of
                                         //  Right
            for (i = 2; i < 18; i++ )    //   Splats
                if ((i - 2) <= j) //  Print a Splat
                    MessageR[i] = '*';
                else              //  Print a Blank
                    MessageR[i] = ' ';

            LCDOut(MessageL); //  Left Graph
            LCDOut(MessageR); //  Right Graph

    }  //  endwhile

}  //  End of Mainline
```

To perform the timing operations, the application uses the interrupt on PORTB change feature of the PICmicro MCU. This feature is often not recommended by PICmicro microcontroller experts because it can be difficult to use.

I would disagree with this recommendation—I have used this feature of the PICmicro MCU since I first started working with it and have found it to be a reliable and easy-to-use method of requesting interrupts from external sources.

This does not mean that there aren't some rules that you should follow. To successfully use the interrupt on PORTB change, you should consider the following rules:

■ The only input pins that should be used with PORTB are the interrupt on PORTB change.

■ There should *never* be any reads of PORTB except when clearing the interrupt request. Do not access any of the bits in PORTB (the instructions that access bits also access the entire register). If necessary, keep a mirror copy of the output contents of PORTB rather than trying to read the current state.

■ To reset the interrupt requesting hardware for interrupt on PORTB change, use the following code:

```
Variable = PORTB; //   Read the current status of PORTB are
                 //   reset
                 //    interrupt requesting hardware
RBIF = 0;        //   Reset interrupt request bit
RBIE = 0;        //   Optional disabling of interrupt
                 //   requests
```

■ Make sure that the circuit connected to the port change interrupt bit does not change before the code is ready for it. This was a problem with light2.c. It required the addition of the 470-ohm current limiting resistors to slow down the flow of current during the capacitor-charging part of the ADC operation.

When the light2.c application executes, it starts a state machine that sequences through the operations of the RC network, charging the capacitor and then changing the pin to an input and using TMR0 to time how long it takes for the charge to pass through the CDS cell.

The ADC state machine only starts executing when the LCD has stopped displaying the current string sent to it. This was deliberate to eliminate timing contention between the operation of the ADC and LCD. To further ensure that the ADC code does not run into any problems with other interfaces taking up interrupt handler cycles that could change the results, the mainline code waits for the ADC state machine to start operating and, once it finishes, sends data to the LCD.

The information received from this interface is much more useful than the two-CDS-cell comparator application, but it has an important difference that can make it difficult to work with in some situations. The ADC operation takes a certain amount of time to charge the capacitor and let it discharge through the CDS cell. This means that the ADC version cannot respond as quickly to changes in input as the comparator-based version.

To mitigate this, change the LeftADC and RightADC values when the new CDS cell data has been received. In this application, the ADC operation executes within a period of 6 milliseconds (giving an operational rate of up to 330 samples per second as 2 samples are taken in each 6-millisecond window). This sample rate is not as fast as many commercial ADCs, but this is one of the trade-offs for implementing an ADC without any external hardware.

■ Sound Sensors

Adding a sound sensor to an electronic circuit has always been easy for me. Many robots have a built-in audio input that operates like a clapper and enables the user to control the robot by clapping his or her hands or yelling at the robot. Some robots use a sound sensor because it can be part of a collision sensor (just listen for the crunch when the robot hits something). Creating a microphone subsystem that detects loud noises is quite easy to do and can be very useful in some robot applications.

One of the most difficult concepts for people to understand about analog signals and sound is that they are made up of many other signals. The basic signal is the *sine wave* (see Figure 4-73) and multiple applications of this signal can be used to produce any analog signal that you can imagine. This is quite a powerful statement and one that requires further explanation.

The sine wave is the result of a drawing of the vertical position of a point on a disk when the disk is rotated for some period of time. There are only a few things that you need to know about sine waves and how they can be used to represent more complex analog signals. The two critical parameters of the sine wave are its amplitude (peak-to-peak measurement described previously) and period.

Sin is the normal abbreviation used for sine and *cos* is the abbreviation used for cosine. The *cosine wave* is identical to the sine wave, but it appears 90 degrees (or π/2 radians, which will be discussed in the following section) ahead of the sine wave. For the information presented in this book, I will work only with a sine wave function, which is described by the following function:

```
Wave Height = A sin(2 * Pi * t / Period)
```

The period of the sine wave is measured in degrees or, for electronics, in radians. Because one full cycle of a wheel covers 2×π×the radius of the wheel, the period of a sine wave is measured in terms of radians, which is multiplied by 2×π to get the current position of the wheel.

In discussions of sines and cosines, you may have heard the period described in degrees (with one full turn of the wheel being 360 degrees). The use of radians

Figure 4-73
Sine and
cosine waves

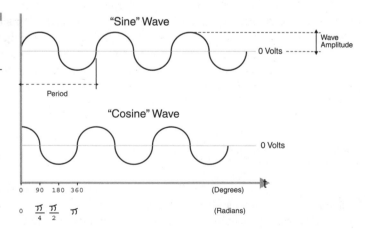

is a convention that is used within electronics and it is recommended that you set your calculator to *rad* when you are working in electronics. To convert from degrees to radians, you can use the following formula:

```
Radians = 2 * Pi * degrees / 360
```

Using this formula, 45 degrees is π/4, or 0.7854.

To convert time into radians to use a sine wave, you should use the following formula:

```
Time in radians = 2 * Pi * t / Period
```

When I produce a clock signal, I always strive to make it as close to a perfect square wave as possible. A perfect square wave has an equal high and low time or a 50 percent duty cycle, and vertically rises and falls when transitioning from low to high and back to low again. It is probably surprising, but the square wave is an excellent example of how a number of different sine waves can be combined together in order to produce a perfect square wave.

In the case of the square wave, numerous sine waves with different periods are added on to each other to produce the actual signal. Figure 4-74 shows how two different sine waves are added together to create a signal that closely approximates a square wave.

This method can be used to produce many other different and arbitrary waveforms.

Note that the frequency of each sine that makes up the square wave is made up of a multiple of the square wave. Each different frequency sine wave is known as a *harmonic*. The sine wave at three times the frequency of the square wave is known as the *third harmonic*, the sine wave at five times the frequency of the square wave is known as the *fifth harmonic*, and so on.

Figure 4-74
Square wave
produced by
sine waves

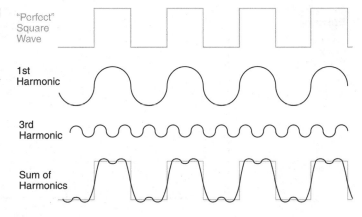

"Perfect"
Square
Wave

1st
Harmonic

3rd
Harmonic

Sum of
Harmonics

Also note that the amplitude of each sine wave changes with the harmonic in a way that is predictable. In fact, the square wave function could be described as

```
squarewave(t) = A * (sin(2 * Pi * t / P) + sin(6 * Pi * t / P)/ 3 +
                     sin(10 * Pi * t / P)/ 5 + ...)
```

or as a sum of terms using the following formula:

```
             ∞
squarewave(t) = A Σsin((2i + 1) * Pi * t/P)/(2i + 1)
             i = 0
```

In this format, the square wave function is known as a *Fourier series*.

I found series fascinating when I learned about them in college. By using a series, you can simplify many mathematical processes or produce arbitrary waveforms like the square wave demonstrated in this example. In fact, many analog signal function generators internally produce a series of sine waves and mix them together to produce an arbitrary signal out.

The obvious question from this introduction is whether or not actual square waves are produced in TTL logic using sine waves. There is a seemingly simple test for this—filter out the harmonics of the signal and see if you are left with a sine wave. Unfortunately, this is very difficult to do practically. The problem isn't with designing and implementing the filter, but with passing a perfect square wave to it.

A filter is simply a device that outputs less of a signal than what is put into it. Your sunglasses are filters because they only let a fraction of the total light through to your eyes. For converting a square wave into its constituent sine waves, a *low-pass filter* is required, which filters out high frequencies, as shown in Figure 4-75.

Figure 4-75
Operation of a
low-pass filter

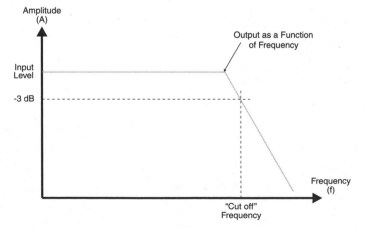

Input frequencies are output from a low-pass filter as long as they are less than the cutoff frequency. At the cutoff frequency, the low-pass filter has attenuated the input signal by 3 dB (the amplitude of the output signal is 0.708 of its input value). As the frequency increases, the amount of attenuation increases.

In the following sample interface, a low-pass filter is implemented using a chip known as an *operational amplifier* (best known as an *op-amp*) with some analog components. One of the most common low-pass filter circuits is the two-pole *Butterworth low-pass filter circuit*, which is shown in Figure 4-76. I like this circuit because the values for the two resistors (R) and two capacitors (C) are the same and the circuit can be used to amplify the resulting filter output signal, although I normally use a second voltage-follower amplifier with a very high gain (amplification factor) for this task.

By amplifying the resulting filter output to as close to infinity as possible, I change the low-frequency sine waves to a chopped wave that resembles a square wave. This is probably confusing because I presented how the low-pass filter works on a square wave—why would I go to all this trouble to convert a square wave back to a square wave?

The answer is that I am not trying to convert a square wave back into a square wave. I am trying to convert an arbitrary sound input wave into a low-frequency sine wave. If the sound input is not present, there should not be any sine wave output from the filter. If a high-frequency input sound from the robot (such as a motor whine) is passed to the filter, it should be filtered out, not passed to the controller.

There are many different types of filters. Along with the low-pass filter, there is the high-pass filter, which only permits signals *above* the cutoff frequency, and the band-pass filter, which only permits frequencies within a specific range to be passed. For robots, the low-pass or band-pass filters are probably the most useful ones to filter out external noises from the internal noises of the robot.

To summarize, there are ways to filter out the noise the robot makes so that sounds external to the robot can be sensed and passed back to the robot. This

Figure 4-76
Two-pole
Butterworth
low-pass filter

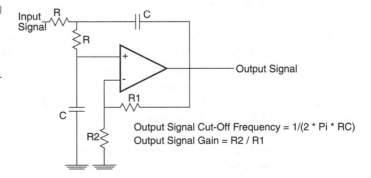

statement also explains why many robot designers do not like to use sound sensors. Often it is very difficult to find frequency ranges where a person's shout or clap or the crunch the robot makes when it hits an object is different from the other sounds going on in the robot.

I have implemented sound sensors in robots using the Butterworth low-pass filter shown previously with a very high gain amplifier successfully with very few problems. To help make your audio input implementation as reliable as possible, you should physically isolate the microphone from the robot as best you can. I use silicon tub caulking to mount the microphone with very thin (28-gauge or higher) wiring that is folded and does not provide a direct mechanical link to the robot. It is also a good idea to build a cardboard or plastic cone to help direct sounds to the microphone that are external to the robot itself.

Modern op-amp chips are very insensitive to electrical noise. Many texts and Internet sites have information regarding the design of op-amp-based filters and amplifiers that you can use. Connecting the output of the op-amp to a microcontroller is nothing more than driving the op-amp output to an input pin of the microcontroller.

When doing this, it is recommended that you drive a counter in the microcontroller rather than a straight I/O pin (or one that can cause an interrupt request). The counter should be polled periodically to see if any signal has come in. You will find that the filter output provides you with many transitions that you have to debounce and will have to wait until the transitions have stopped coming in.

▰▰▰▰ Sound—Recognizing Audio Commands

From the title of this section, I must confess to using a bit of hyperbole. I have probably oversold the capabilities of the interface presented in this section. The circuit shown here gives you the ability to detect loud sounds that are external to the robot such as claps or a frenzied "Stop!"

The circuit shown in Figure 4-77 filters out high frequencies (the ones most likely found in a robot), leaving the low-frequency sounds, which are most likely caused by external sources. The output of Figure 4-77 is passed to the TMR1 input pin (RB6) of a PIC16F627 that has been wired up with the lcd.c application code.

I did not show a breadboard wiring diagram for this application or the PICmicro microcontroller LCD shift register and LCD on a small breadboard because they did not fit—you will have to use a larger breadboard or come up with some other method of building the application circuit.

Table 4-17 lists the bill of materials for the sound circuit. For the PICmicro MCU circuitry with the LCD, refer to the section "LCD—Sample Two-Wire LCD Interface" that describes lcd.c.

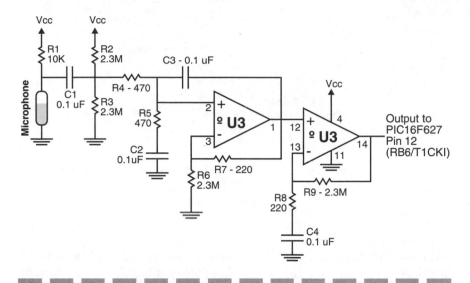

Figure 4-77 Sound application circuit

Table 4-17

Materials for
the Sound
Application
Circuit

Label	Part Number	Comment
U3	LM324	Can use any dual op-amp
Microphone	N/A	Can use any carbon or electret microphone
R1	10K, ¼ watt	
R2, R3, R6, and R9	2.3M, ¼ watt	
R4 and R5	470 ohm, ¼ watt	
R7 and R8	220 ohm, ¼ watt	
C1, C2, C3, and C4	0.1 µF	Uses any type
Misc.		PIC16F627 mounted on a breadboard with an LCD, wiring, and +5-volt power supply

I built the sound circuit with the PIC16F627, shift register, and LCD on a 5-inch-long breadboard. I used the LM324 simply because I had a number on hand —any small signal dual op-amp (such as the 17741) could be used in this application. I find that this circuit is fairly tolerant of power supply noise, which makes it ideal for robot applications.

Figure 4-78
Sound circuit
operation

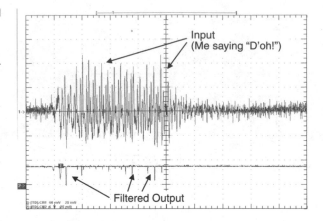

Figure 4-78 shows the circuit in operation. To test it out, I simply said "D'oh!" in a reasonably loud voice. The various high-frequency harmonics have been filtered out from the signal, leaving just a few transient negative spikes that can be input into the PICmicro MCU.

For this application, a PIC16F627 must be used because I used TMR1 to keep track of the transient spikes. The current TMR1 value is polled every 20 milliseconds in the PICC Lite software to see if any sounds have been detected. If they have, a message, along with the current TMR1 count value, is displayed on the LCD for 1 second.

The application code, sound.c, can be found in the robot\sound directory:

```
#include <pic.h>

//   02.01.27 - Sound: Display Message when sound received
//     (and finished) on LCD
//
//   The sound input is filtered and amplified using a 324
//     op-amp.  To remove small transients, the application
//     Checks the TMR1 count once every 20 msecs and compares
//     the current value to see if it has increased by 5 over
//     the previous 20 msec poll.
//
//   This is a two wire interface using a 74LS174 as a shift
//     register and the code called from the mainline (and timed
//     from TMR0 interrupt)
//
//
//   Hardware Notes:
//   PIC16F84 Running at 4 MHz
//   RB1 - CLock Signal to the LCD
//   RB2 - Data Signal to the LCD
//

//   Global Variables
int RTC = 0;                        //  Real Time Clock Counter
```

```
volatile char LCDDlay = 20;        //  Initialization Delay Value
volatile char LCDState = 1;        //  Current LCD State

static volatile bit Clock      @ (unsigned)&PORTB*8+1;
static volatile bit ClockTRIS @ (unsigned)&TRISB*8+1;
static volatile bit Data       @ (unsigned)&PORTB*8+2;
static volatile bit DataTRIS   @ (unsigned)&TRISB*8+2;

char * MessageOut;                 //  Message to be Sent Out
volatile char MessageOuti = 0;       //  Index to current Message
                                     //  Byte

char Message1[18] = "\376\200sound - 0x0xxxx";  //  Message when
sound
//                       1    2    3456789012345678
char Message2[18] = "\376\200                ";  //  Message for
                                                 //  nothing

char soundCounter = 20;            //  Wait 20 msecs for Sound Check
int OldTMR1 = 0;                   //  Timer Values
int CurrentTMR1;
volatile char soundState = 0;  //  Sound State Variable
                               //   0 - No Sound
                               //   1 - Sound Received, Check Next
                               //   2 - Sound Received, Ended


//  Configuration Fuses
#if defined(_16F627)
#warning PIC16F627 with external XT oscillator selected
      __CONFIG(0x03F61);        //  PIC116F627 Configuration Fuses:
                                //   - External "XT" Oscillator
                                //   - RA6/RA7 Digital I/O
                                //   - External Reset
                                //   - 70 msecs Power Up Timer On
                                //   - Watchdog Timer Off
                                //   - Code Protection Off
                                //   - BODEN Enabled
#else
#error Unsupported PICmicro MCU selected
#endif


//  Subroutines
char GetHex(char Value)
{

char returnValue;

     if (Value > 9)
          returnValue = Value + 'A' - 10;
     else
          returnValue = Value + '0';

     return returnValue;

}  //  End GetHex
```

```
Dlay(int msecs)                    //  Delay Specified Number of msecs
{

volatile int variableDlay;     //  1 Second Delay Variable

      variableDlay = RTC + msecs + 1;      //  Get the End Time
      while (variableDlay != RTC);

}  //  End Dlay


LCDNybble(char Nybble, char RS) {    //  Send Nybble to LCD
:


LCDInit()                      //  Initialize the LCD I/O Pins
:


LCDOut(char * const LCDString)      //  Output the Data String
:


//  Interrupt Handler
void interrupt tmr0_int(void) //  TMR0 Interrupt Handler
{

char temp;
int   TMR1Value;

      if (TOIF) {

            TOIF = 0;          //  Reset Interrupt Flag

            RTC++;             //  Increment the Clock

//  Put additional interface code for 1 msec interrupt here

//  Sound Check Software

            if (soundState == 0)     //  Check the Current Sound
                                     //  Value
               if (--soundCounter == 0) {    //  Time to Check?
                   soundCounter = 20;        //  Reset
                   TMR1Value = (TMR1H * 0x0100) + TMR1L;
                   if (TMR1Value >= (OldTMR1 + 2))
                       soundState = 1;
                            //  Indicate Sound Received
                   OldTMR1 = TMR1Value;
                            //  Save Current Counter
               } else;
            else if (soundState == 1)
                            //  Check to see if Sound still
                            //  active
               if (--soundCounter == 0) {        //  Time to
                                                 //  Check?
                   soundCounter = 20;      //  Reset
                   TMR1Value = (TMR1H * 0x0100) + TMR1L;
```

```
                            if (TMR1Value < (OldTMR1 + 2))
                                soundState = 2;
                                    //   Indicate Sound Finished
                            OldTMR1 = TMR1Value;
                                    //   Save Current Counter
                    } else;

//   LCD State Machine
:

        } //   endif

//   Put Different Interrupt Handlers here

}   //   End Interrupt Handler


//   Mainline
void main(void)                      //   Template Mainline
{

        OPTION = 0x0D1;              //   Assign Prescaler to TMR0
                                     //    Prescaler is /4
        TMR0 = 0;                    //   Reset the Timer for Start
        T0IE = 1;                    //   Enable Timer Interrupts
        GIE = 1;                     //   Enable Interrupts

        LCDInit();                   //   Initialize the LCD Port

//   Put in Interface initialization code here

        TMR1H = TMR1L = 0;           //   Initialize Timer 1
        T1CON = 0x003;               //   1:1 Prescaler
                                     //   T1OSCEN Off
                                     //   _T1SYNC Active
                                     //   TMR1CS is Exteral Clock
                                     //   TMR1ON

        while (1 == 1) {             //   Loop forever

//   Put in Robot high level operation code here

            while (soundState != 2);    //   Wait for Sound
                                        //   Received

            CurrentTMR1 = (TMR1H * 0x0100) + TMR1L;

            Message1[13] = GetHex((CurrentTMR1 & 0x0F000) >> 12);
            Message1[14] = GetHex((CurrentTMR1 & 0x0F00) >> 8);
            Message1[15] = GetHex((CurrentTMR1 & 0x0F0) >> 4);
            Message1[16] = GetHex(CurrentTMR1 & 0x0F);

            LCDOut(Message1);        //   Indicate "Sound"

            Dlay(1000);              //   Wait 1 Second

            LCDOut(Message2);        //   Clear the Sound Message

            soundState = 0;          //   Reset and wait for next
```

```
sound

        } // endwhile

} // End of Mainline
```

The execution of this application should be straightforward and should not present any surprises unless a different type of microphone was used from the one that the prototype used. You may have to experiment with R1 in the circuit until the input to the circuit is in the 150 to 200 mV range shown in Figure 4-78.

H-Bridge Motor Control

For most mobile robots, you need a simple, reliable method of turning motors on and off and reversing their direction. In a large robot, you might consider using a transmission, but this is impractical for a smaller robot; a set of four switches arranged as an H-Bridge can be used for this function.

Motors can be controlled by exactly the same hardware as shown in the previous section, but as noted previously, they only run in one direction. A network of switches (transistors) can be used to turn a motor in either direction; this is known as an H-Bridge, the most basic form of which is shown in Figure 4-79.

In this circuit, if all the switches are open, current won't flow and the motor won't turn. If switches 1 and 4 are closed, the motor turns in one direction. If switches 2 and 3 are closed, the motor turns in the other direction. Both switches on one side of the bridge should *never* be closed at the same time as this causes the motor power supply to burn out or a fuse to blow because there is a short circuit directly between the motor power and ground.

Figure 4-79
H-Bridge motor
driver

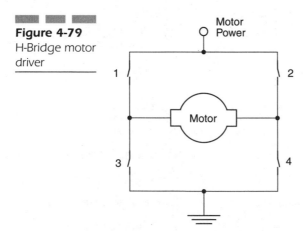

Controlling a motor's speed is normally accomplished using a PWM signal, as shown previously in this chapter (refer to Figure 4-33). This controls the average power delivered to the motors. The higher the ratio of the pulse width to the period, the more power delivered to the motor.

When you are first starting out in robotics, you could create an H-Bridge driver from a set of transistors or you could use a chip with the drivers built in like the 293D (see Figure 4-80).

The 293D chip can control two motors (one on each side) that are connected to the buffer outputs (pins 3, 6, 11, and 14). Pins 2, 7, 10, and 15 are used to control the voltage level (the switches in Figure 4-80) of the buffer outputs. Pin 1 and pin 9 are used to control whether or not the buffers are enabled. These can be PWM inputs, which make control of the motor speed very easy to implement.

Vs is +5 volts used to power the logic in the chip and Vss is the power supplied to the motors, which can be anywhere from 4.5 to 36 volts. A maximum of 500 mA can be supplied to the motors. The 293D contains integral shunt diodes, which means that when a motor is attached to the 293D, no external shunt diodes are required. This reduces the circuitry needed to implement a motor controller to the circuit shown in Figure 4-81.

The resistor and capacitor wired across the brush contacts of the motor help reduce electromagnetic emissions and noise spikes from the motor. If you notice erratic operation from the microcontroller when the motors are running, you may want to put in the 0.1 µF capacitor and 5-ohm (2-watt) resistor snubber across the motor's brushes, as shown in Figure 4-81.

There is an issue with using the 293D motor controller chips—they are bipolar devices with a 0.7-volt drop across each driver (for 1.4 to 1.5 volts for a dual-driver circuit, as shown in Figure 4-81). This drop, with the significant amount of current required for a motor, results in a fairly significant amount of power dissipation within the driver. The 293D is limited to 1 amp of total output. For these circuits to work the best, a substantial amount of heat sinking is required.

Other books in the "Robot DNA" series discuss the intricacies of the discrete component H-Bridge circuit design for both bipolar and MOSFET transistors.

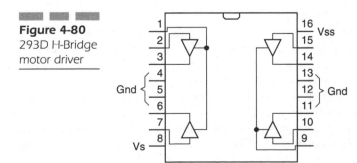

Figure 4-80
293D H-Bridge
motor driver

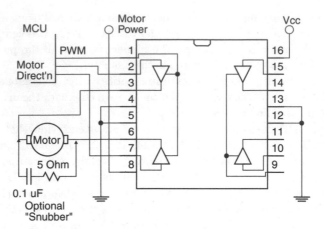

Figure 4-81
Wiring a motor
to the 293D

This is not a trivial topic and is one that can really make or break the operation of your robot. For small, simple robots in which you are going to be investigating how the programming is to be accomplished, it is recommended that you use the 293D chip for controlling your robot's motors.

Wallhug—Using the Tamiya Wall-Hugging Mouse as an Application Base

It can be hard to find a good prototyping base; toys are often used in experimental robots, but generally have a single purpose and can be difficult to work with. There are some commercial robot bases, but these usually start at $100 and require complex motor drivers and power systems. In the middle is the Tamiya *wall-hugging mouse* (Tamiya item number 70068*1300), which is an excellent tool (technically a toy) that can be built into a simple robot base in just a few minutes. I have probably bought a dozen of them over the years to use in my own experimental robots. Figure 4-82 shows a robot I made a few years ago that uses the wall-hugging mouse as a base with a custom PCB, which is small enough for me to put on the mouse body that comes with the kit.

In this section, I will present you with a simple robot platform that you can use for testing out your robot interfaces and software. These platforms are not optimized or have not been tested to work reliably enough for use as a product, but they should be a good starting place for you to learn about robot controller programming.

Figure 4-82
Wall-hugging
mouse robot

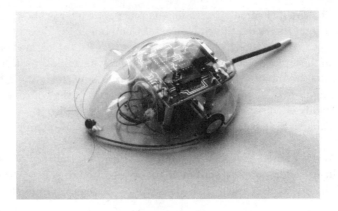

The wall-hugging mouse consists of two DC motors driving differential wheels powered by a C cell and uses a whisker-actuated microswitch that changes which wheel motor is active based on whether the whisker is in contact with a wall. By turning on and off the motors, it follows a wall—or solves a simple maze. At a certain level, the Tamiya wall-hugging mouse can be considered a simple robot.

Before continuing with this application in which you will find a simple robot base to experiment on, build the kit to see how it works. When you assemble the robot, solder the various wires together because twist connections will come loose as the robot runs across the perimeter of the room. Although it is fun to experiment with the original wall-hugging mouse kit, I find that it gets old quick and soon you're ready to use the kit as a robot base.

Before you can use the wall-hugging mouse as a robot base, you must make a few modifications to the kit:

1. Remove the C cell battery holder.
2. Put in a 9-volt battery clip.
3. You may want to remove the whisker that comes with the robot.
4. Drill holes for the prototyping system that you are going to use.
   - I like to use a Radio Shack P/N 276-168B Universal Component PC Board to support the electronics.
   - Use 2-inch #10 bolts with nylon nuts and 1-inch standoffs to support the prototyping system above the motors and battery.

You could use a breadboard with this robot. To do this, you would probably want to attach the breadboard to a PCB with easy prototype wiring.

It is up to you whether or not to keep the whisker that comes with the wall-hugging mouse kit. Ideally, two whiskers should be used to detect collisions from both sides. I have added the whisker and associated microswitch from a second kit to make a robot with two whiskers. I find that this works well, but I prefer using the IR object detection system presented earlier in this chapter.

The circuit I came up with for the robot controller is shown in Figure 4-83. I built it on the Radio Shack Universal Component PC Board (Radio Shack part number 276-168B), which is shown in Figure 4-84. It took me about two hours to build the circuit. Notice that I put in two screw terminals for the battery connector, the two motors, and a switch to turn on and off the battery. I used the screw terminals instead of soldering the wires directly to the board, which enables me to modify or change the circuit when I am finished with it.

Table 4-18 lists the bill of materials for the wall-hugging mouse application circuit.

When I wired my robot, I used a 9-volt alkaline battery source and used the built-in switch (on the bottom of the mouse kit) to control the power. Always remember to place the switch *upstream* of the logic input diode (CR3) and filter

Figure 4-83
Wall-hugging
mouse
application
circuit

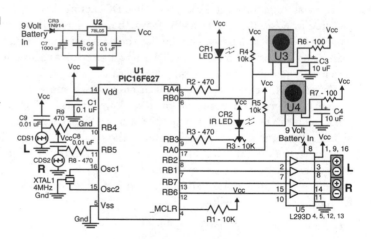

Figure 4-84
Prototype
wallhug PCB

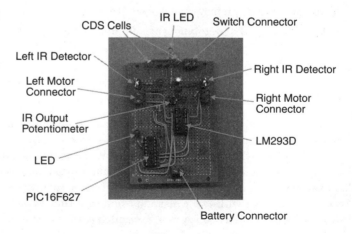

Table 4-18

Materials for
Wall-Hugging
Mouse
Application
Circuit

Label	Part Number	Comment
U1	PIC16F627	
U2	78L05 +5-volt regulator	
U3 and U4	3-pin IR detector	Requires Waitrony PIC-1018SCL or equivalent
U5	L293D	
CR1	LED	Uses any color of visible light LED
CR2	IR LED	Uses IR LED
CR3	1N914	Can use any silicon diode capable of passing 200 mA
R1, R4, and R5	10K, ¼ watt	
R2, R3, R8, and R9	470 ohm, ¼ watt	
R6 and R7	100 ohm, ¼ watt	
CDS1 and CDS2	10K CDS cell	Negative resistance change to light
C1 and C6	0.1 µF	Uses any type
C3, C4, and C6	10 µF, +16 volt	Electrolytic
C7	1000 µF, +16 volt	Electrolytic
XTAL1	4 MHz ceramic resonator with integral capacitors	
Misc.		Requires a Tamiya wall-hugging mouse kit, PCB, wiring, and 9-volt battery clip

capacitors (C7 and C5) to prevent them from being charged and able to shock some unsuspecting passerby.

You are probably surprised at my use of a very large capacitor (1,000 µF) and a diode current block on the 9-volt battery in Figure 4-83. As discussed later in this section, the use of the 9-volt battery, L293D, and 1.5-volt motors of this robot platform is not the optimal solution for a robot's power supply, motor driver, and motors—they are, however, extremely convenient to use.

When the motors are engaged, there is a fairly large voltage drop from the 9-volt battery. This drop can be attenuated by the use of premium alkaline batteries, but you can still have a large enough drop that the PICmicro MCU controlling the robot resets itself. The 1,000 μF capacitor drives the electronics for a reasonable period of time while the motor is engaged without losing any of its input current back to the motors because of the diode. This circuit is somewhat kludgy, but it gets the job done and the robot can be controlled quite effectively by both the remote control and biologic code while the sensors work quite well.

When choosing the locations for the parts, make sure that the PIC16F627 is in a position from which it can be easily pulled from the circuit. I placed it toward the left rear of the robot and in a socket so that it could be easily pried up using a small flat screwdriver. There is nothing more frustrating than building a terrific robot and then discovering that there is no easy way to get the microcontroller out for reprogramming.

When I put in the two screw terminals for the motors, I made sure the + (positive) terminal was at the front of the robot and the + terminal of the motor had a red wire on it (the other has a black). It's easy to confuse the wiring when you are debugging the circuit, which could create a robot that runs backwards or worse. You should also keep the wires reasonably short to prevent them from getting caught up in the wheels, drivetrain, or passing objects.

If you have built all the sample applications up to this point, you should be comfortable with placing the IR object detection circuitry and the CDS cells on the robot.

From the information in Figure 4-83, we can conclude that this robot is made up of the different interfaces that I have already discussed and presented in sample applications in this chapter.

That's really all there is to it.

The application code that I came up for this application is called wallhug.c, which can be found in the robot\wallhug directory:

```
#include <pic.h>

//   02.04.28 - wallhug: Created from "Combine" originally
//
//   A 38 KHz IR Signal to be used as an object detection system
//
//   Take Input from Sony IR Remote Control and Display on LCD.
//
//   IR Remote Control Data Stream:
//
//              Leader/Header    High  "1"         "0"
//   -------+                    +----+  +----+         +---------
//          |                    |    |  |    |         |
//          +----------------+    +-----+  +------------+
//
//          |   2.4 msecs     |540u|660us|540u|  1.2 msecs  |
//
```

```
//                              |1200 msecs|  1.76 msecs    |
//
//                              |   300    |    440         |
//
//
//   Hardware Notes:
//   PIC16F627 Running at 4 MHz with an external ceramic resonator
//   RA0 - IR Detector Input, Poll for Collisions
//   RA4 - LED
//   RB0 - IR Detector Input, Sony IR Input, Use Falling Interrupt
//   RB1 - Left Motor, Negative Connection
//   RB2 - Left Motor, Positive Connection
//   RB3 - IR Light Source Output
//   RB4 - Left ADC
//   RB5 - Right ADC
//   RB6 - Right Motor, Negative Connection
//   RB7 - Right Motor, Positive Connection
//

//   Global Variables
int RTC = 0;                          //   Real Time Clock Counter

#define remote0          0x06EF       //   IR Definitions
#define remote1          0x0FEF
#define remote2          0x07EF
#define remote3          0x0BEF
#define remote4          0x03EF
#define remote5          0x0DEF
#define remote6          0x05EF
#define remote7          0x09EF
#define remote8          0x01EF
#define remote9          0x0EEF
#define remoteVolUp      0x0B6F
#define remoteVolDown    0x036F
#define remoteChanUp     0x0F6F
#define remoteChanDown   0x076F
#define remotePrevChan   0x022F
#define remoteMute       0x0D6F
#define remotePower      0x056F
unsigned int  DataIn;                 //   Data Input
unsigned int  DataTime = 0;
unsigned char DataInCount = 0;        //   Number of Characters to Read
                                      //   In
unsigned int  CurrentRTC;

volatile char LeftCollision = 0;      //   Current Collision
volatile char RightCollision = 0;

volatile char ADCState = 0;           //   Current ADC State Machine
                                      //   Operating State
volatile char ADCDlay = 1;            //   Delay 50 msecs between each
                                      //   ADC Operation
volatile unsigned char LeftADC = 0x0FF;   //   Last Left ADC Value
volatile unsigned char RightADC = 0x0FF;  //   Last Right ADC Value

#define goStop           0x030        //   Stop the Motors
#define goForward        0x072        //   Go Forward
#define goReverse        0x0B4        //   Reverse the Motors
```

```
#define turnLeft        0x074     //   Left Motor Reversed
#define turnRight       0x0B2     //   Right Motor Reversed
                                  //   NOTE: RB4/RB5 Kept High
unsigned char motorState = goStop; //  Motor State
unsigned char PWMDuty = 15;
unsigned char PWMCycle = 0;       //   Start at 1/4 Throttle
unsigned char OpnCount = 1;       //   Want to Monitor the ADC

char ExecuteFlag = 0;             //   Indicating Robot Can
                                  //   Execute


//  Configuration Fuses
#if defined(_16F627)
#warning PIC16F627 with external XT oscillator selected
      __CONFIG(0x03F21);          //   PIC116F627 Configuration
                                  //   Fuses:
                                  //    - External "XT" Oscillator
                                  //    - RA6/RA7 Digital I/O
                                  //    - External Reset
                                  //    - 70 msecs Power Up Timer
                                  //    - Watchdog Timer Off
                                  //    - Code Protection Off
                                  //    - BODEN disabled
#else
#error Unsupported PICmicro MCU selected
#endif


//  Subroutines
//  Interrupt Handler
void interrupt tmr0_int(void)     //   TMR0 Interrupt Handler
{

char temp;

      if (T0IF) {

            T0IF = 0;             //   Reset Interrupt Flag

            RTC++;                //   Increment the Clock

//  Put additional interface code for 1 msec interrupt here

//  Get Motor State and Turn Off Motors to Start

            PORTB = temp = (PORTB & 0x039) | 0x030;
                                  //   Get Current Motor State

//  Check for lost Packet and stop waiting unless packet is active

            if ((!RB0) && (DataInCount == 0)) {
                                  //   Data Packet Coming in?
                  DataInCount = 13; //   12 Bits to Come in/13 Edges
                  INTF = 0;       //   Enable Pin Change Interrupt
                  INTE = 1;

            } else if (DataInCount == 0) {
```

```
                              //  Do Collision Detection

        TRISB3 = 0;         //  Enable PWM Output

        while (TMR0 < 64);
                            //  For a Limited Number of
                            //  Cycles

        if (!RB0)           //  If Low, then Collision
            if (LeftCollision < 3)
                            //  Must Have 3 Collisions in a
                            //  Row
                LeftCollision++;
            else;
        else
            LeftCollision = 0;

        if (!RA0)           //  If Low, then Collision
            if(RightCollision < 3)
                            //  Must Have 3 Collisions in a
                            //  Row
                LeftCollision++;
            else;
        else
            RightCollision = 0;

        TRISB3 = 1;

    }  //  endif

//  Motor PWM Code

        if (PWMDuty == 0)        //  Always Off
            ;
        else if (PWMDuty == 29) //  Always On
            PORTB = temp | motorState;
        else                     //  Some PWM Value
            if (PWMCycle <= PWMDuty)
                PORTB = temp | motorState;
            else;

        if (++PWMCycle == 30)   //  Roll around cycle count?
            PWMCycle = 0;       //  Yes

//  Check Count Operation to Stop Bit

        if ((!ExecuteFlag) && (--OpnCount == 0)) {
                            //  Check the Light Meter
            OpnCount = 1;       //  Reset for Next Time Through
            PORTB = (PORTB & 0x039) | goStop;
            if (LeftADC < RightADC)
                RA4 = 0;
            else
                RA4 = 1;
        }  //  endif

//  Process Remote Control Bit (If Present)
```

```
if (DataTime) {
    if (DataInCount == 0) { //  New Message Coming In
        DataInCount = 12; //  12 Bits to Come in
        DataIn = 0;
    } else {            //  Bit to Process
        DataIn = DataIn << 1;
        if ((DataTime > 225) && (DataTime < 375)) {
                        //  "1" Received
            DataIn++;
            if (--DataInCount == 0) {
                OpnCount = 200;
                //  Timer to Finish
                if (DataIn == remote2) {
                    motorState = goForward;
                //  Go Forward
                } else if (DataIn == remote4) {
                    motorState = turnLeft;
                } else if (DataIn == remote6) {
                    motorState = turnRight;
                } else if (DataIn == remote8) {
                    motorState = goReverse;
                } else if (DataIn ==
                    remoteVolUp) {
                    if (PWMDuty < 29)
                        PWMDuty++;
                    else;
                } else if (DataIn
                    ==remoteVolDown){
                    if (PWMDuty != 0)
                        PWMDuty--;
                    else;
                } else if (DataIn ==
                    remotePower) {
                    ExecuteFlag = 1;
                    motorState = goStop;
                    OpnCount = 1;
                //  Timer to Finish
                } else {     //  Stop or Nothing
                    ExecuteFlag = 0;
                    motorState = goStop;
                    OpnCount = 1;
                //  Timer to Finish
                }  //  endif
            }  //  endif
        } else if ((DataTime > 375) && _
    (DataTime < 500)) {      //  "0" Received
            if (--DataInCount == 0) {
                OpnCount = 200;
                //  Timer to Finish
                if (DataIn == remote2) {
                    motorState = goForward;
                //  Go Forward
                } else if (DataIn == remote4) {
                    motorState = turnLeft;
                } else if (DataIn == remote6) {
                    motorState = turnRight;
                } else if (DataIn == remote8) {
```

```
                                   motorState = goReverse;
                        } else if (DataIn ==
                        remoteVolUp) {
                             if (PWMDuty < 29)
                                   PWMDuty++;
                             else;
                        } else if (DataIn
                        ==remoteVolDown){
                             if (PWMDuty != 0)
                                   PWMDuty--;
                             else;
                        } else if (DataIn ==
                        remotePower) {
                             ExecuteFlag = 1;
                             motorState = goStop;
                             OpnCount = 1;
                        //  Timer to Finish
                        } else {        //  Stop or
                                        //  Nothing
                             ExecuteFlag = 0;
                             motorState = goStop;
                             OpnCount = 1;
                        //  Timer to Finish
                        }  //  endif
                  }  //  endif
            } else if (--DataInCount != 12)
                        //  Invalid Timing
                  DataInCount = 0;
      }  //  endif
      DataTime = 0;      //  Start Over...
}  //  endif

//  Add to the Bit Delay Time

      CurrentRTC += 0x0100;    //  Keep Upcoming

//  Check To see if there is a missing pulse...

      if ((DataInCount) && (CurrentRTC > 0x00900))
            DataInCount = 0;
                        //  Reset and Wait for Next
                        //  Character

//  ADC State Machine:

      if (!DataInCount) {     //  Only Execute if NOT Rx'g IR
            if (ADCState == 0) {          //  ADC Start Delay
                  if (--ADCDlay == 0)
                        ADCState++;
            } else if (ADCState == 1) {
//  Start Charging Left and Right (for 1 msec min.)
                  PORTB = PORTB | 0x030;
                  TRISB4 = 0;
                  TRISB5 = 0;
                  ADCState++; //  Jump to next state in 1
                              //  msec
            } else if (ADCState == 2) {
```

```
                            TRISB4 = 1; //  Turn off Left Pin Driver
                            temp = PORTB;
//  Clear pending Interrupts on Port Change
                            RBIF = 0;
                            RBIE = 1;
                            ADCState++;
                    } else if (ADCState == 3) {   //  Roll Over, No
                                                  //  Interrupt
                            LeftADC = 0x0FF;
                            temp = PORTB;
                            RBIF = 0;
                            RBIE = 0;
                            ADCState++;
                    } else if (ADCState == 4) {
//  Start Charging Right (for 1 msec)
                            TRISB5 = 1; //  Change TRISB to Input
                            temp = PORTB;
                            RBIF = 0;
                            RBIE = 1;
                            ADCState++;
                    } else if (ADCState == 5) {
                            RightADC = 0x0FF;
                            temp = PORTB;
                            RBIF = 0;
                            RBIE = 0;
                            ADCDlay = 50;
                            ADCState = 0;
                    } //  endif
                } //  endif

        } //  endif

//  Put Different Interrupt Handlers here

        if (INTF) {                      //  RB0/INT Pin Interrupt
            DataTime = CurrentRTC + TMR0; //  Get the Bit Timing
            CurrentRTC = 0x0FFFF - TMR0;  //  Get Timing for Next
                                          //  Bit
            INTF = 0;                  //  Reset Interrupt
        } //  endif

        if (RBIF) {                    //  Interrupt on Pin Change
            if (ADCState == 3) {
                    LeftADC = TMR0;    //  Get the ADC Count
                    ADCState++;
            } else {                   //  Step 5 (Right ADC)
                    RightADC = TMR0;   //  Get the ADC Count
                    ADCDlay = 50;      //  Repeat Every 50 msecs
                    ADCState = 0;
            } //  endif
            temp = PORTB;
            RBIF = 0;
            RBIE = 0;
        } //  endif

    } //  End Interrupt Handler
```

```
// User Subroutines
void Dlay(int msecs)              // Delay the Set Number of
msecs
{

int valueDlay;

      valueDlay = RTC + msecs + 1;  // Get the Final Delay Value
      while (valueDlay != RTC);     // Wait for it

} // End Dlay

void LEDOutput(int state )        // Set the LED State
{

      if (state)
            RA4 = 0;                // LED On
      else
            RA4 = 1;                // LED Off

} // End LEDOutput


int GetLeftLight()                // Return Left Light Sensor
Value
{

      return LeftADC;

} // End GetLeftLight


int GetRightLight()               // Return Right Light Sensor
Value
{

      return RightADC;

} // End GetRightLight


int GetLeftWhisker()              // Return State of Left
                                  // Whisker
{

      if (LeftCollision == 3)
            return 1;             // Something Pressing on
                                  // Whisker
      else
            return 0;

} // End GetLeftWhisker


int GetRightWhisker()             // Return State of Right
                                  // Whisker
{
```

```
        if (RightCollision == 3)
              return 1;                    //  Something Pressing on
                                           //  Whisker
        else
              return 0;

}  //  End GetRightWhisker


void LeftMotor(int Movement, int Speed)
{

        switch (Movement) {          //  Set Movement
            case 1:                  //  Forward
                 motorState = (motorState & 0x0F9) + 2;
            case 0:                         //  Stop
                 motorState = motorState & 0x0F9;
            case -1:                        //  Reverse
                 motorState = (motorState & 0x0F9) + 4;
        }  //  endif

        PWMDuty = Speed;

}  //  End LeftMotor


void RightMotor(int Movement, int Speed)
{

        switch (Movement) {          //  Set Movement
            case 1:                  //  Forward
                 motorState = (motorState & 0x0CF) + 0x040;
            case 0:                         //  Stop
                 motorState = motorState & 0x0CF;
            case -1:                        //  Reverse
                 motorState = (motorState & 0x0CF) + 0x080;
        }  //  endif

        PWMDuty = Speed;

}  //  End rightMotor


//  Mainline
void main(void)                      //  Template Mainline
{

        OPTION = 0x0D1;              //  Assign Prescaler to TMR0
                                     //   Prescaler is /4
        TMR0 = 0;                    //  Reset the Timer for Start
        T0IE = 1;                    //  Enable Timer Interrupts
        GIE = 1;                     //  Enable Interrupts

//  Put in Interface initialization code here

        INTEDG = 1;                  //  Interrupt on Rising Edge of
                                     //   RB0/INT IR TX'r
```

```
CCPR1L = 13;                    //  50% Duty Cycle
CCP1CON = 0b000001111;          //  Turn on PWM Mode

PR2 = 26;                       //  26 µsec Cycle for 38 KHz
TMR2 = 0;                       //  Reset TMR2
T2CON = 0b000000100;            //  TMR2 is On, Scalers 1:1

CMCON = 0x007;                  //  PORTA Analog Functions Off

PORTA = 0x010;                  //  RA4 is the LED
TRISA = 0x0EF;

PORTB = 0x000;
TRISB = 0x039;                  //  RB7/RB6 and RB2/RB1 Are
                                //   Motor Drivers
while (1 == 1) {                //  Loop forever

        if (ExecuteFlag) {

//  Put in Robot Biologic Code Here

        } else {

//  Put in Biologic Code Reset Statements Here

        } // endif

    } // endwhile

} // End of Mainline
```

Using this code as a base, I came up with the following set of standard I/Os that are available to the robot:

```
int GetLeftWhisker();    // Return State of Left Collision Sensor
int GetRightWhisker();   // Return State of Right Collision Sensor
// The "Whiskers" can either be a physical sensor or IR

int GetLeftLight();      // Read the Left Light Sensor Value
int GetRightLight();     // Read the Right Light Sensor Value
// Light sensors can either be two CDS cells forming a voltage
// divider or a resistance measuring routine using the CDS cells

void LeftMotor(int Movement[, int Speed]);      // Set Left Motor
                                                // Operation
void RightMotor(int Movement[, int Speed]);     // Set Right Motor
                                                // Operation
// The "Speed" parameter is not available for RC Servos

void LED(int State);              // Turn On/Off LED
void LCDOutput(char * Message);   // Output a Message on an LCD

void Dlay(int msecs);             // Delay the set number of
                                  // msecs
```

Note that I do not include an IR TV remote control in the list—when I implemented the remote control interface, I assumed that the keypad buttons would be used for direct control. I use button 2 to go forward, 4 to turn left, 6 to turn right, and 8 to go in reverse. My default stop key is button 5.

One feature that may be surprising is the ability to change the motor's PWM using the remote control's volume buttons. The default value for the PWM (50 percent) moves the robot quite sedately, but if you change the speed to a 100 percent PWM duty cycle, the robot moves quickly around the room. Unfortunately, moving the robot at a 100 percent duty cycle will also exhaust your battery very quickly—I suggest that you leave the PWM at the default value for most applications.

In the wallhug.c code, the Power On button of the remote control must be pressed to execute the biologic code. To turn off this function, use a direct action command or press 5. The idea behind this convention is that the robot can be moved directly to the desired start location via the remote control and the biologic code can be started to test it out.

When I chose the I/O functions for the wall-hugging mouse robot, I was able to use the software I presented earlier in the book (for the most part). The area where I had the most difficulty (and this is also true for the servo robot base discussed later in the chapter) was with the remote control. You will find that when you are using the remote control, the robot jerks and behaves as if it is not receiving just a few packets.

This operation occurs because it isn't receiving many packets properly. This is due to the electrical noise of the motors on the robot and the extra overhead of the other I/O functions built into the robot. When you create some biologic code for this base (shown in the next chapter), you shouldn't have any problems with the interface execution at all.

To make sure that the sensors are working correctly, you may want to stop the robot's motors and then poll the sensors before moving on. For the light sensors, you should wait at least 60 milliseconds to make sure that you have an updated value, whereas for the collision sensors, you should wait 10 milliseconds.

With a bit of work, you can probably stop the robot's motors for 6 milliseconds each time the ADC operation takes place. During these 6-millisecond interrupt cycles, the ADC operation and collision sense operation could operate without being affected by the operation of the motors.

Electrically, the motor driver and the motors used in this robot base are not optimal. The motors that come with the wall-hugging mouse are designed to be run at 1.5 volts and have a relatively low resistance, causing them to draw excessive current from the 9-volt source. The L293D is a general-purpose H-Bridge driver and is not optimized for this application. Coupled with the 1.5-volt motors of the kit, you will find that there is a surprisingly large voltage drop between the L293D and the motors from the applied battery voltage. This high voltage drop causes the L293D to become quite warm.

If 6-volt-rated (or higher) motors were used in the robot along with properly designed transistor H-Bridges, the robot would operate for a surprisingly long time and the motor drivers would not heat up.

Odometry for Motor Control and Navigation

The task of adding sensors to a robot's drivetrain that enable the robot's controller to measure the movement of the wheels is known as *odometry*. The data from the movement of the wheels can be used to quite accurately calculate the position of the robot from a known starting point and can be used to control the operation of the motors and to keep their speed and response to changing control values within a set range of parameters. This section introduces you to the topic of odometry and how it can be used to navigate a robot. This hardware can also be used to control the operation of the robot's motors, keeping them within a set acceleration and speed profile.

I will not provide you with examples of how odometry is implemented and used in this chapter because it is not practical to implement on the simple robot base designs I have provided in this book. Along with this, I have serious doubts about PIC16F627's capability to be able to implement the algorithms presented in this section.

Measuring the distance that a robot's wheels have turned is actually quite easy—chances are you use a device that provides this function every day. The technology that measures how far you have moved your PC's mouse can be used to measure how much a motor is turned.

If you were to open up your PC's mouse (note I said if), you would discover that the central ball is connected to two wheels with many holes along an edge. At least two opto-isolators (which are actually *optical interrupters* in this application) will be aligned with the holes, as shown in Figure 8-85. When the wheel is turned, one of the opto-isolators passes a signal through the wheel. By placing the opto-isolators at different positions on the wheel, the distance the wheel turns along with the direction it is turning can be calculated.

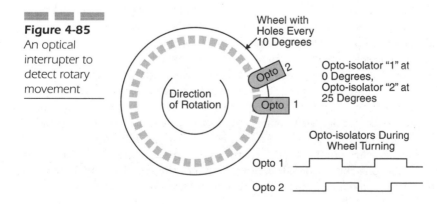

Figure 4-85
An optical interrupter to detect rotary movement

When this system is used within a robot, the holed wheel is connected directly to the robot's wheels. Generally, only one opto-isolator is used in the robot application because the direction of the wheel's motion is known. What is unknown is the speed at which the wheel is turning.

The speed of the turning wheels is very easy to calculate. Using the radius of the robot's wheels and the number of holes in the optical interrupter, the speed can be easily calculated as follows:

```
Speed = 2 * Pi * (Radius) / (# Holes) * (# Holes encountered /
Second)
```

For example, if a 1-inch radius wheel was used with an optical interrupter that had 30 holes in it, the speed of the wheel when it was turning at a rate of 10 holes per second is calculated as follows:

```
Speed = 2 * Pi * (1" / 30 Holes) * 10 Holes/Second
      = 2.094 "/Second
```

You should recognize that although this speed may seem fast, it translates to just 10.5 feet per minute, which is very slow for a robot. When you are calculating the speeds, think in terms of feet per second with 0.5 feet per second (6 inches/second) being acceptable for most robots and 3 feet per second being quite impressive (especially if it doesn't crash into anything).

A single odometer can be added to a robot to monitor the distance it has traveled. By using only one sensor and assuming that both motors (and the wheels they are connected to) turn at the same rate, you can perform a rough calculation on the distance the robot has moved in a given period of time.

If you build the Tamiya wall-hugging mouse example robot base, the robot does not move in a straight line. This is due to differences in the wiring of the robot's motor power distribution, in the motor windings, and in the amount of friction within the motor's drivetrain.

Your response to this statement may be, "Who cares?" Your car only has one odometer and it has never failed to tell you how far you've gone. This is true because the error of measuring the movement of the car from one point is infinitesimally small over the random movement of the car over thousands of miles. This same argument cannot be applied to a robot, where its movement is measured in feet and it tends to go in straight, well-defined lines.

The obvious solution to this problem is to use the odometry sensors to calculate the speed of each robot's wheels and throttle them appropriately so they are both running at the same speed. This is actually much easier said than done—the code that is required for throttling the motors comes under the heading of control theory and it is a very complex subject that is well outside the scope of this book (although it is an interesting topic for a later book in this series).

To give you a brief understanding of the topic, say there are two motor states that must be considered when designing the motor throttle. The first is how the motor reacts to changing inputs. Obviously, a motor cannot start turning at a set rate of speed instantly; it must react to the inertia of itself, the drivetrain, and the mass of the robot. The rate of change of the motor speed is an important control parameter because having two motors that accelerate or decelerate at different rates is just as bad as having motors that run at two different speeds. The second state that must be considered is the steady state. How well do the motors hold a set speed? Both these states have to be considered when the motor controls are implemented.

The most common form of motor control is known as *proportional, integral, and differential* (PID), which continuously keeps track of the motor's odometry and applies simple laws to the measure values and desired output to calculate the appropriate motor setting.

Proportional refers to a constant value multiplied by the motor's speed setting from the controller. *Integral* refers to the number of rotations that have taken place during a set period of time. Finally, *differential* refers to the speed at which the motor is turning. These three parameters are combined together to form an output value to control the motor:

```
Motor Control = (Kp * (Controller Setting)) + (Ki * (#Rotations))
                + (Kd * (Speed))
```

The constant values Kp, Ki, and Kd are dependent on the type of motor, the mass of the robot, and the rate at which the robot controller reads (samples) the odometry and can calculate the new value for the motor control.

Some computer programs calculate the correct PIC controller constant values for your application or you can find them empirically, adjusting one or another until the motor behaves the way you would like it to. The empirical (experimental) method is not that hard to do, but you should have some way of easily updating the constants (such as a remote control); otherwise, you will find the task to be quite onerous.

The first parameter you should adjust is Kp (the proportional constant) for constant motor speeds. Once you have this, you should work on Ki and Kd for when the motor starts and stops to make sure that it doesn't overshoot and go too fast or actually go in the opposite direction of what you want. Motor oscillation at specific speed ranges is another pitfall to watch out for. You may find that the steady state operation at slow speeds is very hard to get working with the constant values and it is advantageous to change the interval in which you sample the odometry data and update the constants.

I can say with almost absolute certainty that it would be impossible to perform the PID calculations on a PIC16F627 within the 1 millisecond available between the TMR0 interrupts. You may be able to perform the calculations in the biologic code section, but this code is clearly not biologic; it is mechalogic, and running it in the mainline of the robot application code prevents the biologic code from running properly. It may be possible to devote a single PICmicro microcontroller (or another microcontroller) to an individual wheel to provide PID operational control over it.

Because it is unlikely that the two wheels will turn exactly the same way in a totally predictable manner, many roboticists look at using the odometry information for predicting the position of the robot within its environment and helping the robot navigate. Using the odometry information is analogous to an airliner using an *Inertial Navigation System* (INS), which consists of a series of gyros that monitor the airliner's motion along with computers that calculate its position based on the accelerations experienced by the aircraft.

As a warning, showing how the odometry information can be used to calculate the position of the robot involves somewhat elaborate mathematics. Nothing that I present here will involve mathematics taught at college, but if it's been a while since you've studied trigonometry, you might want to find an old high-school textbook to help you follow along.

At the outset, navigating a robot probably seems like a pretty simple affair. If the robot is placed at a known initial position, pointing in a known initial direction, then by the use of odometry data, movements to a specific endpoint should be easy to predict.

For example, if a perfect differentially driven robot was at position (0, 0), pointing at angle 0, and you wanted to end up at position (4, 3), you would first have to calculate the angle the robot would have to point in. Knowing that the tangent of an angle is defined as

```
tangent(Angle) = Opposite / Closest
```

Angle can be found using the following equation:

```
Angle = ArcTan(Opposite / Closest)
```

For our example, the angle is 53.13 degrees, or 0.927 radians. This angle should be easy to achieve in the perfect robot and once it has completed this turn to the left, the robot should move five units (according to the Pythagorean theorem).

The problem is that there is no such thing as a perfect robot. Turning to specific angles is very hard to do, and as noted previously, most robots move in a curved path when they are moving forward or backwards, as shown in Figure 4-86. This diagram shows how a robot actually moves about a point—rather than assuming the robot is perfect, it is much better to assume that the robot moves around a central point and if it were allowed to run forward indefinitely, it would make a circle around this point and return to where it started.

In practical terms, this assumption improves the accuracy of odometry-based navigation tremendously. By placing distance sensors on each wheel in the differential robot shown in Figure 4-86, the actual movement of a robot can be calculated to reasonable accuracy using some tricks you learned in high school.

For the purposes of this discussion, the point that is equidistant between the two wheels will be called the robot's *position*. No matter how the wheels turn and move the robot, this point will always be in the center of the motion.

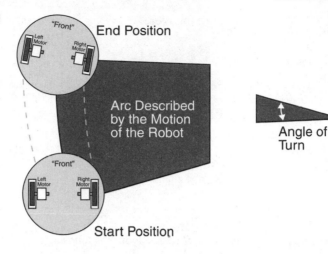

Figure 4-86
Calculating robot position using odometry

Looking at the motion of the wheels, I can describe the motion of the robot moving in a straight line over some period of time, as described in the arc shown in Figure 4-87. The arc in this diagram looks something like a right-angle triangle. We will take advantage of this property when calculating the position of the robot after it has moved some distance.

Along with this, as an angle of an arc becomes smaller, the sine of the angle approximates the height of the opposite side of the triangle. This is a very useful approximation for calculating the position of a robot after it has moved some distance because it means that the position of the inner wheel (after moving distance X) can be described by the equations

```
X = R * sin(Angle)

X / R = sin(Angle)
```

where R is the radius of the circle that the robot is turning in and *Angle* is the angle the robot has moved through.

For the outer wheel (which has moved a distance of Y), a similar set of equations can be written out as

```
Y = (R + Z) * sin(Angle)

Y / (R + Z) = sin(Angle)
```

where Z is the distance between the two wheels.

Combining the second equation for both wheels, the following can be said:

```
X / R = Y /  (R + Z)
```

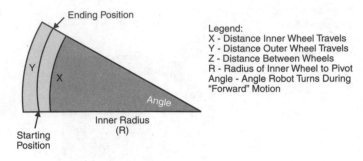

Rearranging this formula, R can be found as follows:

```
R = X * Z / (Y - X)
```

Since X, Y, and Z are all known, the value of R can be calculated and *Angle* can be found as follows:

```
Angle = ArcSin(X / R)
```

To give you an idea of how these formulas are applied, consider the situation where a differentially driven robot that has wheels 4 inches apart is run with the right wheel moving 2.1 inches and the left wheel moving 2 inches. If it starts at position (0, 0), we can calculate the final position by going through the following steps:

1. Identify X, Y, and Z in this example:

```
X = 2"
Y = 2.1"
Z = 4"
```

2. Calculate the inside wheel radius of the turn:

```
R = X * Z / (Y - X)
  = 2" * 4" / (2.1" - 2")
  = 8" / 0.1
  = 80"
```

3. Calculate the angle the robot has turned:

```
Angle = ArcSin(2" / 80")
      = 0.025 radians
```

4. Calculate the final position of the robot assuming that the center of the circle that the robot is turning in is at position (-82", 0"):

```
Ycoord = Radius * sin(Angle)
       = (R + (Z / 2)) * sin(Angle)
       = (80" + (4" / 2)) * sin(0.025)
       = 82" * sin(0.025)
```

```
          = 2.05
Xcoord = Radius * cos(Angle) - 82"
          = (R + Z / 2)) * cos(Angle) - 82"
          = (82" * cos(Angle)) - 82:
          = 82" - 82"
          = 0"
```

The final position of the robot is (0", 2.05") and it is pointing at an angle of 0.025 radians (1.43 degrees). The coordinate value probably makes intuitive sense to you because the position is the point exactly between the left and right wheel. Chances are you might think that the angle is interesting, but that it does not have any bearing on the operation of the robot.

The angle is important to know because as the robot continues to move forward, the angle changes. The next 2-inch step will result in the angle now being 0.050 radians (2.86 degrees) with a final point somewhere around (-0.102, 4.098). At some point, the robot will have to be turned slightly to eliminate this error and point more accurately to the final destination.

With this series of equations, you are probably thinking that you can now navigate your robot to a specific point in a room accurately.

I'm sorry to say that this is not quite true. When you are applying these formulas practically, you must remember that the distance you choose to move before looking at the distance of each wheel affects the accuracy of the formulas provided previously. The longer the distance the robot moves, the less accurate the approximation that I made will be.

The problem with using a shorter distance is twofold. First, the odometry hardware built into the robot must be very accurate for short distances. Secondly, the robot's controller must be able to work with floating point values very quickly and accurately. If these requirements are not met, you will end up with a robot that will take a long time to move inaccurately across the floor.

I also have to point out that in the real world, robots are not started at the absolute correct initial point and they do not point in the absolute direction they think they are. Couple this with wheel slip and you will find that the navigation equations I have presented here really do no more than bring the robot to an approximate point.

Chances are it will be a very accurate approximate point (much more accurate than if you assume that the robot moves in straight lines), but an approximate point just the same.

▆▆▆▆ Radio Control Servos

For many beginner robots, the same servos designed for radio-controlled airplanes, cars, and boats are used to drive the wheels. These servos can be easily modified to rotate continuously in either forward or backward directions.

The output of a radio control (R/C) servo is usually a wheel that can be rotated from 0 to 90 degrees. (There are also servos available that can turn from 0 to 180 degrees as well as servos with very high torque outputs for special applications.) Typically, they only require +5 volts, the ground, and an input signal.

An R/C servo is an analog device; the input is a PWM signal at digital voltage levels. This pulse is between 1.0 and 2.0 milliseconds long and repeats every 20 milliseconds. This is illustrated in Figure 4-88.

The length of the PWM pulse determines the position of the servo's wheel. A 1.0-millisecond pulse causes the wheel to go to 0 degrees, whereas a 2.0-millisecond pulse causes the wheel to go to 90 degrees.

For producing a PWM signal using a microcontroller, you could start with a timer interrupt (set every 20 milliseconds), which outputs a 1.0- to 2.0-millisecond PWM signal using the following pseudocode:

```
Interrupt() {                    //  Interrupt Handler Code

int  i = 0;

  BitOutput( Servo, 1);          //  Output the Signal

  for (i = 0; i < (1 msec + ServoDlay); i++ );

  BitOutput( Servo, 2);

  for (; i < 2 msec; i++ );      //  Delay full 2 msecs

}  //  End Interrupt Handler
```

This code can be easily expanded to control more than one servo (by adding more output lines and ServoDlay variables). This method of controlling servos is also nice because the ServoDlay variables can be updated without affecting the operation of the interrupt handler.

The interrupt handler takes 2 milliseconds out of every 20. This means that there is a 10 percent cycle overhead for providing the PWM function (and this doesn't change even if more servo outputs are added to the device).

Figure 4-88
Servo PWM
waveform

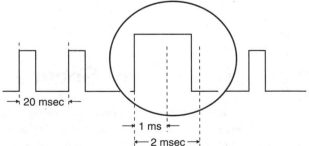

Although this code works very well for a stand-alone servo application, the code does not work very well with the code with the 1-millisecond timer interrupt that has been used so far to implement the different mechalogic and electronic interfaces to the robot. To enable servos to be used with a robot, you must change the operation of the 1-millisecond timer interrupt to a 256-millisecond period and perform the five different steps shown in Table 4-19.

The intermediate-speed (half-speed) timings take advantage of the feature of many servos where the speed the motor turns at is proportional to the time delay's difference between the stop (center) position and full deflection.

To make a servo run continuously, the potentiometer built into the servo for position feedback must be removed and replaced with a simple voltage divider. Along with this, any plastic stops that prevent the servo for turning more than 90 degrees have to be cut out. This voltage divider is actually a trimmer potentiometer that will be calibrated to the actual servo-stopped timing produced by the microcontroller. This operation is discussed in the next section.

Servo2—60-Minute Robot Base Using RC Servos

The most popular type of robot built by beginner (and many expert) robot designers is the *dual radio control servo-driven differential robot*. Radio control servos are quite inexpensive (you should be able to find some inexpensive ball-bearing output servos for $10 or so) and are very easy to interface to, both in terms of hardware and software. Their drawbacks are their expense and the effort required to modify standard servos so they turn continuously instead of stopping once they reach their limits.

Despite the drawbacks, using radio control servos as the drive mechanism for a robot is a reasonable approach to developing your own robot software and you

Table 4-19

Radio Control Servo timings for different robot wheel movements

Step Number	Time Delay	Servo Operation
1	1 msec	Reverse at full speed
2	1.25 msec	Reverse at half speed
3	1.5 msec	Servo stopped
4	1.75 msec	Forward at half speed
5	2 msec	Forward at full speed

can build a robot of your own in less than an hour. It took me less than half an hour to build the robot base (including modifying the servos for continuous rotation)—I called it the *60-minute robot* because it should not require any more than an hour to build, even if you are a beginner.

The robot that I came up with is built on a 8-inch-long, 3.75-inch-wide piece of ³/₈-inch plywood. Before placing the parts on the plywood, I used autobody primer followed by two coats of Tremclad spray paint to create a stable surface that the different parts could be glued or double-sided taped to without any fear of them falling off. The basic design for the robot is shown in Figure 4-89.

The circuitry that I built onto the breadboard is quite simple (see Figure 4-90). Table 4-20 lists the total bill of materials for this robot base.

Figure 4-89
Prototype servo robot

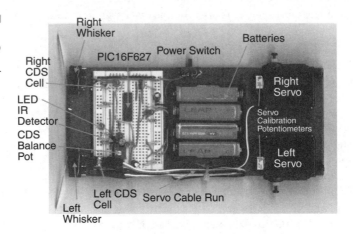

Figure 4-90
Servo-based robot application circuit

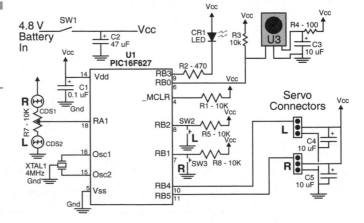

Table 4-20

Materials for
the Servo-
Based Robot
Application
Circuit

Label	Part Number	Comment
U1	PIC16F627	
U3	3-pin IR detector	Requires Waitrony PIC-1018SCL or equivalent
CR1	LED	Uses any color of visible Light LEDR1, R3, R5, R6, and R8 10K, $\frac{1}{4}$ watt
R2	470 ohm, $\frac{1}{4}$ watt	
R4	100 ohm, $\frac{1}{4}$ watt	
R7	10K potentiometer	
CDS1 and CDS2	10K CDS cell	Involves negative resistance change to light
C1	0.1 µF	Uses any type
C2	47 µF, +16 volt	Electrolytic
C3, C4, and C5	10 µF, +16 volt	Electrolytic
SW1	SPST switch	Requires a power switch
SW2, SW3	Momentary on	Requires whiskers, see text
XTAL1	4 MHz ceramic resonator with integral capacitors	
Misc.		Requires a plywood base, 4 AA battery clips, 2 modified servos (see text), small breadboard, miscellaneous wires, 10 mm LED (any color), and 2- to 3-inch diameter model aircraft wheels

When you build the robot, I suggest you take care of the mechanical aspects in the following order:

1. Modify the servos to run continuously.

2. Modify two wheels to be connected to the servo.

3. Using 5-minute epoxy, glue the 10 mm LED to the front bottom of the robot to act as a castor for the robot.

4. Secure the servos, battery pack breadboard, and whiskers (if appropriate) to the robot base using double-sided tape.

5. Add the wiring for the robot.

6. Program the PICmicro MCU and test the robot functions using a Sony remote control.

The six steps to understanding the robot probably seem very simple, but there are points that you should watch for and understand before starting them.

For step 1, modifying radio control servos to run continuously should not be a hard job, but before you go out and buy a couple of servos to practice on, there are a few points to clarify. First, before buying any servos, make sure that there are instructions for modifying them to run continuously on the Internet and download the web page that shows you how to do it. Some servos are very difficult to modify, whereas others are laughably easy. In any case, I recommend getting as cheap a servo as possible to make any mistakes less hard on the pocketbook.

As part of the work in modifying servos to run continuously, you must disconnect the internal potentiometer. I have seen some robots where the designer just connects two resistors in a series to simulate the function of the potentiometer. I always use (and think you should too) multiturn trimmer potentiometers and glue them (using 5-minute epoxy) to the servo case. If you look at Figure 4-89, the potentiometers have been identified (called *calibration potentiometers*) and it should be obvious that they are easily accessible from the top of the robot platform for easy adjustment.

The reason why I am so adamant about how the servo calibration potentiometers are placed and wired is so you can adjust the servos to not turn when the robot is stopped. This operation could be done in software, but it makes the operation of the software I use for my robot applications much more complicated as it now requires some method of changing the servo timing parameters. It is much easier to simply add multiturn potentiometers to the servos and adjust them to stop with a small screwdriver.

Next, you must modify a wheel to work with the servo. The simplest solution is to buy a couple of cheap 2- to 3-inch diameter model aircraft wheels, drill out the center hubs so that the servo's retention screw can pass through it, and then glue (using 5-minute epoxy) a cutdown servo arm (that came with the servo kit) onto the wheel's hub, as shown in Figure 4-91. I apply the 5-minute epoxy liberally until it begins to drip from the other side of the wheel. By going to this

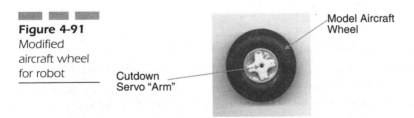

Figure 4-91
Modified
aircraft wheel
for robot

Cutdown
Servo "Arm"

Model Aircraft
Wheel

extreme, you ensure that the servo arm cannot be pulled off from the wheel during handling or operation.

After all of the parts are together, you can place them on the robot body. For servos, battery packs, and breadboards, I use double-sided tape. By using these materials, you will have a very sturdy connection that can be removed with a knife or small screwdriver.

For this robot, I used microswitches that I found with long actuator arms as whiskers. These microswitches were epoxied to the robot's plywood base. This was done at the same time as the servo arms were glued to the wheel hubs and the 10 mm robot castor was glued to the bottom of the plywood base.

When working with 5-minute epoxy, remember to follow the directions given in the manufacturer's instructions and work with it in a well-ventilated area. The fumes from the glue aren't terribly toxic, but they can be nauseating over a long period of time.

At this point, you are ready to start wiring the robot. Note in Figure 4-89, space has been left on either side of the battery pack to mount a simple power switch onto the robot as well as provide a run for the servo cables. To hold down the servo cables, I drilled a hole in the plywood base and used a cable tie to secure them.

The power switch for the robot base is also epoxied to the plywood. If you were planning ahead, you could glue down the whiskers, the LED castor, the switch, the servo calibration potentiometers, and the servo arms to the wheel hubs all at one time. For the first robot base you build, I would recommend not attempting to do all this at the same time.

Going back to Figure 4-90, you should notice that no type of voltage regulator has been included for the robot. For this application, I cheated a bit and decided to use four AA NiCad or NiMH batteries for the robot's power supply. These batteries provide 1.2 volts each and by wiring them in a series, I was able to provide 4.8 volts, which is sufficient to drive the PICmicro MCU, the servos, and all the circuitry on the breadboard.

Wiring the rest of the application is quite simple and will just take a few moments. When I built my prototype, instead of using 10K resistors for R1 and R8, I put in 470 ohm resistors and LEDs that are lit when the whiskers' microswitches are closed. This type of visual feedback from the sensors can be very useful in an application.

Like most of my robot designs, I have included a remote control interface. This interface should not be surprising and works exactly the same way as the Tamiya wall-hugging mouse robot base. Like the wall-hugging mouse robot base, the robot jerks a bit when it is commanded to move forward by the remote control. This is also a function of the remote control signal competing for priority with the other functions being performed in the 1-millisecond interrupt handler.

Finally, note that I included a 10K potentiometer for balancing the left and right CDS cells. This should help the robot sense differences in light around it quite precisely.

With the servos modified and the robot base built and wired, you are now ready to program a PIC16F627 and try out the robot. The basic application code is servo2.c, which can be found in robot\servo:

```c
#include <pic.h>

//   02.04.28 - Updated by Myke Predko by Processing Remote
//    Control Bits in TMR0 Interrupt Handler
//   02.04.26 - Updated with IR Remote Control
//   02.02.23 - Myke Predko
//
//   Servo Test Application.  Run the servos Forwards, Backwards,
//    Together and Apart
//
//   Setup TMR0 to interrupt Mainline once every 512 µsecs
//
//
//   Hardware Notes:
//   PIC16F627 Running at 4 MHz with an External Oscillator
//   RA1 - Comparator1, Tied to Vss
//   RA2 - 1/2 Vdd from Vref module, output to check Vref operation
//   RB0 - IR Detector for Remote Control
static volatile bit IRDetect         @ (unsigned)&PORTB*8+0;
//   RB1 - Right Whisker
static volatile bit RightWhiskerPin @ (unsigned)&PORTB*8+1;
//   RB2 - Left Whisker
static volatile bit LeftWhiskerPin  @ (unsigned)&PORTB*8+2;
//   RB3 - Output LED
static volatile bit LED              @ (unsigned)&PORTB*8+3;
static volatile bit LEDTRIS          @ (unsigned)&TRISB*8+3;
//   RB4 - Left Servo
static volatile bit leftServo        @ (unsigned)&PORTB*8+4;
static volatile bit leftServoTRIS    @ (unsigned)&TRISB*8+4;
//   RB5 - Right Servo
static volatile bit rightServo       @ (unsigned)&PORTB*8+5;
static volatile bit rightServoTRIS   @ (unsigned)&TRISB*8+5;


//   Global Variables
unsigned int  RTC = 0;           //   Real Time Clock Counter
unsigned char RTCDlay = 2;       //   With 512 Cycle Clock, Increment
                                 //   1/2x

//   Servo Variables
#define LeftForward    38        //   Left Motor Values
#define LeftStop       37
#define LeftReverse    36
#define RightForward   36        //   Right Motor Values
#define RightStop      37
#define RightReverse   38
char ServoDlay = 1;              //   Output Servo Every 20 msecs
char leftSpeed = LeftStop;
char rightSpeed = RightStop;

//   Remote Variables
#define remote0         0x06EF   //   IR Definitions
#define remote1         0x0FEF
#define remote2         0x07EF
```

```
#define remote3          0x0BEF
#define remote4          0x03EF
#define remote5          0x0DEF
#define remote6          0x05EF
#define remote7          0x09EF
#define remote8          0x01EF
#define remote9          0x0EEF
#define remoteVolUp      0x0B6F
#define remoteVolDown    0x036F
#define remoteChanUp     0x0F6F
#define remoteChanDown   0x076F
#define remotePrevChan   0x022F
#define remoteMute       0x0D6F
#define remotePower      0x056F
unsigned int  DataIn;            //  Data Input
unsigned char DataInCount = 0;   //   Number of Characters to
                                 //   Read In
unsigned int  DataTime = 0;
unsigned int  CurrentRTC = 0;
int  OpnCount = 1;               //  If Stopped, Check Comparator

volatile char LeftWhisker = 0;   //   Count of the Left Whisker
volatile char RightWhisker = 0;  //   Count of the Right Whisker

char ExecuteFlag = 0;            //  Flag Indicating Robot Can Execute




//  Configuration Fuses
#if defined(_16F627)
#warning PIC16F627 with external XT oscillator selected
    __CONFIG(0x03F21);           //  PIC116F627 Configuration Fuses:
                                 //   - External "XT" Oscillator
                                 //   - RA6/RA7 Digital I/O
                                 //   - External Reset
                                 //   - 70 msecs Power Up Timer On
                                 //   - Watchdog Timer Off
                                 //   - Code Protection Off
                                 //   - BODEN Disabled
#else
#error Unsupported PICmicro MCU selected
#endif


//  Interrupt Handler
void interrupt tmr0_int(void) // TMR0 Interrupt Handler
{

    if (T0IF) {

        T0IF = 0;            //  Reset Interrupt Flag

        if (--ServoDlay == 0) { //  New 20 msec Delay

            ServoDlay = 40;  //  Reset 20 msec Dlay

            leftServo = 1;   //  Make Servos High
            rightServo = 1;
```

```
        } else {          //  Output the Pulses

            if (leftSpeed >= ServoDlay)
                leftServo = 0;    //  Servo Pulse Stop

            if (rightSpeed >= ServoDlay)
                rightServo = 0;   //  Servo Pulse Stop

        }  //  endif

//  Process Remote Control Bit (If Present)

        if (DataTime) {
            if (DataInCount == 0) { //  New Message Coming In
                DataInCount = 12; //  12 Bits to Come in
                DataIn = 0;
            } else {    //  Bit to Process
                DataIn = DataIn << 1;
                if ((DataTime > 450) && (DataTime < 750)) {
                        //  "1" Received
                        DataIn++;
                        if (--DataInCount == 0) {
                            OpnCount = 400;
                            leftSpeed = LeftForward;
                            rightSpeed = RightForward;
                            if (DataIn == remote2)
                                ;  //  Go Forward
                            else if (DataIn == remote4)
                                leftSpeed = LeftReverse;
                            else if (DataIn == remote6)
                                rightSpeed =
                                RightReverse;
                            else if (DataIn == remote8) {
                                leftSpeed = LeftReverse;
                                rightSpeed =
                                RightReverse;
                            } else if (DataIn ==
                                remotePower) {
                                ExecuteFlag = 1;
                                leftSpeed = LeftStop;
                                rightSpeed = RightStop;
                                OpnCount = 1;
                            } else {    //  Stop or Nothing
                                ExecuteFlag = 0;
                                leftSpeed = LeftStop;
                                rightSpeed = RightStop;
                                OpnCount = 1;
                            }  //  endif
                        }  //  Endif
                } else if ((DataTime > 750) && _
            (DataTime < 1000)) {    //  "0" Received
                        if (--DataInCount == 0) {
                            OpnCount = 400;
                            leftSpeed = LeftForward;
                            rightSpeed = RightForward;
                            if (DataIn == remote2)
```

```
                                ;     //  Go Forward
                        else if (DataIn == remote4)
                               leftSpeed = LeftReverse;
                        else if (DataIn == remote6)
                               rightSpeed =
                               RightReverse;
                        else if (DataIn == remote8) {
                               leftSpeed = LeftReverse;
                               rightSpeed =
                               RightReverse;
                        } else if (DataIn ==
                               remotePower) {
                               ExecuteFlag = 1;
                               leftSpeed = LeftStop;
                               rightSpeed = RightStop;
                               OpnCount = 1;
                        } else {        //  Stop or
                                        //  Nothing
                               ExecuteFlag = 0;
                               leftSpeed = LeftStop;
                               rightSpeed = RightStop;
                               OpnCount = 1;
                        }  //  endif
                  }  //  Endif
            } else          //  Invalid - Delete
               DataInCount = 0;
         }  //  endif
         DataTime = 0;       //  Start Over...
      }  //  endif

//  Add to the Bit Delay Time

         CurrentRTC += 0x0100;   //  Keep Upcoming

//  Check To see if there is a missing pulse...

         if ((DataInCount) && (CurrentRTC > 0x01200))
               DataInCount = 0;  //  Reset and Wait for Next
                                 //  Character

//  1 msec Interrupt Code

         if (--RTCDlay == 0) {

            RTCDlay = 2;       //  Reset the Clock
            RTC++;             //  Increment the Clock

//  Increment Counter if Left Whisker Held Down

            if (!LeftWhiskerPin)     //  Left Whisker Pressed
               if (LeftWhisker < 20)   //  debounced
                     LeftWhisker++;
               else; //  Whisker Pressed for 20 msecs
            else          //  Whisker Released
               if (LeftWhisker != 0)
                     //  Whisker Release being debounced
                     LeftWhisker--;
```

```
                          else; //  Whisker released for > 20 msecs

//  Increment Counter if Right Whisker Held Down

                  if (!RightWhiskerPin)   //  Right Whisker Pressed
                      if (RightWhisker < 20)  //  debounced
                              RightWhisker++;
                      else; //  Whisker Pressed for 20 msecs
                  else         //  Whisker Released
                      if (RightWhisker != 0)
                              //  Whisker Release being debounced
                              RightWhisker--;
                      else; //  Whisker released for > 20 msecs

//  Check Count Operation to Stop Bit

                  if ((!ExecuteFlag) && (--OpnCount == 0)) {
                          //  Check the Light Meter
                      OpnCount = 1;
                          //  Reset for Next Time Through
                      leftSpeed = LeftStop;
                      rightSpeed = RightStop;
                      if (C2OUT)
                              LED = 0;
                      else
                              LED = 1;
                  }  //  endif

              }  //  endif

          }  //  endif

//  Put Different Interrupt Handlers here

      if (INTF) {               //  RB0/INT Pin Interrupt
          DataTime = CurrentRTC + TMR0; //  Get the Bit Timing
          CurrentRTC = 0x0FFFF - TMR0;  //  Get Timing for Next
                                        //  Bit
          INTF = 0;             //  Reset Interrupt
      }  //  endif

}  //  End Interrupt Handler



//  User Subroutines
void Dlay(int msecs)          //  Delay the Set Number of msecs
{

int valueDlay;

    valueDlay = RTC + msecs + 1;  //  Get the Final Delay Value
    while (valueDlay != RTC);     //  Wait for it

}  //  End Dlay

void LEDOutput(int state )   //  Set the LED State
{
```

```
         if (state)
                LED = 0;              //  LED On
         else
                LED = 1;              //  LED Off

}  //  End LEDOutput


int GetLeftLight()
{

         if (C2OUT)
                return 0x080;        //  Indicate Left is Brigher
         else
                return 0x0FF;

}  //  End GetLeftLight


int GetRightLight()
{

         if (!C2OUT)
                return 0x080;        //  Indicate Right is Brigher
         else
                return 0x0FF;

}  //  End GetRightLight


int GetLeftWhisker()          //  Return State of Left Whisker
{

         if (LeftWhisker == 20)
                return 1;            //  Something Pressing on Whisker
         else
                return 0;

}  //  End GetLeftWhisker


int GetRightWhisker()         //  Return State of Right Whisker
{

         if (RightWhisker == 20)
                return 1;            //  Something Pressing on Whisker
         else
                return 0;

}  //  End GetRightWhisker


void LeftMotor(int Movement)
{

         leftSpeed = Movement;       //  Use "leftForward", etc. for
                                     //  motions

}  //  End LeftMotor
```

```
void RightMotor(int Movement)
{

     rightSpeed = Movement;   //  Use "rightForward", etc. for
                              //  motions

}  //  End rightMotor


//  Mainline
void main(void)              //  Template Mainline
{

     OPTION = 0x0D0;         //  Assign Prescaler to TMR0 \2
                             //  #### - Note Different from
                             //  Standard
     TMR0 = 0;               //  Reset the Timer for Start
     T0IE = 1;               //  Enable Timer Interrupts
     GIE = 1;                //  Enable Interrupts

//  Put in Interface initialization code here

     INTEDG = 1;             //  Interrupt on Rising Edge of
                             //  RB0/IR
     INTF = 0;               //  Make Sure NO Pending Interrupt
     INTE = 1;               //  Enable RB0/INT Pin Interrupts

     CMCON = 0x002;          //  Enable the Comparator Module
                             //    C1/C2 Normal, not inverted
                             //    CIS = 0
                             //    Mode 2, compare against Vdd
     TRISA = 0x007;          //  Just bits RA0 and RA1 are inputs
     VRCON = 0x0EC;          //  Enable Vref module
                             //    Vref output on RA2
                             //    High Vref Range
                             //    Ladder value of 12

     LED = 1;                //  Enable the LED
     LEDTRIS = 0;

     leftServo = rightServo = 0;   //  Initially Low Output
     leftServoTRIS = rightServoTRIS = 0;

//  Biologic Code Control

     while (1 == 1) {        //  Loop forever

          if (ExecuteFlag) {

//  Put in Robot Biologic Code Here

          } else {

//  Put in Biologic Code Reset Statements Here

          }  //  endif
```

```
}  //  endwhile

}  //  End of Mainline
```

With everything wired, you must perform two calibration operations. The first should not surprise you if you have already programmed the PICmicro MCU with servo2.c and tried it out—you must calibrate the servos. To do this, you should turn on the robots (at startup or after a movement, the robot stops) and use a jeweler's screwdriver to change the calibration potentiometer until the servos stop moving or clicking.

The second calibration operation is to hold the robot up to a light and adjust the CDS cell potentiometer (R7) until the LED flickers on and off when the robot is moved from side to side. This makes the difference between the CDS cells as small as possible and any change in lighting becomes obvious to the robot. Ideally, this should be done in the room where the robot is going to be operating.

This code provides the same biologic interfaces as the wall-hugging mouse robot base with the exception of having the capability to change the speed of the servos and precisely determine the value of the light falling on the CDS cells. For many applications, these functions can be critical. To eliminate any possible issues with them not having the full capabilities of other robots, use the more complete interfaces used in the wall hugging mouse robot base.

Bringing Your Robot to Life

In the example interfaces presented in the previous chapter, I created elelogic and mechalogic interface code around a 1-millisecond timer interrupt request. This code was demonstrated with very simple biologic applications that did little more than just exercise the interfaces and output their current status. Along with discussing the different interfaces, I also discussed how multiple interfaces would be integrated together in a robot. After showing you how mechalogic and elelogic interfaces are created, this chapter examines the biologic code that will literally bring the robot to life.

Several different methodologies are used to program a robot's biologic code and in this chapter I will discuss the most popular ones. These methodologies tend to use nonlinear programming techniques that enable sophisticated robot applications to be created without requiring a large amount of software. This is important for two reasons. The first being that the microcontrollers that tend to control robots have modest program and variable memory, and anything that minimizes the requirements on these parts should be pursued.

The second reason for wanting to use nonlinear programming for robots is a bit more esoteric; programming the applications becomes more of an exercise in setting parameters. This does not mean that there won't be any programming of the robot (there will be as much as you can stand). The parameter programming enables you to tune or tweak the operation of the robot without having to perform a lot of code revisions. Ideally, all you will have to do is modify a few parameters.

When you create your biologic code, remember that you will succeed if the robot behaves "deterministically." This means that the robot's response to input is exactly what was planned and these responses are always repeated. A robot that moves randomly about before getting on track cannot be said to have "artificial intelligence." Most knowledgeable people would say the high-level decision-making code (which I call the biologic code) was not designed well.

Real-Time Operating Systems (RTOS)

One of the most useful application programming environments that can be used in a robot is the *real-time operating system* (RTOS). When properly used, an RTOS can simplify application development; by compartmentalizing the interface software for different features, functions, and peripherals, the opportunity for errors is significantly reduced and the capability to change the interfaces is increased.

The capability to implement an RTOS is available in many microcontrollers; unfortunately, this capability is not built into the mid-range microchip PICmicro MCU's used to demonstrate the different interfaces. When I show how a robot can be implemented using an RTOS in the next section, it will be more of a paper exercise than a practical one.

The best definition I can come up with for a RTOS is as follows:

An RTOS is a background program that controls the execution of a number of application subtasks and facilitates communication between the subtasks.

An important point about subtasks: if more than one can be running at one time, then the operating system is known as a *multitasking operating system* (which I usually refer to as a multitasker). Each subtask is given a short "time slice" of the processor's execution that enables it to execute all or part of its function before control is passed to the next task in line.

Before going on with this chapter, I should point out a few things. The first is that I usually call a subtask a task or process. I will also blur the difference between an RTOS and an operating system.

The reason why I tend to lump both an operating system and an RTOS together is because the central kernel (the part of the operating system that is central to controlling the operation of the executing tasks) is really the same for both cases. The difference comes into play with the processes that are loaded initially along with the kernel. In a PC's operating system, the console *input/output* (I/O), command interpreter, and file system processes are usually loaded with the kernel and everything has been optimized to run in a PC environment (which means the system responds to operator requests). In a robot's RTOS, the initial tasks loaded with the kernel give priority to tasks that are critical to the operation of the robot.

The important features of a robot's RTOS are the minimal resources required to implement it. When I say resources, I am talking about program memory, variable memory, and instruction cycles. For small microcontrollers, the amount of program memory built into the chip is limited. Having a complete RTOS that provides every possible function (including networking) will take up a lot of space and provide functions that are not required by the application.

I generally visualize the operation of a multitasker as a series of hoppers (containers used for dispensing materials such as grain). Each task is put in a different set of hoppers (or queues) based on its priority. Each priority level has several queues, such as ready to execute, waiting for event, or waiting for a message to be sent to the task. Only one task can be taken out of a queue and executed at a time.

The task will be allowed to execute until such time as it is waiting on some other task, on an external event, or until it executes for a set amount of time (time out) and is placed at the end of its ready-to-execute queue. After the current task has stopped, the next available task starts executing until it has to wait on another task, wait on an interrupt or times out. The timeout interval is chosen so that all the waiting tasks will be serviced in a reasonable amount of time. When working with a robot, this timeout interval will have to be determined based on the time required by the different sensor and output tasks, and it will probably require some tuning to get it working properly.

Different tasks within the RTOS are given different priority levels. These priority levels are dependent on how quickly a request or an event has to be responded to.

An interesting irony about RTOSs (or any multitasking operating system) is how different tasks are prioritized to maximize system performance. The tasks that you consider to be of the highest priority (the biologic) code are given lowest priorities in the system. The peripheral interfaces (which most people would consider to be of low priority) are given the highest priorities possible in the RTOS.

This priority arrangement will result in the robot responding to mechalogic and elelogic events quickly. For the biologic code to execute as quickly as possible, all the tasks required for supporting it must have the highest possible priority. Once the support tasks (mechalogic and elelogic) have finished their operations, they will go into a waiting queue until the biologic code places another demand on them. While they are waiting, the biologic code is able to execute until either an external event is detected or until it has to get some information from one of the support tasks.

When determining the priority of tasks, make sure that you have a clear understanding of how the application works. It is very easy to fall into the trap of having one task that is always able to run due to being a higher priority than another task that is a lower priority but is critical to the operation of the robot. In this case, the higher-priority task will be continually executing while the lower-priority task is starved of *central processing unit* (CPU) cycles.

When you first hear the idea of giving the highest-priority task the lowest priority in the system, it probably doesn't make any sense. I agree that it is counterintuitive, but that helps me remember how to prioritize tasks.

An *external event* is a fancy term for *interrupt*. The RTOS must have some methodology built in to enable it to respond to interrupts as well pass execution control to the task that is waiting on it. All the waiting-for-event queues are searched for tasks waiting on this event when an interrupt request is received, and any tasks that are waiting for the event are moved into ready-to-execute queues.

It is important to remember two issues about handling interrupts in an RTOS. The first is that you have to understand how the interrupt request is reset by the RTOS. The RTOS itself should be able to identify the interrupt (event) source and have the code built in to both reset the internal interrupt controller as well as the interrupt request source.

The second issue is more important and will have much more of an impact on the design of your robot software. It is important to remember that interrupts will not necessarily be serviced immediately following the acknowledgement of the interrupt and the reset of the interrupt sources. Interrupt tasks can be given a very high priority within the system, but cases will still occur when many hundreds of microseconds go by before the event is actually processed.

To minimize the effects of the delay on the application code responding to the external event, the use of interrupts should be minimized as much as possible. For robots, this means that many of the sensor and output interfaces that I have shown in the previous chapter should be implemented in hardware rather than software.

A good example of a robot peripheral that will have its operation affected by the response time of the RTOS to a request is the *pulse width modulation* (PWM) motor driver. You cannot implement a timer-based, software PWM and get any kind of consistent timing using an RTOS. In this case, you will need to use a hardware-based PWM generator.

Now, having said this, I should point out that it is a good idea to have a repeating clock interrupt built into the RTOS that is similar to the TMR0 I based the software on in the previous chapter. This interrupt can be used as a mechanism for delaying a task by a set amount of time. A 1-millisecond delay between task switches will enable you to delay events with some degree of precision. Many RTOSs already have this function built in, while in others you will have to create a task that sets up the microcontroller's timer to interrupt it after a set amount of time. The second solution is not as simple as it probably seems; care must be taken to properly write the code to support multiple devices interfacing with the timer simultaneously.

Tasks can communicate with each other via messages. The message-passing mechanism built into the RTOS should be the *only* method of passing data between tasks. Shared global variables could be used, but this avoids the advantages of message passing that include the serialization of messages as well as eliminating the possibility that data is overwritten by another task if multiple messages are sent to a single task before it can respond to the first one.

The message-passing process consists of two parts. The first is sending a message to another task. After the message has been sent, the sending task is taken out of the execution queue and is placed in a waiting-for-acknowledgement queue until the receiver has acknowledged receipt of the message.

On the other side, the receiving task will request the next message waiting for it. This can be done before the sending task sends its message (and the receiving task will be held in a waiting-for-message queue) or after the sending task sends its message. After reading the message, the receiving task will acknowledge the message. Once the receiver has acknowledged the message, the sending task will be placed in a ready to execute queue. It is important to note that both tasks will be marked as ready to execute.

The last operating system resource that is commonly provided is the semaphore flag. This construct consists of a marker within the operating system that can only be owned by one task at a time. A task will request the flag and it will be awarded to the task if the semaphore is available. If it is not available, the task will be placed in a waiting queue until the semaphore is available. When the semaphore becomes available, the semaphore is awarded to the task and the task is placed in a ready-to-execute queue. When the semaphore is finished, it is released, where it may be awarded to the next task waiting for it.

The purpose of the semaphore is to control access to a resource. For example, if the computer is sending status information to the *liquid crystal display* (LCD), the sending tasks may have to be able to complete the message (and get a response) before the next task is allowed to send any characters to the LCD. If the semaphore isn't used and characters are individually sent to the LCD task as a message, then you could end up with a jumbled message on the LCD's screen when multiple tasks write to it at the same time.

A very common beginner's mistake in working with RTOS is to forget to release a semaphore. The task that has it awarded will work fine, but the other tasks in the RTOS that require the semaphore won't work at all.

The basic RTOS requests are listed in Table 5-1 along with the parameters sent and received. Depending on the microcontroller, development tools, and how the RTOS is written, the requests can be a macro of a subroutine call.

The task can be identified using either a number or a label. If a label is used, then messages can be sent using the label rather than a simple number. Unfortunately, using a label-based task reference takes a lot of space and cycles, and can be prohibitive in a small microcontroller. In most microcontroller RTOSs that you will work with, tasks will be referenced by a number.

Note that interrupt handling is not specified in Table 5-1 while the delay task request (which will use the microcontroller's built-in timer and timer interrupt) is. The reason for not including it in the list is the difficulty in defining an interrupt interface that is logical for all situations and intuitive for all developers. I have found that different people have different ideas regarding the best way interrupts are handled in an RTOS.

When the RTOS is booted, it will start something like AllTask, as shown in the following code. This task has two purposes. The first is to start up the application code. To do this, it invokes the StartTask function of the RTOS to start up Task1 at a priority of 1. AllTask has a priority of 0, and once Task1 is executing, if it is allowed to execute, it will just loop as the lowest-priority task in the RTOS, providing the lowest level of functionality for the RTOS.

```
void AllTask()              //  Always Running Task
{

    StartTask(Task1, 1);   //  Start up the first application task

    while(1 == 1);         //  See if something else is executing

}  //  End AllTask
```

The while (1 == 1); loop at the end of the task (which causes the application to loop forever) may not seem like an important function, but it actually is because when most RTOS code is active, interrupts are disabled. If all the other tasks in the RTOS are waiting on interrupt requests, then AllTask will be the only one capable of executing. If you were to look at the instructions that implement this code, it would be Loop: goto Loop. During the time it takes to execute the goto instruction, interrupts are enabled and any incoming interrupt requests will be acknowledged.

Table 5-1

Basic RTOS Requests

RTOS Request	Task Call	Return	Comments
Start task	StartTask (Address, Priority, [TaskName])	Task number or error indicating that the new task could not be created	Start the specified task by pointing to its starting address.
End current task	End()	None/NA	Delete the currently executing task in memory.
Next task	Next()	None	Let other tasks execute before the current one times out.
Send message	Send(Task, Message)	None	Send message to another task. Do not continue until other task acknowledges the message.
Check for message	Check()	Message number or error indicating no message is waiting	Check the message queue to see if any other task has sent a message.
Wait for message	Wait()	Message number	Place task on wait queue until another task sends a message to it.
Read message	Read(Message#)	Message or error indicating Message# does not reference a valid message	Return the specified message.
Acknowledge message	Ack(Message#)	Nothing or error indicating Message# does not reference a valid message	Acknowledge message. Move sending task onto ready-to-execute queue and optionally delete message.
Get semaphore	Get(Semaphore#)	None	Claim the specified semaphore flag or wait until it's free.
Poll semaphore	Poll(Semaphore)	None or error indicating that semaphore is currently claimed by another task	Check semaphore to see if it is available.
Release semaphore	Free(Semaphore#)	None or error indicating the task did not have the semaphore awarded to it	Make the specified semaphore available to other tasks.
Delay task	Dlay(#msecs)	None	Delay the specified number of milliseconds.

When developing tasks for robots, you should always keep in mind the three following rules.

■ The task is of one of three types: a manual control, biologic, or peripheral task. The manual control task enables an operator to take control of the robot. There can be multiple manual control tasks if there are multiple methods of manual control (buttons on the robot, a serial interface, or a remote control). Biologic tasks are responsible for polling the sensors, processing the data, and commanding the outputs of the robot in response to the sensor input. Only one biologic task exists in a robot RTOS application and it should be designed to allow the control tasks to take over the operation of the robot. Peripheral tasks are responsible for interfacing to one (and only one) sensor or output device. Peripheral tasks cannot initiate a message; they only respond when they receive one from the biologic task.

■ If the biologic task isn't active, execute a Dlay request for a few milliseconds to allow the robot to move and for other tasks to execute. It is very easy to think of the different tasks as discrete applications and have them loop infinitely without giving up the processor (except using the timeout function of the RTOS).

■ Always make sure that the task code executes within an endless loop. In some RTOSs, if the task executes to the end of the function code, it may be deleted. In other RTOSs, the task may restart. In a small minority of poorly written RTOSs, the whole application may reset. If you want to clear the task from being executed, don't rely on the operating system to do it for you. Stop it explicitly using the End request.

Roach Example Application Running in an RTOS

In my introduction to RTOSs, I did not really discuss *why* someone would want to use them in the first place. As I will show in this section, one of the biggest advantages an RTOS has is its capability to have new tasks added to it quickly and without detriments to the existing tasks. Although I have tried to design all the example software presented in this book in such a way that it can be added to applications easily and without seriously affecting other tasks, cases will occur when mixing and matching the different pieces of code will be a challenge. This is especially true if they must interact with each other. These problems are greatly minimized when they are used in an RTOS such as the *infra-red* (IR) object detectors and IR remote control.

To illustrate this point, I would like to show how a simple robot application could be implemented in an RTOS and then add some functions to it to demonstrate how easy and nonintrusive it is for the existing tasks.

The initial robot I would like to illustrate is a *roach*, a differentially driven robot that runs away from the light and hides in the darkest part of the room. To implement this robot, I want to create the four tasks: biologic, bump (sensors), light (sensors), and motor (control). The biologic task will be responsible for determining whether or not the robot has collided with a wall and if it hasn't, it will command it to turn away from the lightest source in its field of view.

As part of the process of creating the software, I will draw out the different tasks in oval blocks and any required semaphores in rectangular blocks.

Looking at the list of different task functions in the previous section, you'll see that I did not provide a mechanism for questioning another task. To poll another task, biologic first sends a request to the other task, telling it to send back specific data.

The biologic code would look like the following:

```
void biologic()            // Biologic code for the "Roach"
{

int left, right;           // Saved light sensor values
int msgnum;                // Message number to read
msg message;               // Message passed to/returned from other
                           //   tasks

    while (1 ==1) {        // Run forever

        if (Poll(control_sem) == free) {
                           // Execute if no control task active

            message = biologic_task;
            Send("bump", message);
            msgnum = Wait();
            message = Read(msgnum);
            Ack(msgnum);// Have bump sensor states

            if (message != no_collision) {
                message = motor_stop;
                Send("motor", message);
            } else {    // Move towards the light
                        // Get current Left Light Sensor Value
                message = biologic_task;
                Send("light", message);
                message = get_left_light_sensor;
                Send("light", message);
                Msgnum = Wait();
                left = Read(msgnum);
                Ack(msgnum);

                        // Get current Right Light Sensor Value
                message = biologic_task;
                Send("light", message);
                message = get_right_light_sensor;
                Send("light", message);
                Msgnum = Wait();
                right = Read(msgnum);
                Ack(msgnum);
```

```
                                  //  Turn Away from the light
                         if (right >= left)
                                 message = motor_left;
                         else
                                 message = motor_right;
                         Send("motor", message);

                    }  //  endif

               }  //  endif

               Dlay(100);        //  Delay 100 msecs before polling again

          }  //  endwhile

     }  //  End biologic
```

In the biologic function, I have assumed that the larger the value returned from the light task, the brighter the light the chosen light sensor is exposed to. So, to escape the light, I turn one motor on forward. By doing this, the robot will give the appearance of waddling to the darkest spot in the room. The control_sem is the control task semaphore that will be explained in detail later in this section.

To implement the bump sensors, I have to poll the bump sensors for at least 50 milliseconds without there being a change. The RTOS code to implement this is as follows:

```
void bump()                    //  Poll the Bump Sensors
{

int left, right;               //  Counters for the Bump Sensors
int msgnum;                    //  Message number to read
msg message;                   //  Message passed to/returned from other
msg Sender;                    //   tasks

     left = 0; right = 0;      //  Initialize counters

     while (1 == 1) {

          if (leftsensor == collision)
               if (left < 5)
                    left++;
               else ;          //  Left collision - Increment?
          else                 //  No Left Collision wait for debounce
               if (left > 0)
                    left--;

          if (rightsensor == collision)
               if (right < 5)
                    right++;
               else ;          //  Right collision - Increment?
          else                 //  No Right Collision wait for debounce
               if (right > 0)
                    right--;

          msgnum = Check(); //  Check for message pending
          if (msgnum != no_message) {
```

```
                    Sender = Read(msgnum);
                    Ack(msgnum);//   Read message
                    if (Left == 5)
                            if (Right == 5)
                                    message = "Both";
                            else
                                    message = "Left";
                    else if (Right == 5)
                            message = "Right"
                    else if ((Right != 0) || (Left != 0))
                            message = "Indeterminate"
                    else            //  Both Sensor Counters == 0
                            message = "no_collision";
                    Send(Sender, message);
            }  //  endif

            Dlay(10);           //  Let other tasks execute

      }  //  endwhile

}  //  End bump
```

The bump task will loop continuously, polling the whiskers/bump sensors. Note that I Dlay for 10 milliseconds to avoid the task from running continuously and starving the other tasks. When a message is received, the current state of the sensors is returned; if they have not been in a particular state for 50 milliseconds, then an "indeterminate" message is returned.

For the bump sensor task (as well as the light sensor task), I send the requesting task first to make sure that the bump task knows which task to reply to. When I defined the RTOS functions previously, I did not provide a mechanism for passing the task identifier to the receiver as part of the message. In most RTOSs, a message is made up of a structure with the calling task appended to the message data.

The motor task will run the specified motor control for 100 milliseconds. During this time, it will poll for messages and change the motor operation accordingly. When the motors are moving, the task will claim the motor_running semaphore, which will be used by the light task to make sure that light measurements only take place with the motors stopped and that there is no chance for power upsets due to motor transients or kickback.

```
void motor()             //  Motor Execution Task
{

int i;
int msgnum;              //  Message number to read
msg message;            //  Message passed to/returned from other
                        //    tasks

    i = 0;              //  100 msec running counter reset

    while (1 == 1) {

            if (--i == 0) {   //  100 msecs past - stop motors
```

```
                     motors = stop;
                     i = 1;       //  Reset "i" so next pass == 0
                     Free(motor_sem);
              }  //  endif

         msgnum = Check(); //  Check for message pending
         if (msgnum != no_message) {
                  message = Read(msgnum);
                  Ack(msgnum);
                  motors = message;
                  if (message == stop) {
                          i = 1;//  Indicate Stop Recognized
                          Free(motor_sem);
                  } else {     //  Motors are active
                          i = 10;      //  Run for 100 msecs
                          Get(motor_sem);
                  }  //  endif
         }  //  endif

         Dlay(10);        //  Let other tasks execute

   }  //  endwhile

}  //  End motor
```

The final task is the light sensors. This task waits for a request and reads and returns the value for the specified light sensor.

```
void light()                    //  Poll the Light Sensors
{

int msgnum;                     //  Message number to read
msg message;                    //  Message passed to/returned from other
msg Sender;                     //   tasks

    while (1 == 1) {

         msgnum = Wait();  //  Wait for a message
         Sender = Read(msgnum);
         Ack(msgnum);        //  Save the Sender
         msgnum = Wait();  //  Get the request
         message = Read(msgnum);
         Ack(msgnum)

         if (Poll(motor_sem) != Free) {
                         //  Turn off motors during light measure
             message = motor_stop;
             Send("motor", message);
         }  //  endif

         if (message == get_left_light_sensor)
             message = LeftLightLevel;
         else
             message = RightLightLevel;

         Send(Sender, message);

   }  //  endwhile
```

```
}  //  End light
```

This task does not have an explicit delay; instead it simply waits for a request.

To load all of these tasks together, you will have to create a Task1 that looks something like the following:

```
void Task1()                    //  Load Application Tasks
{

    Start(biologic, 2, "biologic");
    Start(motor, 3, "motor");
    Start(bump, 3, "bump");
    Start(light, 3, "light");

    End()                       //  End this task/free its resources

}  //  End Task1
```

Although the code seems complex, as you read through it, you will discover that it is actually quite straightforward.

This application works quite well, except that it only works once. This is to say that if the robot is started up and allowed to find the darkest spot in the room, it will just sit there until you go over, pick it up, and move it to a different point in the room. Being lazy, I would like to add a control task to the robot that enables me to use a TV remote control to move the robot somewhere else in the room before starting up the roach action again.

The remote control I would like to use has six buttons. Stop and start buttons are used to start or end the roach action, and four buttons manually control the robot's motors: forward, reverse, left, and right. The manual control task could be written as follows:

```
void control()                  //  TV Remote Control Task
{

    Get(control_sem);           //  Claim Control Semaphore to have
                                //    Robot wait for explicit control

msg message;                    //  Message passed to/returned from other
                                //    tasks

    while (1 == 1) {

            if (RemoteControl = ButtonPress) {
                switch(Button) {
                    case Forward:
                        if (Poll(control_sem) == Free)
                            Get(control_sem);
                        Send(motor, Forward);
                        Break;
                    case Reverse:
                        if (Poll(control_sem) == Free)
                            Get(control_sem);
```

```
                        Send(motor, Reverse);
                        Break;
                case TurnLeft:
                        if (Poll(control_sem) == Free)
                                Get(control_sem);
                        Send(motor, TurnLeft);
                        Break;
                case TurnRight:
                        if (Poll(control_sem) == Free)
                                Get(control_sem);
                        Send(motor, TurnRight);
                        Break;
                Case Stop:
                        if (Poll(control_sem) == Free)
                                Get(control_sem);
                        Break;
                Case Start:
                        Free(control_sem);
                        Break;
                } //  endswitch
        } //  endif

        Dlay(10);              // Let other tasks execute

    } //  endwhile

} //  End control
```

Of the original five tasks I presented at the start of this section, I only have to change one, Task1, to start the control. This task becomes as follows:

```
void Task1()                        // Load Application Tasks
{

    Start(control, 1, "control");
    Start(biologic, 2, "biologic");
    Start(motor, 3, "motor");
    Start(bump, 3, "bump");
    Start(light, 3, "light");

    End()                          // End this task/free its resources

} //  End Task1
```

Not having to modify the biologic or peripheral tasks in this application when adding a substantial feature like the remote control is a primary advantage of the RTOS. Along with this, it should be obvious how easy it is to change the different peripheral tasks and the biologic task. In these cases, tasks that are not explicitly related to the changes at hand are not affected.

Another advantage of the RTOS is the minimization and simplification of the variables used in the application. If you look at the various tasks discussed previously, you'll see that to provide the control functions for the roach response to light I only used 32 bytes (assuming that both int and msg types require 2 bytes). This is substantially smaller than what would be required for implementing all these functions using traditional coding techniques.

The biggest advantage to the RTOS is how it simplifies application development. I was able to write the five tasks in just a few hours one afternoon with very little effort. In comparison, I have spent several weeks thinking through the different peripheral interfaces that were presented in the previous chapter to make sure they could be mixed and matched with a minimum of problems. Even with all this work and time spent on the problem, code had to be modified to support the multiple interfaces working together.

State Machines

Creating state machines in software is another technique of nonlinear programming methodology that you can use for your robot biologic code. This enables you to avoid having complex comparison operations and sophisticated decision-making code. Software state machines can be implemented in a variety of different ways, depending on the requirements and capabilities of the application.

State machines were first designed as hardware circuits that perform sequential, decision-based operations without any logic circuits built in, as shown in Figure 5-1. The current state is the address produced by combining the previous state and the external inputs of the circuit.

Software state machines work in a similar manner with a *state variable* used to select the operation to execute. The most common way to describe a state machine is to use a select statement to show how the different states are selected and what happens in them. For a simple randomly moving robot that backs up and starts over when it collides with an object, state-machine-based biologic code could be shown as the following:

```
main()                    //  Randomly moving robot example
{

int  State = 1;           //  Start State at the Beginning

  while(1 == 1)           //  Loop forever
    switch(State) {       //  Execute according to the State
      case 1:             //  Move Forward for 2 Seconds
        Move(Forward);
        Dlay(2000);       //  Use a 1 msec "Dlay" Function
```

Figure 5-1
Hardware state machine implementation

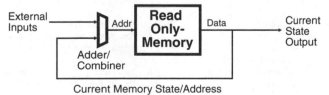

External Inputs → Addr → **Read Only- Memory** → Data → Current State Output

Adder/ Combiner

Current Memory State/Address

```
      if (Collision == "No")
         State = 5;            //  Jump to the Next State
      else                     //  Collision, Back up to avoid
         State = 10;           //   the obstacle
      break;
   case 5:                     //  Turn Left for 3/4 seconds
      Move(TurnLeft);
      Dlay(750);
      if (Collision == "No")
         State = 1;            //  Jump back to moving forward
      else                     //  Collision, Back up to avoid
         State = 10;           //   the obstacle
      break;
   case 10:                    //  Collision, Back up for 1 Second
      Move(Reverse);
      Dlay(1000);
      State = 11;
      break;
   case 15:                    //  Turn right for 1/2 second to
      Move(TurnRight);         //    Point away from object
      Dlay(500);
      State = 0;               //  Resume Random motion
      break;
   }  //  endswitch

}  //  End Randomly moving Robot Example
```

In this example, delays are put into each state and must complete before the next state can execute. This placement and use of the delay code is a problem because, as I discussed in the previous chapter, inputs are ignored until the action has completed.

The solution to the problem requires that the collision input sensors (along with any other sensors in the robot) be continuously polled. This is not hard to implement but requires that the delay subroutine is accessed externally to the switch statements, and the number of times it executes is dependent on the operation within the state machine being carried out.

In the next example application, I have expanded on each state and broken it up into two separate states. The first state starts the motion of the robot and loads a counter with the number of times a common 50-millisecond delay is to count down. The second state decrements the counter and checks the inputs to see if there was a collision and if the collision response must be executed.

```
main()                        //  Randomly moving robot Example 2
{

int  State = 1;               //  Start State at the Beginning
int  DlayCount = 0;           //  Delay Counter

   while(1 == 1) {            //  Loop forever
      switch(State) {         //  Execute according to the State
         case 1:              //  Move Forward for 2 Seconds
            Move(Forward);
            DlayCount = 40;   //  Execute the 50 msec Delay 40x
            State = 2;        //  Execute the Dlay count down and poll
```

```
      break;              //    the collision sensors
   case 2:                //    Moving forward for two seconds
      if (Collision == "No")
        if (--DlayCount)
           State = 2;      //  If not zero, keep with current
        else
           State = 5;      //  If Zero, then Start Turn
      else                 //  Collision, Back up to avoid
         State = 10;       //    the obstacle
      break;
   case 5:                 //  Turn Left for 3/4 seconds
      Move(TurnLeft);
      DlayCount = 15;      //  Turn for 750 msecs
      State = 6;           //  Execute the Delay count down while
      break;               //    Polling the collision sensors
   case 6:                 //  Turning Left for 3/4 seconds
      if (Collision == "No")
        if (--DlayCount)    //  Turned for 750 msecs?
           State = 6;
        else
           State = 1;
      else                 //  Collision, Back up to avoid
         State = 10;       //    the obstacle
      break;
   case 10:                //  Collision, Back up for 1 Second
      Move(Reverse);
      Dlay(1000);
      State = 15;
      break;
   case 15:                //  Turn right for 1/2 second to
      Move(TurnRight);     //    Point away from object
      Dlay(500);
      State = 1;           //  Resume Random motion
      break;
   } // endswitch
   Dlay(50);               //  Common 50 msec Delay
} // endwhile

} // End Randomly moving Robot Example 2
```

I did not put in a collision sense for states 10 and 15 because they were the responses to a collision, and I assumed that the process of backing away from an object will not result in another collision. This is probably an invalid assumption because as the robot moves randomly about the room, it may back into objects.

The fix for the problem of potentially backing into objects is to add collision sensors to the rear of the robot and poll them during the backing up states (states 10 and 15 in the previous applications). Adding the code to support the possibility of a rear collision is quite simple and can be done exactly the same way as handling a collision in the forward direction.

The ease with which new operations can be added to the biologic code is one of the hallmarks of the state machine. Another advantage is how simple the code is for each different state. This means the state machine only executes for a few cycles, leaving more machine cycles for the code in the interrupt handlers.

A downside to the state machine is how confusing its operation can get when the application gets more complex. You will find that as you add more and more functions to the state machine, it becomes difficult to understand exactly what is happening, especially if you are adding multiple input sources. Despite this, using a state machine for your robot's biologic code is a viable alternative and one that you should consider when you are specifying how your robot system should work.

Randomly Moving a Robot Application with IR Remote Control

To demonstrate how a state machine can be implemented to control robots, a simple robot function will be outlined that moves forward for two seconds, turns to the right for three-quarters of a second, and repeats. If, during any part of this operation, the robot collides with an object, it reverses itself for a half-second and then turns away from the direction of the collision for another half-second. Once the collision response has completed, the robot continues executing from the start of the pattern.

Using traditional coding methodology, the biologic code that would implement this robot application would be as follows:

```
while (1 == 1) {

        CollisionFlag = 0;              // Keep Track of Collisions

        LeftMotor(LeftForward);        // Move Forward
        RightMotor(RightForward);
        Dlay = RTC + 2000;             // For 2 Seconds
        while (Dlay != RTC)
                if (GetLeftWhisker())  // Left Collision?
                        CollisionFlag = 1;
                else if (GetRightWhisker())
                        CollisionFlag = 2;

        If (!CollisionFlag) {
                                // Execute Turn if no Collisions
                LeftMotor(LeftForward); // Turn Right
                RightMotor(RightReverse);
                Dlay = RTC + 750;       // For 3/4 Seconds
                while (Dlay != RTC)
                        if (GetLeftWhisker())   // Left Collision?
                                CollisionFlag = 1;
                        else if (GetRightWhisker())
                                CollisionFlag = 2;
        } // endif

        if (CollisionFlag == 1) {      // Left Collision
```

```
                    LeftMotor(LeftReverse);  //  Move away
                    RightMotor(RightReverse);
                    Dlay = RTC + 500;        //  For 1/2 Second
                    while (Dlay != RTC);
                    LeftMotor(LeftForward);  //  Turn Right
                    RightMotor(RightReverse);
                    Dlay = RTC + 500;        //  For 1/2 Second
                    while (Dlay != RTC);
               }  //  endif

          if (CollisionFlag == 2) {       //  Right Collision
                    LeftMotor(LeftReverse);  //  Move away
                    RightMotor(RightReverse);
                    Dlay = RTC + 500;        //  For 1/2 Second
                    while (Dlay != RTC);
                    LeftMotor(LeftReverse);  //  Turn Left
                    RightMotor(RightForward);
                    Dlay = RTC + 500;        //  For 1/2 Second
                    while (Dlay != RTC);
               }  //  endif
     }  //  endwhile
```

This code does work, but it is very awkward. Looking at the code, its function is very hard to ascertain.

To make matters worse, adding functions to the code is just about impossible, and adding the collision checks in the delay while loops changes their functions substantially. If this code were to be used as the basis for a much more sophisticated robot function, adding additional features (such as checking the *cadmium sulfide* [CDS] cell light sensors before turning right) very quickly turns the code into a nightmare of complexity.

The capability to simply add functions to existing code is the hallmark of the state machine. When new functions are added, simple tests are added for new operational parameters, and new states are added to respond to the new operational parameters.

The biologic code that I came up with for this application is as follows:

```
while (1 == 1) {          //  Loop forever
     switch (State) {  //  Process the state command
          case 1:       //  Go forward for 2 Seconds
                    State++;     //  Start on State 2
                    StateDlay = RTC + 2000;
                    LeftMotor(LeftForward);
                    RightMotor(RightForward);
                    break;
          case 2:
          //  Wait for 2 Seconds to be up with no Collisions
                    if (GetLeftWhisker())
                         State = 10;
                    else if (GetRightWhisker())
                         State = 20;
                    else if (StateDlay == RTC)
                         State++;     //  No Collision, Turn
                    break;
          case 3:     //  Turn Right for 3/4 Seconds
                    State++;
```

```
            StateDlay = RTC + 750;
            LeftMotor(LeftForward);
            RightMotor(RightReverse);
            break;
        case 4:
        //  Wait for 3/4 Seconds to be up with no Collisions
            if (GetLeftWhisker())
                State = 10;
            else if (GetRightWhisker())
                State = 20;
            else if (StateDlay == RTC)
                State = 1;  //  No Collision, Start Over
            break;
        case 10:
        //  Left Collision, Back away for 1/2 Second
            State++;
            StateDlay = RTC + 500;
            LeftMotor(LeftReverse);
            RightMotor(RightReverse);
            break;
        case 11:       //  Wait 1/2 Second
            if (StateDlay == RTC)
                State++;
            break;
        case 12:
        //  Turn Away from Collision source for 1/2 Seconds
            State++;
            StateDlay = RTC + 500;
            LeftMotor(LeftForward);
            RightMotor(RightReverse);
            break;
        case 13:    //  Wait 1/2 Second and resume
            if (StateDlay == RTC)
                State = 1;
            break;
        case 20:
        //  Right Collision, Back away for 1/2 Second
            State++;
            StateDlay = RTC + 500;
            LeftMotor(LeftReverse);
            RightMotor(RightReverse);
            break;
        case 21:    //  Wait 1/2 Second
            if (StateDlay == RTC)
                State++;
            break;
        case 22:
        //  Turn Away from Collision source for 1/2 Second
            State++;
            StateDlay = RTC + 500;
            LeftMotor(LeftReverse);
            RightMotor(RightForward);
            break;
        case 23:    //  Wait 1/2 Second and resume
            if (StateDlay == RTC)
                State = 1;
            break;
    } //  endswitch
} //  endwhile
```

The functionality of this biologic application can be enhanced by simply adding more states. As new requirements are added to this application, code is simply added to change the state variable to the new states. The advantage of the state machine is its capability to be enhanced with the complexity and length of the code increasing in a linear fashion. Using traditional programming techniques, you will find that the application's complexity and length will increase geometrically. This ease of enhancement makes state machine programming ideal for robot applications.

This application is demonstrated through the wall-hugging robot and the servo bases by loading and executing the code in robot\statemc. wallhug.c is the wall-hugging mouse robot base code with the previous state machine biologic code. servo.c is the servo robot base code with the state machine biologic code listed previously as well.

If you look at the biologic code for both robots, you will probably be surprised to see that the biologic code is essentially identical. The only difference is in the LeftMotor and RightMotor functions. In the wall-hugging robot base, the PWM duty cycle must be provided (in the servo base, the wheels turn at a constant speed).

Not only does this validate the state machine programming concept, but it also validates how I have been structuring the biologic code and the elelogic and mechalogic interfaces. By keeping the software interfaces consistent, even through the actual interface hardware has changed, the robot designer can mix and match the robot components according to what the application requires, what's available, and what the robot's controller can support.

Behavioral Programming

As unlikely as it seems, robot behavioral (aka subsumption) programming has much in common with the behavior of a bratty kid. A behavior is a set of responses to external events (often referred to as stimulus) that are designed to reach an end goal and have been designed to handle different responses to changing stimulus.

Using the example of the bratty child, a program that dictates this type of behavior could be as follows:

```
GetToyBehavior()              // See Toy, Want it...
{

    printf("Mom, Dad?  Can I have that toy?\n");
    Response = Input.Data;   // Get Parents' response
    while ((Response != "Yes") &&
          (Response != "Okay, Okay, we'll get it)) {
                            // Force Response to "Yes"
          Position(LieDown);
```

```
Crying(Start);
Printf("Why <i>can't</i> I have a toy?\n");

if (Input.Pending) {
                // If No Response, Repeat
        Response = Input.Data;
        if (Response !=
            "Stop it, or you'll get a 'time out'...") {
                // Stop to avoid "Time Out"
            printf("Sorry Mommy and Daddy.")
            Response = "Yes";
        } // endif
    } // endif
} // endwhile

Position(Stand); Crying(Stop);

} // end GetToyBehavior
```

This example program probably seems whimsical, but actual robot behaviors are very similar. As a simple definition, a robot behavior is a simple program that works toward an ultimate goal by responding to its environment.

In the bratty kid example, provisions exist for alternative stimulus. In this example, the end goal was getting the toy, but if the behavior went too far and was eliciting a potentially negative response (time out), the behavior would stop as a safety measure.

Actual robot behaviors can work very similarly but can break away from the previous basic model where there is only one behavior at work at any given time. The idea of there being multiple behaviors at work within a robot is probably confusing, but it can allow for some fascinating and complex operations from the robot without requiring a significant amount of programming.

When multiple behaviors are active, the outputs of the behaviors are combined in some mechanism and the result is passed to the robot's wheels for actions. Many different algorithms can be used for this "arbitration" and you will have to work to understand which one is right for you.

The following example will show how a robot "shadow" application can be built out of some multiple behaviors and then have the outputs of the behaviors combined into motor controls. The shadow behavior consists of having a robot follow a human being three to four feet away. As the human moves, the robot will turn so that it is facing the human and move towards or away depending on the distance to the human and whether or not the human is moving.

As this is a thought experiment, I would like to define a fictional robot: a differentially driven robot with an ultrasonic sensor on a servo that can turn in different directions to determine the position of the human relative to the robot. The left and right wheels can each be moved from -128 (full reverse) to 0 (stop) to 127 (full forward). There are two input functions from the ultrasonic sensor distance(), which returns the distance to the human in inches, and vector(), which returns the relative angle (in degrees) to the human. The distance and angle return simple signed integers and the motor controls consist of signed integers returned in the structure:

```
struct motor {
     int left;
     int right;
} //  end motor sturct
```

In this example, I will assume that the robot is running in an environment that is a flat surface, infinite in size, and just contains the human to shadow. In a real environment, there would be other objects and people to worry about. Although I do not consider these obstacles in this example, adding additional behaviors is relatively easy, as I will show here. The capability to increase the complexity of a robot application, as well as the capability to tune the robot parameters, is an important feature of behavioral programming.

The first behavior to program is the robot turning towards the human. This behavior will be run periodically to keep the human as close to the front of the robot as possible:

```
motor directionBehavior(motor Data)
{                               //  Command motors to turn toward human

int dir;                        //  Relative direction of human to robot

     Data.left = Data.right = 0;
     dir = direction();         //  Get the Direction to the human

     if (dir > 0)               //  Have to Turn Left
          if (dir < tuneAngle) { // slow turn
               Data.left = -slowTurn;
               Data.right = slowTurn;
          } else if (dir < turnAngle) {
               Data.left = -mediumTurn;
               Data.right = mediumTurn;
          } else {              //  Turn fast towards human
               Data.left = -fastTurn;
               Data.right = fastTurn;
          } //  endif
     else if (dir < 0)          //  Have to Turn Right
          if (dir > -tuneAngle) { // slow turn
               Data.left = slowTurn;
               Data.right = -slowTurn;
          } else if (dir > -turnAngle) {
               Data.left = mediumTurn;
               Data.right = -mediumTurn;
          } else {              //  Turn fast towards human
               Data.left = fastTurn;
               Data.right = -fastTurn;
          } //  endif

     return Data;               //  Return Motor Commands

} //  end directionBehavior
```

In this behavior, I have defined all the angles and motor speeds as global parameters that are defined in the primary source file of the application. Looking over the application, it should be obvious that it is designed to turn the robot quickly towards the human if it is a large number of degrees "off beam" and to slow down the robot's turns to face the human. The initialization of the motors to stop is used when the robot is facing the human and does not have to turn.

To follow the robot, I want the wheels to be commanded to move forward when the robot is more than a set distance from the human. The code progressively turns the motors forward, depending on the distance the robot is from the set distance.

```
motor closeBehavior(motor Data)
{                                    //  Move Robot towards the Human

int dist;

        dist = distance();       //  Get the Distance to the human
        Data.left = Data.right = 0;

        if (dist > closeClosest)
             if (dist > closeFurthest)
                  Data.left = Data.right = forwardFull;
             else                    //  Within Range
                  Data.left = Data.right =
                       (forwardFull * (dist - closeClosest)) /
                       (closeFurthest - closeClosest);

        return Data;

}  //  end closeBehavior
```

The farther the robot is away from the human, the more power is applied to the robot to make it move faster towards the human. If the robot is less than the closest programmed distance, this behavior will not command the robot to move closer.

One quick comment must be made about the previous code. Notice that I carry out my multiplications before the divisions. The reason for doing this is to avoid having a value that is less than its divisor, with the result of the integer division being zero. By multiplying by the forwardFull value, I essentially guarantee that there will be a whole value output throughout the range without having to calculate floating point values.

I could have come up with a behavior that checks if the robot is too close and then backs away, but I felt that would make the robot much harder to program (as well as for you to read). Instead I've added the following behavior that is combined with it:

```
motor awayBehavior(motor Data)
{                                    //  Move Robot towards the Human
```

```
int dist;

        dist = distance();          //  Get the Distance to the human
        Data.left = Data.right = 0;

        if (dist < awayFurthest)
              if (dist < awayClosest)
                    Data.left = Data.right = reverseFull;
              else                  //  Within Range
                    Data.left = Data.right =
                          (reverseFull * (awayFurthest - dist)) /
                          (awayFurthest - awayClosest)

        return Data;

}  //  end awayBehavior
```

These three behaviors can be combined into the "main" program:

```
void main(void)             //  Implement a "Shadow" robot applica-
tion
{

motor directionData;        //  Define wheel values for three
motor closeData;            //   Behaviors
motor awayData;
motor CombinedData;         //  Combined wheel commands

        while (1 == 1) {    //  Run forever
              pause(RunDlay);  //  Pause to let robot move as commanded

              directionData = directionBehavior(directionData);
              closeData = closeBehavior(closeData);
              awayData = awayBehavior(awayData);
                          //  Execute Behaviors

                          //  Arbitrate behavior commands
              CombinedData.left =
directionData.left + closeData.left + awayData.left;
              CombinedData.right = directionData.right +
closeData.right + awayData.right;
                  WheelCommand(CombinedData);
                          //  Send the commands to the wheels

              pause(LoopDelay); //  Delay to move robot/new measurements

        }  //  endwhile
}  //  end main
```

As written, I have given each application parameter a label. Table 5-2 lists the different parameters along with the default values for the application.

These parameters can be tweaked or tuned very easily once the application is running in order to find the values that make the robot the most responsive, without jerking back and forth trying to find the perfect position. When the robot first starts and the parameters are at their default values, you will

Table 5-2

Parameters
and Their
Default
Values

Parameter	Default Value	Function
tuneAngle	5 (degrees)	Largest angle for slow turn.
turnAngle	20 (degrees)	Normal turning angle.
slowTurn	5 (motor control)	Slow turning speed.
mediumTurn	15 (motor control)	Standard turning speed.
fastTurn	40 (motor control)	Fast turning speed.
closeClosest	36 (inches)	Minimum distance for the robot to approach the human.
closeFurthest	48 (inches)	Maximum distance for the robot to move at full speed towards the human.
forwardFull	50 (motor control)	Maximum speed the robot moves towards the human.
awayClosest	36 (inches)	Minimum distance for the robot to move away from the human. Less than this distance, the robot moves away from the human at full speed.
awayFurthest	48 (inches)	Maximum distance from which the robot pushes away from the human.
reverseFull	-50 (motor control)	Maximum speed the robot moves away from the human.

probably find that the robot is very nonresponsive and it does not try very hard to shadow the human.

When the parameters are correct, the operation of the robot is quite simply amazing. For example, if the robot were 47 inches away from a person and 15 degrees to the left, the three behaviors would return the following values:

	DirectionBehavior	closeBehavior	awayBehavior	Combined
Left	-15	45	-4	26
Right	15	45	-4	56

In this case, the robot will be making a sweeping left turn as it moves towards the human. As it approaches the human, the awayBehavior component of the combined motor value will increase until it is in equilibrium with the closeBehavior and the robot will stop moving towards or away from the human. This motion will be reminiscent of a robot responding with PID or fuzzy logic control. The difference is that this case is much easier to program and it is simpler to tune the behaviors to get the most out of the robot.

Note that in the previous code, addition, subtraction, multiplication, and division are all the mathematical functions and no "floating-point" variables are required. It isn't unusual to see a tenfold reduction or more in required code space for a behavior-programmed robot over one that performs the same functions using classic control theory.

Neural Networks and Artificial Intelligence

It is virtually impossible to discuss robotics without the topic of artificial intelligence coming up. Many people think that artificial intelligence means that a robot (or computer) will be able to operate as if it were an organism at least as intelligent as a human being. Along with this thinking, many people consider that for a robot or computer to be truly intelligent, it must be able to pass the "turning test."

It is appropriate to end a discussion on how biologic code can be implemented with a look at artificial intelligence. If you look at the examples in this chapter, you will see that I have taken the basic robot interfaces presented in the previous chapter and used them to implement some fairly intelligent applications with the robots working in a very lifelike way. In saying this, I mean that they imitate an actual living entity, not that they are living or have some form of artificial intelligence.

If I were asked to define the term artificial intelligence, I would define it as follows:

> The capability of a computer or robot to carry out a specific task, navigating through a set of natural obstructions and learning from its experience so that it can accomplish the task more efficiently in the future.

Nowhere in this definition do I discuss communications, processing visual information, or destroying villages, the actions that are generally assumed when the term artificial intelligence is used. The point I am trying to make is that artificial intelligence is not as extensive as is generally perceived.

Trying to develop code that meets the previous definition of artificial intelligence is difficult, and it is hampered by our inability to define the terms knowledge and learning as applied to computer technology.

To illustrate how difficult it is, consider the famous experiment of training a worm to move through a maze. For this experiment, the maze that was used was a T maze. The worm would start at the bottom of the T and when it reached the wall at the end of the corridor, it would have the choice of turning left or right to continue. To train the worm, a series of electrodes were placed in one of the corridors, shocking the worm. In the other corridor, food was waiting for the worm. Eventually, the worm learned that it should go down the corridor without electrodes, and every time its way was blocked it decided to turn in the direction that did not have electrodes in the maze.

This experiment became famous because after training a certain number of worms to turn in one direction after encountering an obstacle, they were ground up and fed to worms that had not received this training. When these worms were presented with an obstacle ahead of them, a statistically significant number of them always turned away from the electrodes and towards the food without any kind of training. About 56 percent of the time, the worm would turn away from the electrodes when encountering an obstacle, as opposed to 50 percent of the time for the worms without "training or eating worms that had been trained."

Ignoring the ramifications of the experimental evidence that suggests knowledge can be passed along as food, consider the difficulty in developing a computer program that would train a robot to turn only in one direction when it encounters an obstacle. Using the paradigms from the worm experiment, how would you represent the pain the worm felt going over the electrodes? What kind of sensation or emotion takes place in the worm when it finds the food after going down the other corridor? How would a robot respond when it finds itself at the far end of the corridor with the electrodes? A very basic question is *why* would a worm move down the first corridor in the first place?

When you use a human or biological animal paradigm, all of these questions are understood at a very basic level. The problem is, nobody knows how a computer is programmed with concepts like pain, pleasure, hunger, or memory and have them all work together. Personally, I believe that in a human being these concepts are never defined; at birth there are just genetically programmed reflexes. From these basic reflexes, we learn these different concepts and they are intimately interrelated in ways we can't understand.

For example, I remember being a young boy and touching my mother's iron with the inside of one of my arms and getting a fairly large burn. Now, when I think of pain, burns, clothes, irons, or arms, I am always reminded of this incident and I unconsciously take steps to protect the insides of my arms. I've been burned many times on other parts of my arms and hands, but these incidents, all of which have been very painful, are never the ones that I remember and I certainly don't try to protect myself in the same manner as I described with my arms. I'm sure that you also have incidents in your lives that have trained you to respond a certain way when some kind of stimulus or a threat of this stimulus is being received. These incidents create knowledge that is remembered and from this stored knowledge we learn how to act in a certain way to a specific stimulus.

This discussion sounds like I think of people (and animals) like computers and, to a certain extent, this is correct. Going back to my definition of artificial intelligence, this description of how we think fits in with the initial points but does not address how we learn facts and store them; it does not answer how we look for ways to do things more efficiently.

I believe that some time in the very near future there will be a computer program that passes the turning test with ease. You will be able to talk to it just as if it were a human and its responses will be completely normal. The reason why I do not believe that this computer is artificially intelligent is that it is still incapable of learning (a human programmed all the facts that the computer needs to respond like a human).

I believe that it will be many years until computer scientists understand how knowledge is entered, processed, and stored within an autonomous being. The term autonomous being is my label for a computer program that executes a task on its own or is prompted in some way to carry out the task and learn from it for the next time it must be carried out.

When I look at the current "state of the art" of artificial intelligence, I believe that the first steps towards understanding how the brain works and how intelligence takes place will be by studying and simulating the building blocks of the brain: the neurons. By simulating neurons in "neural nets," scientists will learn how our brains are initialized with basic reflexes, how these reflexes are programmed out, and how the neurons are loaded with knowledge and learned behaviors.

Since the 1940s, many different efforts have been made to model how a neuron (nerve) cell works, interacts with other neurons, and provides logic functions within our brains. Depending on your understanding of digital logic, you may be under the impression that neurons provide very simple digital logic (AND, OR, NOT, and so on) functions. In actuality, the logic functions within the neuron are often not very simple; they often perform analog functions that control different parts of the body while monitoring how the body is responding.

Despite the decades of work on understanding neural networks, I haven't found a single description of what a neural network is that is universally agreed upon. Many different descriptions are used to help model the operation of a specific neural network product. The best one that I have found from the DARPA Neural Network Study (1988, AFCEA Internationl Press, p. 60):

> A neural network is a system composed of one or more simple processing elements operating in parallel whose function is determined by network structure, connection strengths, and processing performed at different computing elements or nodes.

Simply stated, a neural network is a collection of interconnected processing elements that weigh different inputs and provide different outputs based on the inputs. Figure 5-2 shows an example that consists of a robot moth using bump and light sensors interfaced with the motor controllers.

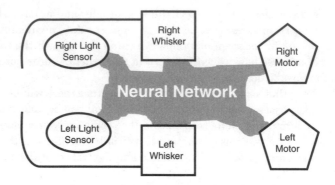

Figure 5-2
Robot moth
using a neutral
network

The neural network is the amphorous blob linking the various parts of the robot together. I usually visualize a single-celled neural network as having the following programming form:

```
Outputs = f(Inputs);
```

where multiple inputs are processed together using some function(s) to produce multiple outputs.

It is very important to note that there can be multiple outputs from the neural network function. The operations performed within the neural network could be a truth table, a state machine, or even straight programming, but, however they are implemented, the parameters used by the decision-making code are "learned" and not pre-programmed.

Designing the neural network topology (how it connects to different inputs and outputs) is very important. When defining the neural network, you have to define all the possible links that the neural network will connect to. If you err on the side of linking too many inputs into your neural network, you will find that the training operation will minimize or eliminate the unimportant inputs from the output response calculations. The downside of having too many inputs is that it will result in a longer training period than if just the primary inputs were used in defining the neural network.

I have never seen a calculation in the literature I have read about neural networks, but I suspect that there is some kind of power law with regards to increased training due to increased inputs. Assuming a square law, it could take four times as long to train a neural network with twice as many I/Os as was required to train a base neural network. This extra time and effort has to be calculated when estimates are given for how long it will take to create (or program) a neural network application.

Along with initial learning, a true neural network will continue to keep learning, trying to find a better way to perform the necessary task. In the robot moth example, the neural network may discover that if one light sensor is receiving a bright light, by taking into account the light at the other sensor the rate of turn can be tuned to find the most efficient path to the light source.

Although the robot moth example may be trivial, a much more important example would be using a robot gripper to pick up an arbitrary object (or a known object at an arbitrary orientation). Our hands do this naturally, but programming a robot's gripper to perform the same task is extremely difficult. Situations like this are really meant to take advantage of neural network control.

As you would expect, the major issue with neural networks is training. Ideally, we would want to train a neural network by showing the robot how to respond to a given set of inputs. Many different methods exist for doing this, but virtually all of them come down to the concept that after the neural network starts performing the task, it is stopped when it responds incorrectly to the inputs. The wrong response is noted, the neural network changes the input response parameters, and it tries the task again to see if it can perform it better. This is somewhat like housetraining a puppy.

Robot neural network software can appear to be similar to behavioral programming, but actual logic implementation will be different. The code that I have seen places a very high emphasis on the learning aspects of the neuron's operating parameters.

Instead of training a neural network, you may want to create the necessary parameters for the neural network using a genetic algorithm. This is almost exactly the same as the neural network, but the final goal is programmed into the software and it is allowed to execute by changing these parameters systematically (or randomly to achieve leaps in the evolution of the application). For the robot moth, different parameters would be tested in the neural network until the final destination is reached (the brightest spot in the room).

Once these parameters have been found, the process will be repeated with simple variations on the parameters until the optimal response has been found. In some applications, the parameters are never frozen and are continually modified to keep attempting to find the best possible set of responses. This continual operation is often used in applications when depending solely on evolution to find the best solution would take an unreasonable amount of time and would cost valuable production time.

With the different methods of training and finding parameters, you are probably asking yourself if neural networks provide a path to true artificial intelligence. Many people would say yes because the resulting action produced by the robot from its inputs is based on a decision process that simulates life. I would not consider it to be true intelligence but rather an artificial reflex. It is a pre-programmed response, just as your leg jumping after being hit by a hammer.

Designing the Robot System

It might be somewhat surprising that this chapter is at the end of the book. The reason for placing this chapter here is due to the need to present some new concepts that had to be explained before the discussion on how to design the system could take place. My goal for this chapter is to give you the knowledge required to integrate the different parts of a robot and design a software application that will use each of these parts effectively.

In the "Microcontroller Connections" Chapter 4, I designed the different interfaces to work together around a common 1-millisecond interrupt. Having the software peripheral interface code that is designed to work with others makes the task of creating the complete robot software easier (as was shown in Chapter 5, "Bringing Your Robot to Life"), but these functions are really similar to the different mechanical and electrical parts that make up the robot. Work is still required to create the final application.

Before you can start defining the requirements and working through to the final software application for the robot, you must have a clear vision of what the robot is to be. A vision is not a statement like "I want to build a robot to learn more about them." It is something much more specific like "I want to design a line-following robot using a two-wheel differential drive robot." With a concrete vision in place, you will find that developing a plan that meets your requirements is much easier. Having a concrete vision statement does not mean that you cannot enhance the robot later; with a clear vision statement you will find that you can efficiently specify which modifications to the software (and robot structure) are required.

A lot can be said about visions that are totally unrealistic. Many hobbyist robots are started with the vision of doing a task that professionals have worked on for years and have spent literally millions of dollars trying to get working. A classic example of a hobbyist robot that has extremely aggressive requirements is one that will get a beer from the designer's refrigerator on call. Problems with this type of robot include answering how does the robot open the door (it will be easier or harder depending on the mass of the door), where will it find the beer, and how does it locate the recipient. This does not include such questions as what if the refrigerator and the recipient are on different floors. These are just questions off the top of my head; I'm sure if you were to look at the application in more detail you would discover a lot of other questions have to be answered before you could start building the robot.

I am not exaggerating when I say that scientists have spent many person-years and millions of dollars trying to solve these problems. Consider the potential difficulty in designing a robot that goes upstairs. The robot will probably have to be facing the stairs exactly perpendicularly to the staircase (a curved staircase makes this very difficult) and it will have to be able to go up stairs that are of differing heights and may have different materials. Once these problems are worked out, you will have to work on the problem of going downstairs; this is a completely different problem with some possible solutions.

I have discussed how different mechalogic and elelogic interfaces are integrated together and provided to the biologic code, but this is a major consideration when designing a robot. Without proper planning of how these interfaces are written, you will find that you may not be able to get the interfaces you want integrated into your robot, and you may find that there are not enough cycles for the different interfaces to execute or for the biologic code. In the "Microcontroller Connections" chapter, I designed the code interfaces to be as robust as possible with an eye towards being integrated with other interfaces, but in some cases there will be problems getting them all to work together. It is easy to resolve issues with different mechalogic and elelogic interfaces not working together before you start building your robot; it is much more difficult when you are halfway through the project and two interfaces that you have chosen are not working together.

A big problem for many people designing robots is what I call *magazine envy*. Often when reading hobbyist magazines, you will find robot designs or circuits that seem perfect for your requirements. The problem is that the article will use a part that is only available from the author. Chances are the part is very expensive, if you can find it anywhere else. I emphasize designing your robot with parts that are available in popular catalogs and web sites. Not only will using readily available parts make it easier for you to build your robot, but you should also be able to get the best price for it and if you decide to publish the robot's design, other people will appreciate being able to use commonly available parts.

Lastly, I want to point out that it is easy to build a robot. It is much more difficult to get it to work properly or as expected. In this chapter, I will go through some of the steps that I use to debug applications, circuits, and systems, and I recommend that you follow them or at least try to understand what I am trying to say. The cornerstone to the information here about fixing problems is making sure you understand *exactly* what is happening. I discourage people from making changes based on "feel" and just making changes for the sake of seeing if they can find the source of the problem and decide what is the best course of action for fixing it. Instead of simply debugging problems, I want you to go through the process of failure analysis on every problem you encounter.

Defining the Requirements for the Robot

Before going through the steps that I use to define and develop robot applications, I want to give you a few words of advice. First of all, document everything. Get into the habit of carrying a notebook around with you so that if you get an idea on how to do something, you can record it. Human memory is a pretty fallible storage device and I've spent many hours trying to remember that great idea I had the day before at lunch.

It is also important that you start small. There are a lot of half-finished robots out there that were just too much work for their designers. Your ultimate goal may be a robot caddy that follows you around on the links and hands you the club that you need. But if this is your first robot, I recommend that you start with a small platform with a lot of space for electronics to figure out the different technologies that will be required.

When defining the robot, remember to plan for the worst case. When I say worst case, I mean you should expect that different parts may not work. Don't expect that you will settle on the first method you come up with. There's always more than one way to do something.

Take, for example, the first practical example application I presented in this book: a flashing *light-emitting diode* (LED). The problem of using a PICmicro MCU to flash an LED could be solved by one of the following methods:

- Delay by code

- Delay by timer and polling the TMR0

- Counting TMR0 interrupts (the method used)

- TMR2 *pulse width modulation* (PWM) output

Which method is best in your application? Spending a few minutes thinking of options can make the application development much easier later on. This is an extremely important question because you will often find that your planned way of doing something will work great on a workbench, but when you implement it in the electrically noisy environment of a robot, the circuit may be unreliable or not work at all.

It is unfortunate, but I usually discover better methods for carrying out something *after* I feel like I have finished with the robot. If I figure out a better way, I'll usually keep it in my notebook for the next time rather than go back and recreate the application. I may go back and change the application, but that is only if the change offers a substantial improvement to what was already there. The basic rule here is "if it ain't broke, don't fix it."

Lastly, steal other people's ideas and methods. This does not mean literally stealing their mechanical designs, code, or circuit diagrams, but instead understanding how their ideas work and recreate them as they work best for you. This will not only help you avoid getting "blocked," but chances are you will learn from others and be able to use their ideas in ways that aren't readily apparent.

The Internet has a plethora of robots, circuits, and code at various web sites. I just want to caution you not to believe everything that you read on the Internet, as the information can be inaccurate (as it is for any topic). I've also found a number of web pages that offer a few "free" circuits as an inducement for you to buy design information from them or use them as consultants.

The most important thing you can do before starting to design and implement a robot is to create a set of requirements. These requirements will help you to plan the steps needed to end up at the desired end point and allow you to check whether or not you are meeting the original vision of your robot. You should be very rigorous in defining the requirements of your robot, as this will allow you to keep track of what you are doing and not get overwhelmed by the task in front of you.

I find it useful to first write out a simple, one-sentence statement specifying what the robot is supposed to do and how it is going to be implemented. Here are a couple of examples:

"I want to create a robot that is driven by radio control servos and controlled by a BASIC Stamp 2 in order to learn more about robot programming."
"I want to make a tracked robot that can be built into a model tank shell and controlled by an Atmel AVR 8515."

Next, the physical requirements for the robot must be determined with ideas on how the robot is to be implemented. A basic list would consist of answers to the following questions:

- What type of robot platform is to be used?
- What is the expected mass and size of the robot, and how much payload can it carry?
- What is the operating speed of the robot?
- What types of motors and controllers will be used?
- What types of surfaces is the robot expected to run on?
- What is the power source?
- What are the safety concerns?
- What types of controllers are to be used with the robot?
- What language will be used for writing the robot's application code?
- What are the sensors that are to be used with the robot?
- Are there any output devices?
- What are the methods for manually controlling the robot?
- What are the user interfaces?
- What is the qualification plan for the robot?
- How much should the project cost?

None of these answers should have the perfunctory answers "none" or "no restrictions." You may consider the robot to be completely safe (because it is small), but there will probably be environmental concerns about batteries if lead/acid/NiCad batteries are used. Along with this, you should consider potential problems because of the mass of the robot (if the robot weighs 100 pounds, what happens if the controller software causes the robot to run amok?).

With the requirements of the robot specified, prepare a list of requirements for the software used to control the robot. These requirements include the following:

- Language application software written in
- Simulator availability
- Programmer availability
- Interrupts to be used
- Built-in application interfaces to be used
- Data structures to be used

I should point out that all the lists of requirements are interconnected. You may initially assume that a specific microcontroller is to be used only to discover that it is not appropriate for other requirements, or it may have capabilities you don't require.

You may note that I don't include an expected application code size, which would help select the microcontroller that will be used. When you are first starting out, it can be very difficult to predict what will be the final size of an application. I've been doing it for years and I still can be off by 50 percent or more with my best guess. Often, I will specify a device with double the program memory that I think I will need.

Nothing gives an observer a feeling that the robot is sophisticated than the speed it runs at. The faster the robot runs, the more sophisticated it seems. In light of this, many robots are designed to run at a speed that would make a NASCAR driver envious. Note that to achieve these speeds, a powerful motor and batteries will be required (which have larger masses), and the sensors must be much faster. Not thinking through all of the ramifications could result in serious safety concerns with the robot in the future.

Later in this chapter, I will address the issue of qualifying the design of the robot. If you are a hobbyist, this probably sounds like a lot of extra work for something that you're just building to learn about robots. I disagree because, as will be discussed later, the qualification plan helps give you an organized approach to finding and resolving problems with the robot.

As you develop your requirement list, you may feel that the different requirements are mutually exclusive; you may want to use a certain microcontroller part number, but it doesn't support the features required in software. You may find that your suppliers cannot reliably provide you with the parts that are required. If this is the case, you may find you have to go back and change your original defining statement. This is not a serious problem.

If you are willing to work objectively with your requirements and the assumptions they're based on, you will find that you can get a solid set of requirements that will lead you to efficiently create a robot that performs according to your original vision statement.

■ Choosing the Peripherals to Be Used in the Robot

After specifying the desired function of the robot, you will have to select the type and implementation of the interface peripherals that will be used in the robot. Poor choices will result in the robot not working well or making unexpected movements. The root cause of the problems will be difficult to find, and you will spend a lot of time trying to figure out what the problem is in the biologic code when in actuality the interface peripherals are not working as you would expect.

The list of potential problems from poorly chosen mechalogic and elelogic interfaces includes the following:

■ The interfaces using up more instruction cycles than are comfortably available between the interrupts and the biologic code

■ The different interfaces used in the robot have resource conflicts that result in missed input events

In these cases, the biologic code cannot respond correctly to the input coming into the robot.

To avoid these problems, I advocate an elelogic/mechalogic base architecture like the 1-millisecond timer interrupt that I introduced with the example interface code. I have taken quite a few pains to ensure that the sample interfaces can be used as interchangeably as possible. When you design your robot, you must make sure the elelogic and mechalogic interfaces work independently of each other and give the biologic code enough cycles to work effectively.

It is important to remember that the 1-millisecond interrupt handler enables the biologic code to treat the elelogic and mechalogic interfaces like they were hardware built into the controller. For each of the interfaces, they must either set variables that can be read by the biologic code or poll a variable that will start a series of operations that are controlled or sequenced by the interrupt handler. This enables the biologic code to perform its primary task, deciding what the robot should do next.

When looking at the mechalogic and elelogic interface code, make sure that there are as few sophisticated operations as possible. Complex data types with similarly complex operations should be implemented in the biologic code where long operations (like a floating point divide) can execute without affecting any peripheral operations.

When writing your elelogic and mechalogic code, there are a few guidelines that help make sure the opportunities for any of the code execution in the robot are not compromised:

■ Use state machines for interface execution. This may slow down the execution of the interface function but will not affect the overall robot operation.

■ Do not use floating point or array variables.

■ Do not carry out complex mathematical operations such as multiplication, division, trigonometry, and logarithms.

■ Choose a controller that has as many built-in functions that will help you as possible. I realize that there really isn't a microcontroller available that has the built-in functions required for *your* robot (most likely requiring multiple PWM outputs along with multiple timers and interrupt vectors), but there are many chips out there that will come close.

■ Selecting the Parts and Assembly Techniques

When I have decided upon the robot I want to create and have chosen its features, I decide upon the hardware that is appropriate for the project. This list includes the following:

■ What is the desired microcontroller to be used?

■ What type of clocking will be used?

■ What kind of power supply?

■ Sensors/input devices?

■ Motors?

■ Output devices?

■ Storage devices?

■ Part availability?

These questions probably seem very straightforward, except for the last one. As I write these books, I am discovering more and more that what I think will be easy to find often isn't. For example, when I first designed the El Cheapo PICmicro MCU programmer, I used a P-Channel MOSFET that I could find easily from a variety of sources. Unfortunately, this wasn't true for many other people. I changed the transistor to a 2N3906, a very common PNP transistor, and the complaints stopped and acceptance of the programmer increased.

When I design anything, I try to choose electronic parts that are listed in standard distributor catalogs (such as Digi-Key, Jameo, and, ideally, Radio Shack). Most robots are one-offs and part availability is more important than cost. This is especially true if you are going to publish your design on the Internet or in a magazine article. Your "customers" will not appreciate spending a long time trying to find a part that is only available from a surplus store that is local to you. You will find that with the Internet, your designs could be built anywhere around the world.

Sometimes you will find that parts you have used for years have reached "end of life" or are obsolete. This is almost always a problem, although in many cases you can come up with parts that can be directly substituted for them. In other cases, parts no longer being available will be a big problem, and circuitry and software will have to be redesigned. The most important thing that I can say is that if you are in this situation, update your design as quickly as possible and make sure you publish the changes to make it easier for other people using your robot design.

To minimize the possibility that parts will be difficult to find or become obsolete, use standard (banal) parts as much as possible. The list of parts that I recommend includes the following:

- 1N4001, 1N4192 (1N915) diodes
- 2N3904/2N3906 or BC547/BC557 bipolar transistors
- IRF510/IRF9510 MOSFET transistors
- ECS/Panasonic ceramic resonators
- Watrony *infra-red* (IR) LEDs and detectors
- 74LS *Transistor-to-Tansistor Logic* (TTL)
- 78(L)xx regulators
- (7)555 timers

Discretes (resistors and capacitors) are generally not a problem, although you may find that different parts may be difficult to find at different times. As I write this, there is a world-wide shortage of tantalum capacitors. In most of my designs, I like to have a 0.1 uF, 16-volt tantalum capacitor for decoupling power in logic chips. Due to the shortage, I have been using ceramic or monolithic 0.01-to-0.1 uF capacitors.

The most significant part that you have to choose for your robot's electronics is the microcontroller. People generally choose microcontrollers according to their personal likes and dislikes, as well as their experiences and what kind of equipment they have access to. This is a good thing for the most part, although certain issues must be considered, including what kind of software development tools are available (and what are their costs), what kind of programmers are required, and what kind of support components are required. When I select a microcontroller, I believe that the only support it should require is just +5 volts of power, a crystal or ceramic resonator for clocking, and a pull-up resistor for reset. If the microcontroller requires more than this, then you should consider another microcontroller.

Note that I do not consider microcontroller operating/execution speed to be a significant consideration. Due to different "spectrums" of code execution, problems due to slow microcontroller operating speed can be minimized. I recommend keeping the speed down for minimal microcontroller current consumption and maximize the noise immunity. The most important consideration that I would make when choosing a microcontroller is how well it works in the noisy robot environment.

Once you have chosen the parts to use, the assembly of the circuitry in the robot is the next major consideration. A robot is a pretty tough environment for electronics and microcontrollers with issues to be considered, including the following:

- Electrical noise
- Magnetic device kickback
- Vibration
- Acoustic noise
- Vunerability to weather
- Moving parts that can grab wires

When building or integrating the electronic circuits for your robot, I recommend that you follow these assembly guidelines:

- Use large ground planes or good grounding with motor cases.
- Use soldered connections.
- Wires should be glued down every 0.75 inches (2 cm).
- Minimal and standard connector wiring harnesses should be used.
- Use screws in terminal blocks for wires or tied or screwed-down connectors for harnesses.

The ideal platform for robot circuitry is a customized *printed circuit board* (PCB). This will be the most reliable method of integrating the electronic parts over the long term. You may not consider a custom PCB to be cost effective or see it as being very difficult to modify. I once did a cost analysis between a prototype PCB and a wire-wrapped board and found that the cost difference was surprisingly minimal. The customized PCB cost 20 percent more than the wire-wrapped circuitry before the build time was taken into account. If the time difference between assembling a circuit on a PCB versus wire wrapping were considered, the customized PCB would be much less expensive.

As well as being surprisingly cost effective, if additional circuits are to be built, you will not have to invest again in laying out the circuit and additional copies will be quite inexpensive. Actually, a customized PCB will be the least expensive method of building the robot circuit if more than one prototype is required.

As for modifying the PCB circuit, I always make sure I place an area on the PCB that has an array of holes where additional circuits can be added. This enables me to add peripherals to the robot as well as correct any serious mistakes in the original circuitry.

Wire wrapping and point-to-point wiring are possible methods that you can use for building your robot circuitry, but you will find that they are very labor intensive as well as very costly (for sockets and specialized connectors). However, they generally have the reliability required for a robot application.

The worst method of building a robot circuit is using the "breadboard" (which was used to demonstrate many of the peripherals in Chapter 4, "Microcontroller Connections") in terms of long-term reliability, cost, and ease of assembly. As the circuitry on the breadboard becomes more complex, you will find that the breadboard circuitry is also extremely difficult to follow and modify. I would only recommend using a breadboard for example applications that are simple and will be torn down and built back up fairly quickly.

When you are designing your first robots, look for a compromise solution like a PCB with some temporary prototyping capabilities. I recommend that the microcontroller should be socketed (for reprogramming) or soldered in with the programmer circuit built into the board. The microcontroller's power regulator, reset mechanism, and oscillator should be soldered in. I also recommend having a soldered-in motor driver with provisions for screw terminals to reliably connect the motors to the PCB. In this PCB, you might want to have sockets with a standard bus or pinout for different sensors.

Validating the Operation of the Robot

The requirements that you define for your robot should include an application qualification plan. This plan consists of a list of tests that the application must pass before you can consider it "done" and ready for use or presentation to others. I recommend creating a list, similar in format to the list of definitions and how the robot is implemented.

A typical robot qualification plan would consist of the following tests:

- Test for the power supply to supply 4.75 to 5.25 volts of regulated power at the required current load under all conditions. This includes stalling the robot's motors.

- Test for the microcontroller to command the robot to stop when the battery voltage or current drops below a set value.

- The microcontroller runs when the motors are started and stopped (high-current/large-voltage transients).

- User interface output functions are at the correct initial state on power-up.

- User interface output functions respond correctly to user input.

- Sensors work (on the robot) as specified.

- Reliability calculations for different robot components have been made.

- *Federal Communications Commission* (FCC), and other emissions and regulatory testing (if a commercial product) are performed.

■ User documentation is complete and correct (if a commercial product or design is going to be published on your web page).

■ Manufacturer documentation is complete and correct (if a commercial product).

The power range that I specified in this list probably seems quite restricted, especially with the wide input voltage ranges available to different microcontrollers, but you should understand what the operating conditions of other chips of the circuit are and find out what their tolerance of input voltages is. Many active integrated circuits only work reliably within a plus or minus five percent input power voltage (Vcc or Vdd) "window."

I have tried to order the qualification tests to ensure your safety first and make sure that the robot will be under positive control at all times. This may seem unimportant for a small robot, based upon the order in which the application would be debugged, but there's nothing as upsetting as watching your robot start up spontaneously and run off the edge of the table, smashing itself to bits on the floor.

For this reason, I always recommend having some method of keeping the driving wheels (or feet) "jacked up" above the flat surface it normally sits on. Even a fully debugged and qualified robot can unexpectedly run away from you when an unexpected sensor input is received or when you forget to turn the power off after a test. It's even happened to me when I've changed some code that I thought had nothing to do with the motor controller code.

Note that for many different kinds of inputs, a simple digital multimeter or logic probe will not be sufficient for making sure the circuitry works correctly. Ideally, an oscilloscope or logic analyzer should be used to look at the actual *input/output* (I/O) and confirm that it is correct. Once you understand how the robot works, you should rigorously simulate the microcontroller code to make sure that it will respond appropriately in all conditions.

Coming up with appropriate tests for the robot's operation is something that is near and dear to my heart (mostly because I have spent almost all of my professional life ensuring that products work correctly). The previous list of qualification items should test for all functional aspects of an application. If you've worked around electronic circuits for any length of time, you will know that this is almost always impossible because there are so many different ways in which a circuit can fail.

The important point to make about testing is that you must ensure that the application will respond correctly to the expected inputs. There will be cases when unexpected inputs will be received (such as when the application connection is jostled or the circuit is zapped by static electricity) and these are hard to plan for. You should have a good idea of how the application works under normal operating conditions and be able to test these conditions.

If abnormal inputs are received or if internal variables are outside expected limits, the robot control software should stop and indicate that there is a problem. Along with cheking the internal variables, you may want to implement a "watchdog timer" function to your application. The watchdog timer will reset the MCU if it itself is not reset periodically.

Reliability qualification is not something that many companies worry about and, with the ruggedness of modern components, it is not perceived as being a critical factor to a product's success. Personally, I believe that testing the reliability of a product is the most important indicator of the quality of the design and of the components used. In the stressful robot environment, reliability testing can add some costs to a product, but it can keep you from unhappy customers and lawsuits if your products fail prematurely.

Environmental qualification is simply ensuring that the product will run reliably in the location where it is expected to operate. Your robot should be designed with good electrical practices to minimize the possibility that it creates excessive noise. For products to be sold commercially, they must attain FCC Class B (or appropriate according to country) certification.

Lastly, make sure you have properly documented your application. I consider this aspect of project development to be part of the qualification plan because the documentation for users, as well as manufacturers, is critical to having your product accepted. User interfaces should be designed to make the product easy and intuitive to use. Any extra hardware required by the user, as well as instructions on how to connect and power the application, must be clearly listed and detailed.

Documentation should be written using an electronic tool that is widely available to any potential users. If you do not feel comfortable with one single format (for example, your product can be used with PCs, Macintoshes, or Unix workstations), then you should document how to use the product in straight ASCII text files.

If straight ASCII text files don't seem appropriate because formatted text is required with graphics, then I would suggest either Adobe .pdf format or using *HyperText Markup Language* (HTML) for your documentation. Both data formats are widely available on different computer systems along with development tools that can be found fairly cheaply and easily.

All these requirements can be summarized in an ISO9000 registration. ISO9000 (the generic term) is a certification required by the European Union and is used to document development and manufacturing processes. Even if you are a one-person company, ISO9000 certification can be achieved for a modest cost and will add enforced rigor to your development process.

▆▆ Finding and Resolving Problems

At work, I try to discourage people from using the term *debug*, as it seems to denote a quick fix. Instead, I prefer that the term *failure analysis* be used. It's interesting how changing a simple word to a technical phrase can change an operation that is "quick and dirty" into a structured process that enables you to understand exactly what is happening and allows you to find the most expedient method of fixing the problem.

You'll find that following the steps and procedures I present to you will allow you to fix problems faster than if you just look for the first problem and try to "nail it." This will seem like a paradox because I will be pushing for you to characterize and understand exactly what the symptoms of the problem are. By using this information, hypothesizing about what the defect is, and testing the hypothesis, you will understand exactly what is happening and it will be obvious what the best fix is for the problem.

So, instead of blindly going ahead and removing components, the process you should use consists of the following three steps:

1. Characterizing the problem.
2. Hypothesizing what the cause of the problem is.
3. Developing a fix for the problem.

Even if your robot just seems to be sitting there and doesn't seem to be working at all, you can still get information out of it to help you figure out what the problem is. At the other end of the spectrum, if the robot works "mostly okay" or fails occasionally, you should still be able to isolate where the errors are based on understanding what is happening. This process of looking at a failing application and trying to detail what its environment is, along with its response, is known as characterization and is the most critical aspect of performing failure analysis.

When I have any kind of application that isn't running properly, the first thing I do is examine the circuit to look at all the input and output voltage levels. I normally use a logic probe and I record all of my results on a sheet of paper. Digital logic levels can be checked with a logic probe, but not voltage levels. When you first work with your applications, you may want to check the signal pins with both a logic probe and a voltmeter until you get a good idea of what is happening and what you expect to see.

When checking a microcontroller's built-in oscillator, check both pins with a logic probe to make sure the capacitance of the probe doesn't inadvertently set the application running. In some cases, you might want to go back program in and execute an application like ledflash to make sure the microcontroller can still execute in the robot.

When this is done, compare your notes with what the expected values are. This is an important point because I can't tell you how often I've thought I've seen a problem, only to discover I miscounted and was looking at the wrong pin.

Any discrepancies you find can be put in one of two *buckets*: the ones that are caused by an external device driving the pin to the invalid state or the application code driving the pin incorrectly. To discover which is the case, an Ohm meter check is performed before and after isolating from the circuit. This is known as *floating* the pin and can be done by desoldering or removing a component from its socket.

When you are looking around the circuit, it is very important to not make any assumptions about what you think the problem is. I call this state of mind as being "totally naïve," and looking at what is there without prejudice or expectation is the only way you will see the problem. If you don't try to look at everything objectively, you will probably miss the shorted pins, incorrect wiring, or incorrectly typed instructions. Being totally naïve is a skill and I've only met one or two people over the course of my life that have truly mastered it. As humans, we tend to remember what we expect and tend to see that first and miss what is actually in front of us.

From this check of the robot circuitry, you will have either found a problem with the circuit or are going to have to look at the software running in the application. The first thing I do is to pull the microcontroller out of the circuit and check it on a programmer against the latest application code. The problems that are typically found here are programming errors (often they were not programmed at all—somehow a step was skipped) or the configuration fuses are set wrong (wrong oscillator type, *watchdog timer* [WDT] incorrectly enabled, and so on).

If the circuit and microcontroller programming are good, you will then have to check your application code, but with the knowledge that the circuit is good and the problem must lie in the software.

The important aspect of the characterization process is to document your problems. When I'm setting up a failure analysis process for a new product at work, I issue notebooks to all the technicians and instruct them in how to fill them out. This notebook is to be kept forever as it is a database of problems and what was done to find and fix them. This is true even if the failures are recorded on a computer database.

In cases where the application starts up and then fails, the importance of keeping notes detailing what has happened is critical. For example, if you have an RS-232 interface that locked up, you would have to figure out if the problem was based around time, a data value, or data volume (the number of incoming characters). Once this determination is made, then you could go ahead and look for the problem by simulating the application code with the data input that seems to cause the failure. The goal of characterizing the problem is to document exactly how the circuit is operating and what its inputs are so you can reproduce these conditions (to make sure the problem is fixed once you have made changes).

Once you have your problem characterized, you should be able to make a hypothesis on what the cause of the problem is. Before trying to fix the application, you should try to test the hypothesis out, either as a "thought experiment" or on actual hardware to see if you can reproduce the problem exactly as it was on the failing board.

Part of making your hypothesis is deciding what the problem actually is. This is done using the observations you have made of the problem circuit. I am saying this because it is not unusual to have multiple problems with an application. In some of my microcontroller applications, I have identified more than five separate problems that required debugging before I could expect the application to work. When making your hypothesis, you should first list all the observations that seem to be pertinent to the problem at hand.

In trying to find a problem, don't be afraid of creating a small application to force the problem to happen. If you can come up with an automated way of causing the error, then you have correctly characterized the hypothesized situation and have translated this knowledge to a simulator to reproduce the problem, which will give you the opportunity to test out different fixes.

With the problem characterized and with a good hypothesis of what the root cause is, you can now go about coming up with a fix. If you have done a thorough and unbiased characterization of the problem and your hypothesis fits what was observed, chances are the repair action will be obvious. What you will often find, more often than not, is that you have a poorly soldered joint or an instruction that was entered into the source code incorrectly.

Occasionally, you will find a problem where the design or the software just isn't right. If you have followed the steps in designing the application I outlined previously and have a problem, then you will probably have a situation where you incorrectly applied technology or have designed your circuitry incorrectly. In these cases, the answer will be obvious and you will be able to redesign the circuitry or rewrite the code that is causing the problem. When you are redesigning the functions that do not work correctly, you will find that the time spent on characterizing the problem will be invaluable, as it will help you understand exactly what is happening and what you can do to fix the problem.

On some rare occasions, you will find interaction problems, like two sensors that interfere with each other (such as the IR collision detection circuitry and the IR remote control receiver). In these cases, you will have to go back and look at how you have created your application code (as well as look to see if any problems exist in your circuitry). When you work out the fix for this situation, you will discover that both the characterization as well as the hypothesis that you came up with earlier will be invaluable in helping you decide upon a fix that will solve the problem and enable the multiple functions to coexist.

Note that in all of these cases, you should go through the qualification plan again that you created for the robot. The reason for doing this is to ensure that the fix you have made will not affect any other functions of the robot. There's nothing so frustrating as to discover that once you have fixed one problem, two others pop up in its place. If something like this happens, you will have to go back and work at understanding where the difficulty lies and come up with either a new fix or a new hypothesis as to what the problem actually is.

Adding Features, Functions, and Peripherals

Once you have your robot working, I'm sure you will want to add or try out different functions or peripherals to your robot. Chances are that mechanically or electronically you will not have any problems, or if you do, they will be easily resolvable. The biggest problem that most robot designers have to come to grips with when they are enhancing their robot is how the microcontroller will handle the changes. The issues here consist of the software that is running in the robot and the hardware resources built into the microcontroller.

Most developers just look at the microcontroller's pins when the term resources is discussed, but there are also the resources inside the controller as well. These include program memory, variable memory, timers, and interrupt handlers. Further complicating the problem is that many of these resources are already used within the microcontroller, and for the enhancements to work, the allocation of the internal microcontroller resources will have to be changed.

This problem is fairly unique to robots because the designs tend to be very dynamic and experimentally based. It isn't unusual for somebody to get the robot working, only to decide that there is a better way of doing something.

Further complicating the task of upgrading a robot application is the use of other people's code and circuits that were picked up from the Internet. The code source may have widely varying authors and styles, and may require substantial rewrites for even simple additions to the robot. This is why I discussed the importance of thinking through the operation of creating the robot's software.

A well-designed robot software architecture will enable the addition of new functions and features quite easily. At the other end of the scale, a poorly designed robot architecture will require a complete rewrite if new features are added or changed in any way. This task becomes almost impossible if the code came from a variety of sources and the source may not be available.

To avoid these problems, I try to create the robot control software right from the start to allow for changes. What works best for me is a central, timer-based interrupt that can be used to sequence the various events. The 1-millisecond timer interrupt that was demonstrated in the example code seems to be a good compromise in terms of providing a fast interrupt for sequencing mechalogic interfaces. If virtually all hardware interfacing is done in the 1-millisecond interrupt timing routine and the number of instructions or cycles used for each interface is kept to a minimum, you will find that you can add new interfaces with very little danger of having to rewrite the application.

Even with complex biologic code, you can use a majority of the instruction cycles available in the interrupt handler for interfacing. To avoid any possibility of problems, I recommend that you have at least 25 percent of the cycles in between timer interrupts. This spacing will enable the biologic code to work without any serious delays as well as enable nontimer-based interrupts to execute without resulting in missed interrupts. Remember that the interrupt handler is just for executing the sequenced interface code; complex operations and data processing should be done outside the interrupt handler as either biologic or elelogic code.

I used this philosophy when designing the code for the two robot bases and it worked out very well, except for the IR remote control. This interface used a hardware input that was used by another peripheral that complicated the development effort. Ideally, all interface functions should use their own hardware.

One comment on using code from other sources must be made. You should only use modules from other people if you can get the source code that goes with them. I've seen a number of cases where authors try to protect "their assets" and only release object code to their peripheral interfaces. The problem with taking the object code is that down the road when you are modifying or upgrading your robot, chances are you will discover that the code does not work well with the new devices. If you do not have the source code to the function, then you will have to try and figure out why it is a problem and modify *your* code to implement a fix. If you have the source code to the interface module, then you will find that it is quite a bit easier to modify it to work with your enhanced robot. It may even be advantageous for you to have the source code when you first create the robot, because it will give you more flexibility when you are making your initial specifications.

It is important to remember that you can always add more chips when you are modifying your robot. To allow for the addition of other microcontrollers, it is a good idea to leave two pins free as a synchronous bus. The advantage of a synchronous bus is the elimination of relying on precise timing required for *nonreturn to zero* (NRZ) serial or Manchester encoding.

When building multiple microcontrollers in your robot application, you should just have one microcontroller responsible for the biologic functions. Having multiple controllers executing high-level functions will make the software much more complex and very difficult to debug when the perceived inputs by the two microcontrollers do not agree and the two microcontrollers respond in conflicting ways.

One last point when enhancing your robot application: always remember that sometimes drive motors should be turned off during peripheral operation. IR light- and sound-level sensors should only be polled when the motors are off to ensure that the power is as clean as possible and that motor or gear noise is at a minimal level.

■ APPENDIX A ■

Glossary

I've tried to give a complete list of all the acronyms, terms, and expressions that may be unfamiliar to you. Acronyms are expanded before they are described.

active components   Generally, integrated circuits and transistors. Active components require external power to operate.

ADC (analog-to-digital converter)   Hardware devoted to converting the value of a DC voltage into a digital representation. *See* DAC.

address   The location of a register, RAM byte, or instruction word within a processor's memory space.

alkaline   Alkaline technology radio batteries. 1.75 to 2.25 volts per cell output.

amps   A measure of current. One amp is the movement of 1 coulomb of electrons in 1 second.

analog   A quantity at a fractional value rather than a binary value, 1 or 0. Analog voltages are the quantity most often measured.

AND   A logic gate that outputs a 1 when all inputs are 1.

anode   Positive connection of an electronic device. *See* Cathode.

array   A collection of variables that can be addressed by an arbitrary index.

ASCII (American Standard Character Interchange Interface)   The bit-to-character representation standard most used in computer systems.

ASCIIZ   A string of ASCII characters that ends in a null (0x000) byte.

assembler   A computer program that converts assembly language source to object code. *See* cross assembler.

assembly language   A set of word symbols used to represent the instructions of a processor. Along with a primary instruction, there are parameters that are used to specify values, registers, or addresses.

asynchronous serial   Data sent serially to a receiver without clocking information. Instead, data synching information for the receiver is available inside the data packet or as part of each bit.

bare board   *See* raw card.

BCD (Binary Code Decimal)   The use of four bits to represent a decimal number (0 to 9).

binary numbers   Numbers represented as powers of 2. Each digit is 2 raised to a specific power. For example, 37 decimal is $32 + 4 + 1 = 2^4 + 2^2 + 2^0 = 00010101$ binary. Binary can be represented in the following forms: 0b0nnnn, B'nnnn,' or %nnnn where nnnn is a multidigit binary number comprised of 1's and 0's.

bipolar logic Logic circuits made from bipolar transistors (either discrete devices or integrated onto a chip). *See* NPN transistor and PNP transistor.

bit banging Simulating interface functions with code.

bit mask A bit pattern that is ANDed with a value to turn off specific bits.

bounce Spurious signals in a changing line. Most often found in mechanical switch openings and closings.

burning *See* programming.

bus An electrical connection between multiple devices, each using the connection for passing data.

capacitor A device used for storing electrical charge. Often used in microcontroller circuits for filtering signals and input power by reducing transient voltages. The different types include ceramic disk, polyester, tantalum, and electrolytic. Tantalum and electrolytic capacitors are polarized. The shorter lead of a polarized capacitor is the cathode.

cathode Negative connection of an electronic device. *See* anode.

ceramic resonator A device used to provide timing signals to a microcontroller. It is more physically robust than a crystal, but with poorer frequency accuracy.

character A series of bits used to designate an alphabetic, numeric, control, or other symbol or representation. *See* ASCII.

clock A repeating signal used to run a processor's instruction sequence.

clock cycle The operation of a microcontroller's primary oscillator going from a low voltage to a high voltage, and back again. This is normally referenced as the speed at which the device runs. Multiple clock cycles may be used to make up one instruction cycle.

CMOS logic Logic circuits made from N-channel and P-channel Metal Oxide Silicon Field Effect Transistors (MOSFET) devices, either discrete devices or integrated onto a chip.

comparator A device that compares two voltages and returns a logic 1 or 0 based on the relative values.

compiler A program that takes a high-level language source file and converts it to either assembly language code or object code for a microcontroller. *See* cross compiler.

concatenate Joining two pieces of data together to form a single, contiguous piece of data.

constant A numeric value used as a parameter for an operation or instruction. This differs from a variable value that is stored in a RAM or register memory location.

contiguous When a set amount of data is placed together in memory and can be accessed sequentially using an index pointer, it is said to be contiguous. Data is noncontiguous if it is placed in different locations that cannot be accessed sequentially.

control store *See* program memory.

coulomb The charge produced by 1.6×10^{19} electrons.

CPU (central processing unit) What I refer to as the microcontroller's processor.

cross assembler A program written to take assembly language code for one processor and convert it to object code while working on an unrelated processor and operating system. *See* assembler.

cross compiler A program written to convert high-level code to object code while working on an unrelated processor and operating system. *See* compiler.

crystal A device used for precisely timing the operation of a microcontroller.

current The measurement of the number of electrons that pass by a point each second. The units are called amps, which represent coulombs per second.

DAC (digital-to-analog converter) Hardware designed to convert a digital representation of an analog DC voltage into that analog voltage. *See* ADC.

DCE (Data Communications Equipment) The RS-232 standard by which modems are usually wired. *See* DTE.

debounce Removing spurious signals in a "noisy" input.

debugger A program used by an application programmer to find the problems in the application. It is normally used for the target system on which the application will run.

decimal numbers Base 10 numbers used for constants. These values are normally converted into hex or binary numbers for the microcontroller.

decoupling capacitor A capacitor placed across the Vcc and ground of a chip to reduce the effects of increased/decreased current draws from the chip.

demultiplexor A circuit for passing data from one source to an output that can be specified from a number of sources. *See* demux, multiplexor, and mux.

demux An abbreviation for demultiplexor.

digital A term used to describe a variety of logic families where values are either high (1) or low (0). For traditional logic families, the voltage levels are either approximately 0 volts or approximately 5 volts with a switching level somewhere between 1.4 and 2.5 volts.

DIP Acronym for "Dual In-line Package." Proper term for a pin through hole electronic chip.

driver Any device that can force a signal onto a net. *See* receiver.

D-Shell connectors A style of connector often used for RS-232 serial connections as well as other protocols. The connector is D shaped to provide a method of polarizing the pins and ensuring that they are connected correctly.

DTE (Data Terminal Equipment) The RS-232 pinout standard used by a PC's serial port. *See* DCE.

duty cycle In a pulse-wave-modulated digital signal, the duty cycle is the fraction of time that the signal is high over the total time of the repeating signal.

edge triggered Logic that changes based on the shift in a digital logic level. *See* level sensitive.

Editor A program located on your development system that is used to modify application source code.

EEPROM (electrically erasable programmable read-only memory [also known as Flash]) Nonvolatile memory that can be erased and reprogrammed electrically. (That is, it doesn't require the ultraviolet light of EPROM.)

emulator An electrical circuit connected to the development system that allows the application to be executed under the developer's control, enabling observation of how well it's working and the testing of changes to the application.

EPROM (erasable programmable read-only memory) Nonvolatile memory that can be electrically programmed and erased using ultraviolet light.

event-driven programming Application code that waits for external events to process before executing.

FIFO (first-in—first-out) Memory that will retrieve data in the order in which it was stored.

Flash A type of EEPROM. Memory that can be electrically erased in blocks, instead of as individual memory locations. Flash is very unusual in microcontrollers; many manufacturers describe their devices as having Flash when, in actuality, they use EEPROM.

flip flop A basic memory cell that can be loaded with a specific logic state and read back. The logic state will be stored as long as power is applied to the cell.

floating The term used to describe a pin that has been left unconnected and is floating relative to the ground.

frequency The number of signal repetitions that can take place in a given period of time (typically, 1 second). *See* period and hertz.

FTP (File Transfer protocol) A method of transferring files to/from the Internet.

function A subroutine that returns a parameter to the caller.

fuzzy logic A branch of computer science in which decisions are made on partially correct data rather than correct-or-incorrect data, which is used by digital logic. These decisions are often made for controlling physical and electronic systems. *See* PID.

ground Negative voltage to the microcontroller/circuit, also referred to as Vss for CMOS logic.

GUI (graphical user interface) Often pronounced "gooey," a GUI is used, along with a graphical operating system (such as Microsoft Windows), to provide a simple, consistent computer interface for users that includes a screen, keyboard, and mouse.

hertz A unit of measurement for frequency. One hertz (or Hz) means that an incoming signal is oscillating once per second.

hex number A value from 0 to 15 that is represented using four bits, or the characters 0 through 9 and A through F.

high-level language A set of English (or other human language) statements that have been formatted for use as instructions for a computer. Some popular high-level languages used for robot microcontrollers include C, BASIC, Pascal, and Forth.

hysteresis The characteristic response to input that causes the output response to change, based on that input. It is typically used in microcontroller projects to debounce input signals.

Hz *See* hertz.

I2C (Inter-Intercomputer Communication) A synchronous serial network protocol enabling microcontrollers to communicate with peripheral devices and each other. Only common lines are required for the network.

index register An 8- or 16-bit register whose contents can be used to point to a location in variable storage, control store, or the microcontroller's register space. *See* stack pointer.

inductor Wire wrapped around some kind of form (metal or plastic) to provide a magnetic field for storing energy. Inductors are often used in oscillator and filtering circuits.

infrared (IR) A wavelength of light (760 nanometers or longer) that is invisible to the human eye. It is often used for short-distance communications or control applications.

interpreter A program that reads application source code and executes it directly, rather than compiling it.

interrupt An event that causes the microcontroller's processor to stop what it is doing and respond.

instruction A series of bits that is executed by the microcontroller's processor to perform a basic function.

instruction cycle The minimum amount of time needed to execute a basic function in a microcontroller. One instruction cycle typically takes several clock cycles. *See* clock cycle.

in-system programming The capability to program a microcontroller's control store while the device is in the final application's circuit without having to remove it.

I/O space   An address space totally devoted to providing access to I/O device control registers.

KHz   This is an abbreviation for kilohertz (one thousand cycles per second).

label   An identifier used within a program to denote the location of a control store or register address. *See* variable.

latency   The time or number of cycles required for hardware to respond to a change in input.

LCD (liquid crystal display)   A device used for outputting information from a microcontroller. It is typically controlled by a Hitachi 44780 controller, although some microcontrollers contain circuitry for interfacing to an LCD directly without an intermediate controller circuit.

LED (light-emitting diode)   A diode (rectifier) device that will emit light of a specific frequency when current is passed through it. When used with microcontrollers, LEDs are usually wired with the anode (positive pin) that is connected to Vcc and the microcontroller I/O pin sinking current (using a series 200 Ohm to 1K resistor), enabling the LED to turn on. For typical LEDs in hemispherical plastic packages, the flat side (which has the shorter lead) is the cathode.

level conversion   The process of converting logic signals from one family to another.

level sensitive   Logic that changes based on the state of a digital logic signal. *See* edge triggered.

LIFO (last-in—first-out)   A type of memory in which the most recently stored data will be the first retrieved.

linker   A software product that combines object files into a final program file that can then be loaded into a microcontroller.

list server   An Internet server used to distribute common interest mail to a number of individuals (also known as a ListServ).

logic analyzer   A tool that will graphically show the relationship of the waveforms for a number of different pins.

logic gate   A circuit that outputs a logic signal based on input logic conditions.

logic probe   A simple device used to test a line for being in a high, low, transitioning, or high-impedance state.

macro   A programming construct that replaces a string of characters (and parameters) with a previously specified block of code or information.

Manchester encoding   A method for serially sending data that does not require a common (or particularly accurate) clock.

mask-programmable ROM   A method of programming memory that takes place during the final assembly of a microcontroller. When the final set of aluminum traces of a chip are laid down, a special photographic mask is made to create wiring that will result in a specific program being read from a microcontroller's control store.

master In microcontroller and external device networking, a master is a device that initiates and can control the transfer of data. *See* multimaster and slave.

MCU An abbreviation for microcontroller.

memory A circuit designed to store instructions or data.

memory array A collection of flip flops arranged in a matrix format that enables consistent addressing.

memory-mapped I/O A method of placing peripheral registers in the same memory space as RAM or variable registers.

MHz This is an abbreviation for megahertz (one million cycles per second).

microwire A synchronous serial communications protocol.

MIPS (millions of instructions per second) This acronym should really stand for misleading indicator of performance and should not be a major consideration when deciding which microcontroller to use for a robot application.

monitor A program used to control the execution of an application inside a processor.

MPU The acronym/abbreviation for microprocessor.

msec An abbreviation for millisecond (one thousandth of a second, or 0.001 seconds). *See* nsec and usec.

multimaster A microcontroller networking philosophy that enables multiple masters on the network bus to initiate data transfers.

multiplexor A device for selecting and outputting a single stream of data from a number of incoming data sources. *See* demultiplexor, demux, and mux.

mux An abbreviation for multiplexor.

Negative Active logic A type of logic where the digital signal is said to be asserted if it is at a low (0) value. *See* Positive Active logic.

nesting Placing subroutine or interrupt execution within the execution of other subroutines or interrupts.

net A technical term for the connection of device pins in a circuit. Each net consists of all the connections to one device pin in a circuit.

Net, the A colloquial term for the Internet.

NiCad An abbreviation for nickel-cadmium batteries. These batteries are rechargeable, although typically provide only 1.2 volts per cell output, compared to 1.5 to 2.0 volts for standard "dry" or alkaline radio batteries.

NiMH An abbreviation for nickel-metal hydride batteries. These batteries have characteristics similar to NiCads, but the chemicals used are not as toxic.

NPN transistor Electronic switch which is turned on when current is injected into its "base."

NMOS (negative-channel metal-oxide semiconductor) logic A digital logic where only N-channel MOSFET transistors are used.

noise High-frequency variances in a signal line that are caused by switch bounce or electrical signals picked up from other sources.

NOT A logic gate that inverts the state of the input signal (for example, 1 NOT is 0).

nsec One billionth of a second, or 0.000000001 seconds. *See* usec and msec.

object file After assembly or high-level language compilation, a file is produced with the hex values (op codes) that make up a processor's instructions. An object file can be loaded directly into a microcontroller, or multiple object files can be linked together to form an executable file that is loaded into a microcontroller's control store. *See* linker.

octal numbers Numbers represented as the digits from 0 through 7. This method of representing numbers is not widely used, although some high-level languages, such as C, have made it available to programmers.

one's complement The result of XORing a byte with 0x0FF that will invert each bit of a number. *See* two's complement.

op codes The hex values that make up the processor instructions in an application.

open collector/drain output An output circuit consisting of a single transistor that can pull the net to which it is connected to ground, but not to source current.

OR A basic logic gate where if any input is set to 1, a 1 is output.

oscillator A circuit used to provide a constant frequency repeating signal for a microcontroller. This circuit can consist of a crystal, ceramic resonator, or resistor-capacitor network, providing the delay between edge transitions. The term is also used for a device that can be wired to a microcontroller to provide clocking signals without having to provide a crystal, caps, and other components to the device.

oscilloscope An instrument that is used to observe the waveform of an electrical signal. The two primary types of oscilloscopes in use today are the analog oscilloscope, which writes the current signal onto the phosphers of a cathode ray tube (CRT), and the digital storage oscilloscope, which saves the analog values of an incoming signal in RAM for replaying on either a built-in CRT or a computer connected to the device.

OTP (one-time programmable) This term generally refers to a device with EPROM memory encased in a plastic package that does not allow the chip to be exposed to ultraviolet light. Note that EEPROM devices in a plastic package may also be described as OTP when they can be electrically erased and reprogrammed.

parallel This term describes passing data between devices with all the data bits being sent at the same time on multiple lines. This is typically much faster, but more electrically complex than transferring data serially.

parameter A user-specified value for a subroutine or macro. A parameter can be a numeric value, a string, or a pointer depending on the application.

passive components Generally, resistors, capacitors, inductors, and diodes. These are components that do not require a separate power source to operate.

PCA (printed-circuit assembly) A bare board with components (both active and passive) soldered onto it.

PCB (printed-circuit board) *See* raw card.

PDF files Files suitable for viewing with Adobe Postscript.

period The length of time that a repeating signal takes to go through one full cycle (the reciprocal of frequency).

PID (parallel integrating differential) A classical method of controlling physical and electronic systems. *See* fuzzy logic.

ping The operation of sending a message to a device to see if it is operating properly.

PNP transistor Electronic switch which is turned on when current is drawn from its "base."

pointer A variable that indexes to another variable in memory that contains data.

poll A programming technique in which a bit (or byte) is repeatedly checked until a specific value is found.

pop The operation of taking data off a stack memory.

port The task of moving code from one processor or programming language to another.

Positive Active logic Logic that becomes active when a signal becomes high (1). *See* Negative Active logic.

PPM An abbreviation for parts per million. An easy way to calculate the PPM of a value is to divide the value by the total number of samples or opportunities, and then multiply by 1 million. One percent is equal to 10,000 PPM; 10 percent is equal to 100,000 PPM.

Princeton architecture A computer processor architecture that uses one memory subsystem for instructions (control store) memory, variable memory, and I/O registers. *See* Harvard architecture and Von Neumann.

program counter A counter within a computer processor that keeps track of the current program execution location. This counter can be updated by the current application code and can have its contents saved/restored on a stack.

program memory   Also known as program storage or program store, this is memory (usually nonvolatile) devoted to saving the application program for when the microcontroller is powered down.

programming   Loading a program into a microcontroller control store, also referred to as burning.

PROM (programmable read-only memory)   This was originally an array of fuses that were blown to load a program. Now, PROM can refer to EPROM in an OTP package.

pull-down   A resistor (typically 100 to 500 Ohms) that is wired between a microcontroller pin and a ground. This method is not recommended to hold an input at a specific state. Instead, a pull-up driving an inverter should be used. *See* pull-up.

pull-up   A resistor (typically 1 to 10K Ohms) that is wired between a microcontroller pin and a Vcc. A switch pulling the signal to ground the microprocessor pin can be used to provide user input. *See* pull-down.

push   The operation of putting data onto a stack memory.

PWB (printed-wiring board)   *See* raw card.

PWM (pulse-width modulation)   A digital output technique where a single line is used to output analog information by varying the length of time that a pulse is active on the line.

RAM (random access memory)   Memory that you can write to and read from. In microcontrollers, virtually all RAM is static RAM (SRAM), which means that data is stored within it as long as power is supplied to the circuit. Dynamic RAM (DRAM) is very rarely used in microcontroller applications. EEPROM can be used for nonvolatile RAM storage.

raw card   A fiberglass board with copper traces attached to it enabling components to be interconnected (also known as PCB, PWA, and bare board).

R/C   An abbreviation for Remote Control. In this book, the term "R/C" is used to represent parts (such as servos) designed for remote control cars, airplanes and boats.

RC   An abbreviation for resistor/capacitor. This is a network used to provide a specific delay for a built-in oscillator or reset circuit.

receiver   A device that senses the logic level in a circuit. A receiver cannot drive a signal. *See* driver.

recursion   A programming technique where a subroutine calls itself with modified parameters to carry out a task. This technique is not recommended for microcontrollers that might have a limited stack.

reenterent   A subroutine/API/function/interrupt handler that is written in such a way that it can be called by another routine while it is active, executing a request from another.

register A memory address devoted to saving a value (like RAM) or providing a hardware interface for the processor.

relocatable Code that is written or compiled in such a way that it can be placed anywhere in the control store memory map after assembly, and run without any problems.

resistor A device used to limit current in a circuit.

reset The operation of placing a microcontroller in a known state before allowing it to execute.

ROM (read-only memory) This type of memory is typically used for control store because it cannot be changed by a processor during the execution of an application. Mask-programmable ROM is specified by the chip manufacturer to build devices with certain software as part of the device and cannot be programmed in the field.

rotate A method of moving bits within single or multiple registers. No matter how many times a rotate operation or instruction is carried out, the data in the registers will not be lost. *See* shift.

RS-232 An asynchronous serial communications standard. The normal logic level for a 1 is −12 volts, and for a 0 it is +12 volts.

RS-485 A differential pair, TTL voltage-level communications system.

RTOS (real-time operating system) A program that controls the operation of an application.

scan The act of reading through a row of matrix information for data rather than interpreting the data as individual units.

serial Passing multiple bits using a serial line, one at a time. *See* parallel.

servo A device that converts an electrical signal into mechanical movement. A radio-control modeler's servos are often interfaced to microcontrollers. In these devices, the position is specified by a 1- to 2-millisecond pulse every 20 milliseconds.

shift A method of moving bits within single or multiple registers. After a shift operation, bits are lost. *See* rotate.

simulator A program used to debug applications by simulating the operation of the microcontroller.

slave In microcontroller networking, a device that does not initiate communications but does respond to the instructions of a master.

software The term used for the application code that is stored in the microcontroller's program memory. Some references might use the term firmware for this code.

source code Human-readable instructions used to develop an application. Source code is converted by a compiler or assembler into instructions that the processor can execute and is stored in a .hex file.

SPI   Acronym for "Synchronous Peripheral Interface." SPI is a synchronous serial communications protocol.

splat   Another name for asterisk (*). It's easier to say and spell, and funnier than asterisk.

SRAM (static random access memory)   A memory array that will not lose its contents while power is applied.

stack   LIFO memory used to store program counter and other context register information.

stack pointer   An index register available within a processor that is used for storing data and updating itself, enabling the next operation to be carried out with the index pointing to a new location.

state analyzer   A tool used to store and display state data on several lines. Rather than requiring a separate instrument, this is often an option available in many logic analyzers.

state machine   A programming technique that uses external conditions and state variables to determine how a program will execute.

string   A series of ASCII characters saved sequentially in memory. When ended with 0x000 to indicate the end of the string, it is known as an ASCIIZ string.

subroutine   A small application program devoted to carrying out one task or operation. It is usually called repeatedly by other subroutines or the application mainline.

synchronous serial   Data transmitted serially along with a clocking signal that is used by the receiver to indicate when the incoming data is valid.

task   A small, autonomous application that is similar in operation to a subroutine, but that can execute autonomously to other application tasks or to mainline.

timer   A counter incremented by either an internal or external source. Often used to time events, rather than counting instruction cycles.

traces   Electrical signal paths etched in copper in a printed circuit card.

transistor   An electronic device by which current flow can be controlled.

TTY   Serial Teletype interface protocol. Simple serial communications interface with no text or graphic formatting capabilities.

two's complement   A method for representing positive and negative numbers in a digital system. To convert a number to a two's complement negative, it is complemented (converted to one's complement) and incremented.

UART (universal asynchronous receiver/transmitter)   Peripheral hardware inside a microcontroller used to asynchronously communicate with external devices. *See* USART and asynchronous serial.

ultrasonic   Sound at frequencies above human hearing (20 KHz and above).

USART (universal synchronous/asynchronous receiver/transmitter)
Peripheral hardware inside a microcontroller used to synchronously (using a clock signal either produced by the microcontroller or provided externally) or asynchronously communicate with external devices. *See* UART and synchronous serial.

μsec One millionth of a second, or 0.000001 seconds. *See* nsec and msec.

UV light (ultraviolet light) Light at shorter wavelengths than the human eye can see. UV light sources are often used with windowed microcontrollers with EPROM control store for erasing the contents of the control store.

variable A label used in an application program that represents an address containing the actual value to be used by the operation or instruction. Variables are normally located in RAM and can be read from or written to by a program.

Vcc Positive power voltage applied to a microcontroller/circuit. Generally, 2.0 to 6.0 volts, depending on the application (also known as Vdd).

Vdd *See* Vcc.

vias Holes in a printed circuit card.

volatile RAM is considered to be volatile because when power is removed, the contents are lost. EPROM, EEPROM, and PROM are considered to be nonvolatile because the values stored in the memory are saved, even if power is removed.

voltage The amount of electrical force placed on a charge.

voltage regulator A circuit used to convert a supply voltage into a level useful for a circuit or microcontroller.

volts Units of voltage.

Vss *See* ground.

watchdog timer (WDT) A timer used to reset a microcontroller upon overflow. The purpose of the watchdog timer is to return the microcontroller to a known state if the program begins to run errantly (that is, amok).

wattage The measure of power consumed. If a device requires one amp of current with a one-volt drop, one watt of power is being consumed.

word The basic data size used by a processor. In all the PICmicro microcontroller families, the word size is eight bits.

XOR A logic gate that outputs a 1 when the inputs are at different logic levels.

ZIF (zero insertion force) ZIF sockets will allow the plugging/unplugging of devices without placing stress on the device's pins.

.ZIP files Files combined and compressed into a single file using the PKZIP program by PKWARE, Inc.

Useful Tables and Data

■■■■ **Physical Constants**

Symbol	Value	Description
AU	149.59787x(10^6) km 92,955,628 miles	Astronomical unit (distance from the Sun to the earth)
c	2.99792458x(10^8) m/s 186,282 miles/s	Speed of light in a vacuum
e	2.7182818285	
Epsilon-o	8.854187817x(10^{-12}) F/m	Authorization of free space
Ev	1.60217733x(10^{-19}) J	Electron volt value
g	32.174 ft/sec^2 9.807 m/sec^2	Acceleration due to gravity
h	6.626x(10^{34}) Js	Planck constant
k	1.380658x(10^{-23}) J/K	Boltzmann entropy constant
me	9.1093897x(10^{-31}) kg	Electron rest mass
mn	1.67493x*10^{-27}) kg	Neutron rest mass
mp	1.67263x(10^{-27}) kg	Proton rest mass
pc	2.06246x(10^5) AU	Parsec
pi	3.1415926535898	Ratio of circumference to diameter of a circle
R	8.314510 J/(K * mole)	Gas constant
sigma	5.67051x(10^{-8}) W/ (m**2 * K^4)	Stefan-Boltzmann constant
u	1.66054x(10^{-27}) grams	Atomic mass unit
mu-o	1.25664x(10^{-7}) N/A^2	Permeability of vacuum
Mach 1	331.45 m/s 1087.4 ft/s	Speed of sound at sea level in dry air at 20C
None	1480 m/s 4856 ft/s	Speed of sound in water at 20C

▓▓▓ Audio Notes

The following table outlines notes around middle C. Note that an octave above is twice the note frequency and an octave below is one half-note frequency.

Note	Frequency (Hz)
G	392 Hz
G#	415.3 Hz
A	440 Hz
A#	466.2 Hz
B	493.9 Hz
C	523.3 Hz
C#	554.4 Hz
D	587.3 Hz
D#	622.3 Hz
E	659.3 Hz
F	698.5 Hz
F#	740.0 Hz
G	784.0 Hz
G#	830.6 Hz
A	880.0 Hz
A#	932.3 Hz
B	987.8 Hz

Touch-Tone Telephone Frequencies

Frequency	1209 Hz	1336 Hz	1477 Hz
697 Hz	1	2	3
770 Hz	4	5	6
852 Hz	7	8	9
941 Hz	*	0	#

Electrical Engineering Formulas

This section's formulas include the following abbreviations:

V = Voltage

I = Current

R = Resistance

C = Capacitance

L = Inductance

Ohm's Law:

$V = IR$

Power:

$P = VI$

Series Resistance:

Rt = R1 + R2 ...

Parallel Resistance:

Rt = 1/((1/R1) + (1/R2) ...)

Two Resistors in Parallel:

Rt = (R1 * R2)/(R1 + R2)

Series Capacitance:

Ct = 1/((1/C1) + (1/C2) ...)

Parallel Capacitance:

Ct = C1 + C2 ...

Wheatstone Bridge:

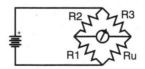

Figure B-1
Wheatstone
bridge
operation

Ru = R1 * R3 / R2

When NoCurrent Flow
In the Meter

Resonance:

Frequency = 1/(2 * pi * SQRT(L * C))

RC Time Constant:

Tau = R * C

RL Time Constant:

Tau = L/R

RC Charging:

$V(t) = Vf * (1 - e^{(-t/Tau)})$
$i(t) = if * (1 - e^{(-t/Tau)})$

RC Discharging:

$V(t) = Vi * e^{(-t/Tau)}$
$i(t) = ii * e^{(-t/Tau)}$

Transformer Current/Voltage:

Turns ratio = Number of turns on primary ("p") side/number of turns on secondary ("s") side

Turns ratio = Vs/Vp = Ip/Is

Transmission Line Characteristic Impedance:

Zo = SQRT(L/C)

Mathematical Formulas

Frequency = Speed/wavelength

For Electromagnetic Waves:

Frequency = c/wavelength

Perfect Gas Law:

PV = nRT

Boolean Arithmetic

Identify Functions:

A AND 1 = A
A OR 0 = A

Output Set/Reset:

A AND 0 = 0
A OR 1 = 1

Identity Law:

A = A

Double Negation Law:

NOT(NOT(A)) = A

Complementary Law:

A AND NOT(A) = 0
A OR NOT(A) = 1

Idempotent Law:

A AND A = A
A OR A = A

Commutative Law:

A AND B = B AND A
A OR B = B OR A

Associative Law:

(A AND B) AND C = A AND (B AND C) = A AND B AND C
(A OR B) OR C = A OR (B OR C) = A OR B OR C

Distributive Law

A AND (B OR C) = (A AND B) OR (A AND C)
A OR (B AND C) = (A OR B) AND (A OR C)

De Morgan's Theorem

NOT(A OR B) = NOT(A) AND NOT(B)
NOT(A AND B) = NOT(A) OR NOT(B)

NOTE:

- AND is often represented as multiplication, with nothing between the terms, or the period (.) or asterisk (*) characters between them.
- OR is often represented as addition with a plus sign (+) between terms.
- NOT is indicated with a hyphen (-) or an exclamation point (!) character before the term. A tilde (~) is usually used to indicate a multibit, bitwise inversion.

Mathematical Conversions

1 inch = 2.54 centimeters

1 mile = 1.609 kilometers

1 ounce = 29.57 grams

1 U.S. gallon = 3.78 liters

1 atmosphere = 29.9213 inches of mercury

= 14.6960 pounds per square inch

= 101.325 kilopascals

10,000,000,000 angstroms = 1 meter

1,000,000 microns = 1 meter

Tera = 1,000 giga

Giga = 1,000 mega

Mega = 1,000 kilo

kilo = 1,000 units

Unit = 100 centi

Unit = 1,000 milli

1 hour = 3,600 seconds

1 year = 8,760 hours

ASCII

The ASCII definition uses the seven bits of each ASCII character.

3-0	6-4 ->	Control Characters			Data Characters					
		000	001		010	011	100	101	110	111
V										
0000		NUL	DLE		Space	0	@	P	`	p
0001		SOH	DC1		!	1	A	Q	a	q
0010		STX	DC2		"	2	B	R	b	r
0011		ETX	DC3		#	3	C	S	c	s
0100		EOT	DC4		$	4	D	T	d	t
0101		ENQ	NAK		%	5	E	U	e	u
0110		ACK	SYN		&	6	F	V	f	v
0111		BEL	ETB		'	7	G	W	g	w
1000		BS	CAN		(	8	H	X	h	x
1001		HT	EM		)	9	I	Y	I	y
1010		LF	SUB		*	:	J	Z	j	z
1011		VT	ESC		+	;	K	[	k	{
1100		FF	FS		,	<	L	\	l		
1101		CR	GS		-	=	M	]	m	}
1110		SO	RS		.	>	N	^	n	~
1111		SI	US		/	?	O	_	o	DEL

ASCII Control Characters

The ASCII control characters were specified as a means of allowing a computer to communicate and control how data is displayed. Normally, a carriage return or line feed is used to indicate the start of a line. Null is used to indicate the end of an ASCII string. Backspace will move the cursor back one column to the start of the line. The Bell character, when sent to MS-DOS, will cause the PC's speaker to beep. Horizontal tab is used to move the cursor to the start of the next column that is evenly distributed by eight. Form feed is used to clear the screen.

Hex	Mnemonic	Definition
00	NUL	Null, which is used to indicate the end of a string.
01	SOH	Message's start of header.
02	STX	Message's start of text.
03	ETX	Message's end of text.
04	EOT	End of transmission.
05	ENQ	Enquiry for identification or information.
06	ACK	Acknowledges the previous transmission.
07	BEL	Rings the bell.
08	BS	Backspace, which moves the cursor to the left.
09	HT	Horizontal tab, which moves the cursor to the right to the next tab stop (normally a column evenly divisible by eight).
0A	LF	Line feed, which moves the cursor down one line.
0B	VT	Vertical tab, which moves the cursor down to the next tab line.
0C	FF	Form feed up to the start of the new page. For CRT displays, this is often used to clear the screen.
0D	CR	Carriage return, which moves the cursor to the leftmost column.
0E	SO	The next group of characters does not follow ASCII control conventions so they are *shifted out* (SO).
0F	SI	The following characters follow the ASCII control conventions and are *shifted in* (SI).
10	DLE	Data link escape, the ASCII control character start of an escape sequence. In most modern applications, escape (0x01B) is used for this function.
11	DC1	Not defined, normally application-specific.
12	DC2	Not defined, normally application-specific.
13	DC3	Not defined, normally application-specific.

Hex	Mnemonic	Definition
14	DC4	Not defined, normally application-specific.
15	NAK	Negative acknowledge, which means the previous transmission was not properly received.
16	SYN	Synchronous idle. If the serial transmission uses a synchronous protocol, this character is sent to ensure the transmitter and receiver remain synched.
17	ETB	End of transmission block.
18	CAN	Cancel and disregard the previous transmission.
19	EM	End of medium, which indicates the end of a file. For MS-DOS files, 0x01A is often used instead.
1A	SUB	Substitute the following character with an incorrect one.
1B	ESC	Escape, which is used to temporarily halt the execution or put an application into a mode to receive information.
1C	FS	Marker for file separation of data being sent.
1D	GS	Marker for group separation of data being sent.
1E	RS	Marker for record separation of data being sent.
1F	US	Marker for unit separation of data being sent.

APPENDIX C

Microcontroller Selection Table

The following table lists some of the most common microcontrollers used for robots. The vast majority of these parts are 8-bit microcontrollers that can be bought for just a few dollars.

The information presented here lists the information that I consider critical to choosing a microcontroller for use in a robot. At the start of each different manufacturer's product, I have included the manufacturer's name and do not repeat it. Note that there is a wide variability in some devices of the same family.

MCU	Number of Pins	PM Size	Var Size	Base Tool	Programmer	Comments
Microchip PIC12C5xx	6	512–1K Instr'ns	25– 41bytes	Microchip MPLAB	Serial ICSP (many home-build designs available)	Excellent for peripheral controllers. Has a built-in 4 MHz oscillator.
PIC16C505	12	512 Instr'ns	72 bytes	Microchip MPLAB	ICSP	Excellent peripheral/small robot controller. Has a built-in 4 MHz oscillator.
PIC16HV540	12	512 Instr'ns	25 bytes	Microchip MPLAB	Parallel algorithm	Has built-in voltage regulator.
PIC16C62x	13	512–2K Instr'ns	80–96 bytes	Microchip MPLAB	ICSP	Has built-in voltage comparators.
PIC16F628	16	2K Instr'ns	224 bytes	Microchip MPLAB	ICSP	Has a built-in 4 MHz oscillator.
PIC16F84(A)	13	1K Instr'ns	68 bytes	Microchip MPLAB	ICSP	Flash program memory.
PIC16C71x	13	512–2K Instr'ns	36–128 bytes	Microchip MPLAB	ICSP	Includes four ADC pins.
PIC16F87x	22, 33	8K Instr'ns	368 bytes	Microchip MPLAB	ICSP	Built-in ADC, flash program memory, and internal debug hardware (ICD).
PIC17C4x	33	8K Instr'ns	232–454 bytes	Microchip MPLAB	Parallel algorithm	Able to access external 8/16-bit bus devices.
PIC18Fxx2	23–68	16K Instr'ns	768–3840 bytes	Microchip MPLAB	ICSP	Flash program memory, built-in ADC, and internal debug hardware.

Device	I/O	Program memory	Data memory	Programming	Interface	Notes
Parallax BASIC Stamp 1	8	253 bytes	14 bytes	Parallax stampw	Parallel port cable	Flash program memory and PBASIC.
BASIC Stamp 2	16 (plus 2 serial I/O)	2K	26 bytes	Parallax stampw	RS-232	Flash program memory and PBASIC.
BASIC Stamp 2e	16 (plus 2 serial I/O)	16K	26 bytes plus 63 byte scratch	Parallax stampw	RS-232	Enhanced PBASIC.
BASIC Stamp 2sx	16 (plus 2 serial I/O)	16K	26 bytes plus 63 byte scratch	Parallax stampw	RS-232	Enhanced PBASIC/very fast MCU.
BASIC Stamp 2p	16/30 (plus 2 serial I/O)	16K	26 bytes plus 63 Byte scratch	Parallax stampw	RS-232	Enhanced PBASIC.
Motorola 68HC08	13–51	0–32K	128–1K bytes	Third party	Proprietary	Many features and packages. Optional external program memory.
Motorola 68HC11	13–50	0–32K	64–2K bytes	Third party	Proprietary	Many features and packages. Optional external program memory.
Handyboard (68HC11-based)	20	32K battery-backed RAM	Use 32K battery-backed RAM	Free/third party	RS-232	Designed for robots/good support.
OOPIC2	40	95 objects	172 bytes	OOPic language	I2C	Available with many different interfacing objects.
Atmel AT90S1200 (AVR)	15	1K bytes	64 bytes	AVR studio	Serial or parallel (home-build designs available)	Suitable for I/O subsystem controller and with Flash program memory.

MCU	Number of Pins	PM Size	Var Size	Base Tool	Programmer	Comments
AT90S8515 (AVR)	32	8K bytes	512 bytes	AVR studio	Parallel (home-build designs available)	Voltage comparators and Flash program memory.
AT90S8535 (AVR)	32	8K bytes	512 bytes	AVR studio	Parallel (home-build designs available)	ADC on I/O pins and with Flash program memory.
Intel or other 87C51	32	4K bytes	128 bytes	Third party	Third party	Basic 8051 with EPROM program memory, built-in serial port, and can access external 8-bit memory.
87C52	32	8K bytes	256 bytes	Third party	Third party	Basic 8052 with EPROM program memory, enhanced serial port, and can access 8-bit memory.
Atmel AT89Cx051 (8051)	15	1–4K bytes	128 bytes	Third party	Third party (home-build versions available)	20-pin 8051 that includes voltage comparator and USART.
Kg Systems BASIC-Tiger®	38	128K	128K	BASIC-Tiger® development kits	Serial	Analog I/O, I2C, serial, and LCD interfaces. Built-in multitasking.
PC/104	N/A	N/A	N/A	Various	Serial/Flash/EEPROM	104-pin I/O bus, standard form factor with PC-compatible processor. Memory, I/O, and other features added via I/O cards.

APPENDIX D

PICC Lite C Language and Library Reference

Radix Formats

Radix	Format	Comments
Binary	0b*number* or 0B*number*	MPASM binary is in format B'*number*'.
Octal	0*number* or *number*	Inadvertently putting a 0 in front of a decimal number will convert it to octal for the compiler, potentially resulting in errors that are very hard to find.
Decimal	*number*	MPASM default is hexadecimal.
Hexadecimal	0x*number* or 0X*number*	MPASM also enables X'*number*'.

Comments

Comments are lines of text used for notes and information that are not meant to be compiled. There are two formats for comments in C:

```
//   Everything to the right of the two "slash" characters on a
//     line is ignored

/*   Everything inside the "slash-splat" and "splat-slash" is
       ignored  */
```

Declarations

The following is a single variable declaration

```
[qualifier] type [* [qualifier]] Label [= Value];
```

where "type" can be one of the following options:

Type	Size (in Bits)	Arithmetic Type
bit	1	Boolean
char	8	Signed integer
unsigned char	8	Unsigned integer
short	16	Signed integer
unsigned short	16	Unsigned integer
int	16	Signed integer
unsigned int	16	Unsigned integer
long	32	Signed integer
unsigned long	32	Unsigned integer
float	24	Real (modified IEEE 754)
double	Default 24	Real and can be 32 bits by using the -D32 command-line option

The signed qualifier can optionally be used with the different integer variable types. For example, the variable declaration

```
signed char i;
```

is equivalent to

```
char i;
```

In some cases, data that is different from the type declaration of the variable will have to be stored in the variable. The variable's type can be overridden by placing the new type in front of the variable in brackets:

```
(unsigned long) StringAddr = 0x012345678;
```

Value is an optional initialization constant.
An array variable declaration is as follows:

```
type Label[Array Size] [= {value1, value2, ...}]
```

Strings are defined as single dimensional ASCIIZ arrays

```
char String[ 17 ] = "This is a String";
```

where the last character is an ASCII NUL.

The optional splat (*) indicates that the variable is a pointer. In this case, if the variable is initialized, it is given the address of another object. In the PIC16F84 MCU, variable registers can only be in the address range 0x0C to 0x04F with the addresses in bank 1 (0x08C to 0x0CF) being "shadowed" onto the addresses in bank 0. In the PIC16F627 MCU, variable registers can be in either bank 0 or bank 1 and are not shadowed between the two.

In PICC Lite, pointers cannot be set to the addresses of other variables (or array elements) using the & character. Instead, data to be shared between variables should be defined using unions.

Strings can be defined as pointers to characters:

```
char * String = "This is a String";       // Define ASCIIZ String
```

To access individual characters in the string, the array offset can be used:

```
    b = String[n];                         // Load "B" with String
                                           // Element
```

Specific registers can be accessed in the PICmicro MCU directly in PICC Lite. The declaration format is

```
static unsigned char RegisterName @ 0xAddress;
```

Individual bits of a register can be accessed using the following format:

```
static bit BitName @ (unsigned)&RegisterName*8+BitNumber;
```

The standard registers and bits declarations are loaded when pic.h is included in the application.

If you want to access a register at a specific address, the following declaration format should be used:

```
volatile unsigned RegisterName @ 0x0Address;
```

When the *RegisterName* label is accessed, the data in the register at the specified address will be accessed without regard as to whether or not another label has already been assigned to this register's address.

It is critical that any variables that are modified by an interrupt handler and polled by mainline code are declared as volatile to make sure that loops polling on the value changing continually read the variable. For example, if you have the mainline code

```
    while (!flag);          // Wait for interrupt to happen
```

and if flag is declared as a byte, the compiler generates code that would load the PICmicro MCU's accumulator with flag once and continually test that value without loading in the most recent value of flag. By declaring flag as volatile, each time the while loop executes, a new value of flag is loaded for the test.

When declaring variables, pointers, and structures in PICC Lite, a number of qualifiers can be used to explicitly specify the placement and operation of the different variables.

The const qualifier will place the initial value in *read-only memory* (ROM) and will not allow it to be modified. For example, the statement

```
const int Label = Value;
```

declares Label as a constant returning value and can never be modified.

The const qualifier can be confusing when it is used with pointers. For example, the declaration

```
const char * ptr;
```

creates a pointer (ptr) to a character. The value being pointed to can be modified, but the pointer itself cannot.

In the following declaration, the pointer can be modified, but the value being pointed to cannot:

```
char * const ptr;
```

Lastly, the declaration

```
const char * const ptr;
```

creates a pointer that cannot be modified, pointing to a value that cannot be modified.

In the PIC16F627, you can specify variables to be placed in bank 1 of the register space (normally, variable data is assigned to addresses in bank 0) using the bank1 qualifier. This option is not available when the PIC16F84 is used.

The bank1 qualifier can be used similarly to the const qualifier to specify where pointers and the value that they are pointing to are placed.

The volatile qualifier indicates to the compiler (and its optimizers) whether or not the data will change between accesses. If the variable is defined as volatile, then the compiler will not use a temporarily saved value (such as in the processor's accumulator) and instead reads the register to get its most current value.

Normally, variables that are not declared with initial values are cleared (set to zero) on startup. If the variable should not be cleared on startup, the persistent qualifier should be used.

Structures and Unions

The traditional format for defining structures and unions is

```
struct | union {
     type Label;
     :
} StructUnionLabel [VariableLabel];
```

To access data elements within a structure, the -> operator is used. For the example of defining the result of a division operation

```
struct {
     int dividend;
     int remainder;
} div_result result;
```

setting dividend and remainder in the result structure could be implemented as

```
     result -> dividend = quotient / divisor;
     result -> remainder = quotient % divisor;
```

PICC Lite also provides the capability to build bit structures at specified registers. The following example shows how the PCLATH register of a generic PICmicro MCU could be defined:

```
struct {
     unsigned          : 3;  //  Upper 3 Bits not used
     unsigned PageSel  : 2;  //  2K Instruction Page Selection
     unsigned OffSel   : 3;  //  256 Instruction group within Page
} PCLATH @ 0x00A;
```

In this structure, the three bits that select the 256 Instruction group within the page are separated from the two bits that select the active page. The structure also isolates the three bits within the PCLATH register that are not used.

Functions

The application start is as follows:

```
void main(void)
{  //  Application Code

   :                      //  Application Code

}  //  End Application
```

The function format is as follows:

```
Return_Type Function([ Type Parameter [, Type Parameter...] ])
{  //  Function Start

   :                          //  Function Code

   return value;

}  //  End Function
```

The function prototype is as follows:

```
Return_Type Function([ Type Parameter [, Type Parameter...] ]);
```

In PIC16F627 (and other midrange PICmicro MCUs), only one interrupt vector is built into the processor. To differentiate between different interrupt request sources, the INTCON bits ending in F are polled to see which one is set. The base interrupt handler definition is

```
void interrupt Handler Name(void)
{

}  //  End Interrupt Handler
```

NOTE: *Variables defined outside of all functions are known as global variables and can be accessed by all the functions in the application. Variables defined within functions can only be accessed within these functions and are known as local variables. The labels assigned local variables can be used within other functions without other local variables being affected.*

Statements

Statements are built from expressions, which are in the format

```
[(..] Variable | Constant [Operator [(..] Variable | Constant
][)..]]
```

An assignment statement is structured as follows:

```
Variable = Expression;
```

C conditional statements consist of if, while, do, for, and switch. The if statement is defined as

```
if ( Statement )
   [Statement]; | { Statement...}
[else [Statement]; | { Statement...}]
```

The while statement is added to the application following the definition:

```
while ( Statement ) [Statement]; | { Statement...}
```

The for statement is defined as

```
for ([initialization (Assignment) Statement]; [Conditional
Statement[, Conditional Statement . . . ]]; [Loop Expression
(Increment) Statement[, Loop Expression Statement . . . ]])
[Statement]; | { Statement . . . }
```

To jump out of a currently executing loop, the break statement

```
break;
```

is used.

The continue statement skips over remaining code in a loop and jumps directly to the loop condition (for use with while, for, and do/while loops). The format of the statement is

```
continue;
```

For looping until a condition is true, the do/while statement is used:

```
do Statement; | {Statement...}
while ( Expression );
```

To conditionally execute according to a value, the switch statement is used:

```
switch( Expression ) {
   case Value:              //  Execute if "Statement" == "Value"
     [Statement...]
     [break;]
   :
   default:                 //  If no "case" Statements are True
     [Statement...]
} // End switch
```

The goto Label statement is used to jump to a specific label within the current function (it cannot jump to a label outside of the current function):

```
goto Label;

Label:
```

To return a value from a function, the return statement is used:

```
return Statement;
```

Operators

Unary statement operators are summarized in the following table:

Operator	Operation
~	Bitwise complement (only to the left of the expression to complement)
++	Increment (to the left, before the rest of the statement, or to the right, after the rest of the statement)
-	Two's complement negation (only to the left of the expression to negate)
—	Decrement (to the left, before the rest of the statement, or to the right, after the rest of the statement)

The binary statement operators are as follows:

Operator	Operation
!	Logical negation
&&	Logical AND
&	Bitwise AND or address
||	Logical OR
|	Bitwise OR
^	Bitwise XOR
+	Addition

Operator	Operation
-	Subtraction
*	Multiplication or indirection
/	Division
%	Modulus
==	Equals (returns 0 if not true or Not Zero if true)
!=	Not equals (returns 0 if not true or Not Zero if true)
<	Less than (returns 0 if not true or Not Zero if true)
<=	Less than or equals to (returns 0 if not true or Not Zero if true)
<<	Shift left
>	Greater than (returns 0 if not true or Not Zero if true)
>=	Greater than or equals to (returns 0 if not true or Not Zero if true)
>>	Shift right

The compound assignment operators are as follows:

Operator	Operation
&=	AND value with the destination
|=	OR value with the destination
^=	XOR value with the destination
+=	Add value to the destination
-=	Subtract value from the destination
*=	Multiply value with the destination
/=	Divide value from the destination
%=	Get the modulus of destination with respect to value
<<=	Shift destination left value times
>>=	Shift destination right value times

The order of operations is outlined in the following table:

Operators	Priority	Type
() [] . ->	Highest	Expression evaluation
- ~ ! & * ++ —		Unary operators
* / %		Multiplicative
+ -		Additive
<< >>		Shifting
< <= >= >		Comparison
== !=		Comparison
&		Bitwise AND
^		Bitwise XOR
|		Bitwise OR
&&		Logical AND
||		Logical OR
?:		Conditional execution
= &= |= ^= += -= *= /= %= >>= << =		Assignments
,	Lowest	Sequential evaluation

Reserved Words

The following words cannot be used in C applications as labels:

auto
bank1
break
case
char
const
continue
default
do
else
extern

fastcall

for

goto

if

int

interrupt

persistent

return

signed

static

struct

switch

union

unsigned

void

volatile

while

Backslash Characters

String	ASCII	Character
\r	0x00D	Carriage return (CR)
\n	0x00A	Line feed (LF)
\f	0x00C	Form feed (FF)
\b	0x008	Backspace (BS)
\t	0x009	Horizontal tab (HT)
\v	0x00B	Vertical tab (VT)
\a	0x007	Bell (BEL)
\'	0x027	Single quote (')
\"	0x022	Double quote (")
\\	0x05C	Backslash (\)
\ddd	N/A	Octal number
\xddd	0x0dd	Hexadecimal character

◼ Directives

All directives start with # and are executed before the code is compiled.

Directive	Function
#	Null directive.
#asm ... #endasm	Place PICmicro assembly language code within the two directives.
#assert *Condition*	Generates error if Condition is false.
#define *Label*[(*Parameters*)] *Text*	Defines a label that will be replaced with text when it is found in the code. If parameters are specified, then replace them in the code, similar to a macro.
#include "*File*" | <*File*>	Loads the specified file in line to the text. When < and > enclose the filename, the file is found using the INCLUDE environment path variable. If the filename is enclosed by " and ", then the file in the current directory is searched before checking the INCLUDE path.
#error *Text*	Forces the error listed in text.
#if *Condition*	If the condition is true, then compile the following code to #elif, #else, or #endif. If the condition is false, then ignore the following code to #elif, #else, or #endif.
#ifdef *Label*	If the #define label exists, then compile the following code. #elif, #else, and #endif work as expected with #if.
#ifndef *Label*	If the #define label does *not* exist, then compile the following code. #elif, #else, and #endif work as expected with #if.
#elif *Condition*	This directive works as an #else #if to avoid lengthy nested #ifs. If the previous condition was false, it checks the condition.
#else	Placed after #if or #elif, this toggles the current compile condition. If the current compile condition was false, after #else it will be true. If the current compile condition was true, after #else it will be false.
#endif	Used to end an #if, #elif, #else, #ifdef, or #ifndef directive.
#line *number filename*	Specifies the line number and file name for the listing.
#number	Specifies the line number for the listing.

Directive	Function
#pragma interrupt_level 1	The following interrupt function will be calling a function; a suppress error is normally generated.
#pragma jis	Enables JIS (two-byte national character set) strings.
#pragma nojis	Disables JIS (two-byte national character set) strings.
#pragma printf_check *type*	Specifies that a following function is to accept string data in printf format followed by a number of parameters.
#pragma psect *psectlabel=label*	Specifies code or text put into a specific segment.
#pragma regused *register* . . .	Specifies registers to be saved in an interrupt handler.
#undef *Label*	Deletes (undefines) the label.
#warning *Text*	Outputs a warning message.

▉ Command-Line Parameters

If PICC Lite is invoked from the MS-DOS Prompt (or command line), the following parameters can be specified.

Option	Meaning
-processor	Defines the processor (only 16C84, 16F84, 16F84A, or 16F627).
-A*spec*	Specifies offset for read-only memory (ROM).
-A-*option*	Specifies that *-option* is to be passed to the assembler.
-AAHEX	Generates an American automation symbolic HEX file.
-ASMLIST	Generates an assembler .LST file for each compilation.
-BIN	Generates a binary output file.
-C	Compiles to object files only.
-Ck*file*	Makes OBJTOHEX use a checksum file.
-CR*file*	Generates a cross-reference listing and stores it in *file*.

Option	Meaning
-D24	Uses the truncated, 24-bit floating point format for doubles.
-D32	Uses the IEEE754 32-bit floating point format for doubles.
-Dmacro	Defines the preprocessor macro (same function as the #define directive).
-E	Defines the format for compiler errors in a human-readable format.
-Efile	Redirects compiler errors to a file.
-E+*file*	Appends errors to a file.
-FAKELOCAL	Produces MPLAB-specific debug information (use with -Gfile parameter).
-FDOUBLE	Enables the use of faster 32-bit floating point mathematic routines.
-Gfile	Generates an enhanced source-level symbol table.
-HELP	Prints a summary of options.
-ICD	Compiles code for the MPLAB-ICD debugger.
-Ipath	Specifies a directory pathname for include files.
-INTEL	Generates an Intel HEX format output file (default).
-Llibrary	Specifies a library to be scanned by the linker.
-L-option	Specifies *-option* to be passed directly to the linker.
-Mfile	Requests the generation of a MAP file.
-MOT	Generates a Motorola S1/S9 HEX format output file.
-Nsize	Specifies the identifier length (default is 31 characters).
-NORT	Does not link standard runtime module.
-O	Enables post-pass optimization.
-Ofile	Specifies the output filename.
-P	Preprocesses assembler files.
-PRE	Produces preprocessed source files.
-PROTO	Generates function prototype information.
-PSECTMAP	Displays complete memory segment usage after linking.
-Q	Specifies quiet mode (must be the first option).
-RESRAM*ranges*	Reserves the specified RAM address ranges.

Option	Meaning
-RESROM*ranges*	Reserves the specified ROM address ranges.
-ROM*ranges*	Specifies the external ROM memory range.
-S	Compiles to assembler source files only.
-SIGNED_CHAR	Makes the default char signed unless otherwise specified in variable declarations.
-STRICT	Enables strict ANSI keyword conformance. Note if -STRICT is specified on the command line, bit variables are not available.
-TEK	Generates a Tektronix HEX format output file.
-*Usymbol*	Undefines a predefined preprocessor symbol.
-UBROF	Generates an UBROF format output file.
-V	Meaning Verbose, displays compiler pass command lines.
-*Wlevel*	Sets compiler warning level.
-X	Strips local symbols.
-Z*glevel*	Specifies the optimization level (for PICL, *level* 3 is considered the maximum.

■ Built-in Standard C Functions

The following functions are available in PICC Lite:

Function	Include	Operation
int abs(int)	stdlib.h	Calculates the absolute value of the parameter.
double acos(double)	math.h	Finds the Arccosine of the value (returned in radians).
double asin(double)	math.h	Finds the Arcsine of the value (returned in radians).
double atan(double)	math.h	Finds the Arctangent of the value (returned in radians).

Function	Include	Operation
double atan2(double, double)	math.h	Finds the Arctangent of the value (returned in radians).
double atof(const char *)	math.h	Converts a string to a real number.
int atoi(const char *)	stdlib.h	Converts an ASCII string to an integer.
int atol(const char *)	stdlib.h	Converts an ASCII string to a long integer.
double ceil(double)	math.h	Returns the floating point value representing the largest integer greater than or equal to the parameter.
char *cgets(char *)	conio.h	Gets a string of characters.
double cos(double)	math.h	Finds the cosine of the angle (in radians).
double cosh(double)	math.h	Finds the hyperbolic cosine of the parameter.
void cputs(char *)	conio.h	Outputs string of characters.
void eeprom_write(unsigned char, unsigned char)	pic1684.h	Writes byte to data EEPROM.
unsigned char eeprom_read(unsigned char)	pic1684.h	Reads byte from data EEPROM.
double exp(double)	math.h	Calculates the natural exponent.
double fabs(double)	math.h	Calculates the absolute value of the parameter.
double floor(double)	math.h	Returns the floating point value representing the largest integer less than or equal to the parameter.
double frexp(double, int *)	math.h	Returns the power of the floating parameter to the integer pointer.
char getch(void)	conio.h	Gets one character from standard input (the serial port). If no character is available, then wait for it. It is written by the user who may have to create a getche function as well.
Char getche(void)	conio.h	Gets a two-byte character from standard input. Two calls are required to return the full 16 bits. Written by user.
void init_uart(void)	conio.h	Initializes the UART function. In PIC16F84, this is to set I/O pin operation.

Function	Include	Operation
bit isalnum(char)	ctype.h	Returns true if alphanumeric.
bit isalpha(char)	ctype.h	Returns true if alphabetic.
bit iscntrl(char)	ctype.h	Returns true if control (0x000-0x01F or 0x07F).
bit isdigit(char)	ctype.h	Returns true if digit/number.
bit isgraph(char)	ctype.h	Returns true if printable, but not blank (0x021-0x07E).
bit islower(char)	ctype.h	Returns true if lower case.
bit ispunct(char)	ctype.h	Returns true if punctuated.
bit isprint(char)	ctype.h	Returns true if printable (0x020-0x07E).
bit isspace(char)	ctype.h	Returns true if 0x09-0x00D or " ".
bit isupper(char)	ctype.h	Returns true if upper case.
bit isxdigit(char)	ctype.h	Returns true if hex digit.
char kbhit(void)	conio.h	Returns pending state of the input function. Written by user.
double ldexp(double, int)	math.h	Returns the power of the floating parameter to the integer.
double log(double)	math.h	Calculates the natural logarithm.
double log10(double)	math.h	Calculates the base 10 logarithm.
const void *memchr(const void *, int, unsigned int)	string.h	Searches buffer for the first occurrence of the specified character.
int memcmp(const void *, const void *, unsigned int)	string.h	Compares two strings and it returns less than zero if the first is less than the second, zero if the two are equal, or greater than zero if the first is greater than the second.
void *memcpy(void *, const void *, unsigned int)	string.h	Copies the specified number of bytes.
void *memmove(void *, const void *, unsigned int)	string.h	Copies the specified number of bytes from the source to destination. If there is overlap, no data is lost.
void *memset(void *, int, unsigned int)	string.h	Writes the character a specified number of times.

Function	Include	Operation
double modf(double, double *)	math.h	Breaks the floating point number into integer and fractional parts.
double pow(double, double)	math.h	Calculates the power of parameters.
int printf(const char *, . . .)	stdio.h	Outputs the Const string text. Escape sequence characters for output are embedded in the Const string text. Different data outputs are defined using the following conversion characters: %d, %i: Decimal integer %o: Octal integer %x, %X: Hex integer (with upper- or lowercase values). No leading "0x" character string output. %u: Unsigned integer %c: Single ASCII character %s: ASCIIZ string %f: Floating point %#e, %#E: Floating point with the precision specified by # %g, %G: Floating point %p: Pointer %%: Print % character
void putch(char)	conio.h	Outputs one character to standard output (the serial port). Written by user.
int rand(void)	stdlib.h	Returns a random number. See srand.
double sin(double)	math.h	Finds the sine of the angle (in radians).
double sinh(double)	math.h	Finds the hyperbolic sine of the parameter.
double sqrt(double)	math.h	Returns the square root of the parameter.
void srand(void)	stdlib.h	Loads a random number generator with a seed. See rand.
char *strcat(char *, const char *)	string.h	Puts ASCIIZ append string on the end of the existing ASCIIZ string.
const char *strchr(const char *, int)	string.h	Returns the position of the first char in the ASCIIZ string.
int strcmp(const char *, const char *)	string.h	Compares two ASCIIZ strings. Zero is returned for a match, negative when String1 is less than String2, and positive when String1 is greater than String2.

Function	Include	Operation
char *strcpy(char *, const char *)	string.h	Copies the contents of ASCIIZ String2 into String1.
unsigned int strcspn(const char *, const char *)	string.h	Finds the number of characters in the second string not in the first string.
char *strdup(const char *)	string.h	Duplicates string.
const char *strchr(const char *, int)	string.h	Returns the position of the first character in the ASCIIZ string without regard for case.
int stricmp(const char *, const char *)	string.h	Compares strings without regard to case.
const char *stristr(const char *, const char *)	string.h	Returns a pointer in the first string of the first occurrence of the second without regard to case.
char *strncat(char *, const char *, unsigned int)	string.h	Puts the number of characters from append on the end of the existing ASCIIZ string.
int strncmp(const char *, const char *, unsigned int)	string.h	Compares two ASCIIZ strings for the number of characters. Zero is returned for a match, negative when String1 is less than String2, and positive when String1 is greater than String2.
int strnicmp(const char *, const char *, unsigned int)	string.h	Compares two ASCIIZ strings for the number of characters without regard for case. Zero is returned for a match, negative when String1 is less than String2, and positive when String1 is greater than String2.
unsigned int strlen(const char *)	string.h	Returns the length of the passed string.
char *strncpy(char *, const char *, unsigned int)	string.h	Copies the number of characters from String2 into String1.
const char *strpbrk(const char *, const char *)	string.h	Returns the first occurrence in the first string of any character from the second string.
const char *strrchr(const char *, int)	string.h	Returns the position of the last character in the ASCIIZ string.

Function	Include	Operation
unsigned int strspn(const char *, const char *)	string.h	Returns the index of the first character in the first string that does not belong in the second.
const char *strstr(const char *, const char *)	string.h	Returns the pointer in the first string of the first occurrence of the second string.
double tan(double)	math.h	Finds the tangent of the angle (in radians).
double tanh(double)	math.h	Finds the hyperbolic tangent of the parameter.
char tolower(char)	ctype.h	Converts a character to lower case.
char toupper(char)	ctype.h	Converts a character to upper case.
unsigned int xtoi(const char *)	stdlib.h	Converts an ASCII string to hexadecimal integers.

■■ APPENDIX E ■■ ■■ ■■

Resources

■■ Contacting the Author

I can be contacted by sending an e-mail to myke@passport.ca or visiting my web site at www.myke.com.

■■ Microchip

Microchip Technology, Inc.

2355 W. Chandler Blvd.

Chandler, AZ 85224

Phone: (480) 786-7200

Fax: (480) 917-4150

Web: www.microchip.com

The web site contains a complete set of data sheets in .pdf format for download as well as the latest versions of MPLAB-IDE.

▰ Hi-Tech Software

Hi-Tech Software
PO Box 103
Alderley, QLD 4051
AUSTRALIA
Phone: +61-7-3552-7777
Fax: +61-7-3552-7778

The address in the United States is

Hi-Tech Software, LLC
6600 Silacci Way
Gilroy, CA 95020
Phone: (800) 735-5715
Fax: (866) 898-8329
Web: www.htsoft.com

The web site contains information for downloading the latest version of PICC Lite and ordering their line of compilers for the Microchip PICmicro MCU and other microcontrollers and processors.

▰ Other Compiler Vendors

CCS C Compilers

PICmicro Microcontroller Compilers

Custom Computer Services, Inc.
PO Box 2452
Brookfield, WI 53008
Phone: (414) 797-0455
Fax: (414) 797-1459
Web: www.ccsinfo.com
E-mail: ccs@ccsinfo.com

PicBasic and PicBasic Pro

PICmicro Microcontroller Compilers

microEngineering Labs, Inc.

PO Box 7532

Colorado Springs, CO 80933

Phone: (719) 520-5323

Fax: (719) 520-1867

Web: www.melabs.com

E-mail: info@melabs.com

CC5X C Compiler

PICmicro Microcontroller Compilers

B Knudsen Data

Stubbanv. 33B

7037 Trondheim

NORWAY

Fax: (47) 73-96-51-84

Web: www.bknd.com

E-mail: sales@bknd.com

MPC Code Development System

Compilers Available for a Wide Range of Microcontrollers

Byte Craft Limited

421 King Street North

Waterloo, Ontario N2J 4E4

CANADA

Phone: (519) 888-6911

Web: www.bytecraft.com/impc.html

E-mail: info@bytecraft.com

Keil C Compilers

Compilers Available for a Wide Range of Microcontrollers

Keil Software

1501 10th Street, Suite 110

Plano, TX 75074

Phone: (972) 312-1107

Toll free: (800) 348-8051

Fax: (972) 312-1159

Web: www.keil.com

E-mail: sales.us@keil.com

Embedded Software Development Tool Suites

Compilers and IDEs for a Wide Range of Microcontrollers

Tasking Software Development from Altium

17140 Bernardo Center Drive, Suite 100

San Diego, CA 92128

Phone: (858) 521-4280

Fax: (858) 521-4282

Toll-free sales: (877) TAS-KING

Toll-free support: (800) 458-8276

Web: www.tasking.com

E-mail: tasking.sales.na@altium.com

Useful Books

Here are a collection of books that I have found useful over the years when developing electronics and software for robot applications. I have started out with listing different books that are specific to microcontrollers and robots, and then gone on to list some books that would be useful references for you to have.

Some of these are hard to find, but definitely worth it when you do find them in a used bookstore.

Robot Builder's Bonanza, 2E
Author: Gordon McComb
ISBN: 0-07-136296-7

Robot Building for Beginners
Author: David Cook
ISBN: 1-89-311544-5

Mobile Robots: Inspiration to Implementation
Author: Joseph Jones, et al.
ISBN: 1-56-881097-0

Build Your Own Robot!
Author: Karl Lundt
ISBN: 1-56-881102-0

Robot Building for Beginners
Author: David Cook
ISBN: 1-89-311544-5

Applied Robotics
Author: Edwin Wise
ISBN: 0-79-061184-8

Handbook of Microcontrollers (1998)
Aukthor: Myke Predko
ISBN: 0-07-913716-4

Programming and Customizing the PICmicro® Microcontroller, 2E
Author: Myke Predko
ISBN: 0-07-136172-3

Programming and Customizing the 8051 Microcontroller
Author: Myke Predko
ISBN: 0-07-134192-7

Programming and Customizing the Basic Stamp, 2E
Author: Scott Edwards
ISBN: 0-07-137192-3

Programming and Customizing the AVR Microcontroller
Author: Dhananjay V. Gadre
ISBN: 0-071-34666-X

Programming and Customizing the HC11 Microcontroller
Author: Tom Fox
ISBN: 0-07-134406-3

Design with PIC Microcontrollers
Author: J.B. Peatman
ISBN: 0-13-759259-0

The Art of Electronics (1989)
Author: Paul Horowitz and Winfield Hill
ISBN: 0-521-37095-7

Bebop to the Boolean Boogie (1995)
Author: Clive "Max" Maxfield
ISBN: 1-878707-22-1

The Encyclopedia of Electronic Circuits, Volumes 1 through 7
Volume 1 ISBN: 0-8306-1938-0
Volume 2 ISBN: 0-8306-3138-0
Volume 3 ISBN: 0-8306-3348-0
Volume 4 ISBN: 0-8306-3895-4
Volume 5 ISBN: 0-07-011077-8
Volume 6 ISBN: 0-07-011276-2
Volume 7 ISBN: 0-07-015116-4

CMOS Cookbook (Revised 1988)
Author: Don Lancaster
ISBN: 0-7506-9943-4

The TTL Data Book for Design Engineers
From: Texas Instruments
ISBN: N/A

PC PhD (1999)
Author: Myke Predko
ISBN: 0-07-134186-2

The Embedded PC's ISA Bus: Firmware, Gadgets and Practical Tricks (1997)
Author: Ed Nisley
ISBN: 1-5739-8017-X

The C Programming Language, 2E (1988)
Authors: Brian W. Kernighan and Dennis M. Ritchie
ISBN: 0-13110-362-8

▬ Periodicals

Here are a number of magazines that give a lot of information on robots and project ideas. Every month, each magazine will have examples of circuits and software that can be applied to a robot:

Circuit Cellar Ink
Subscriptions
PO Box 698
Holmes, PA 19043-9613
Phone: (800) 269-6301
BBS: (860) 871-1988
Web: www.circcellar.com/

Poptronics
Subscription Department
PO Box 459
Mt. Morris, IL 61054-7629
Phone: (800) 827-0383
Fax: (631) 592-6723

Micro Control Journal (This is published on the Web.)
Web: www.mcjournal.com/

Nuts & Volts
Subscriptions
430 Princeland Court
Corona, CA 91719
Phone: (800) 783-4624
Web: www.nutsvolts.com

Everyday Practical Electronics
EPE Subscription Department
Allen House, East Borough
Wimborne, Dorset
BH21 1PF
UNITED KINGDOM
Phone: +44 (0) 1202-881749
Web: www.epemag.wimborne.co.uk

Robot Web Sites

Seattle Robotics Society

www.seattlerobotics.org/

One of the most famous robot sites on the Internet. The Seattle Robotics Society has lots of information on interfacing digital devices to such real-world devices as motors, sensors, and servos. They also do a lot of exciting things in the automation arena.

Silicon Valley HomeBrew Robotics Club

www.wildrice.com/HBRobotics/HBRCBuildersBook.html

This is an excellent introduction to robotics. Included on the page are a considerable number of links and instructions for building your own robot.

robots.net

http://robots.net

A nice site with pointers to resources as well as a presentation of many different robots that have been designed and built by others.

The Robot Menu

www.robotics.com/robomenu/index.html

A web site devoted to showing off different people's robots.

Robotics FAQ List

www.frc.ri.cmu.edu/robotics-faq/

This covers just about everything to do with robotics.

Innovatus

www.innovatus.com

Innovatus has made available PICBots, an interesting PICmicro microcontroller simulator that enables programs to be written for virtual robots that fight among themselves. Also available from Innovatus is MazeBots, which is a robot simulation designed to teach programming and robotics. These products are an excellent introduction to developing biologic robot code.

Part Suppliers

The following companies supplied components that are used in this book. I am listing them because they all provide excellent customer service and are able to ship parts anywhere you need them.

Digi-Key

Digi-Key is an excellent source for a wide range of electronic parts. They are reasonably priced and most orders will be delivered the next day. They are real lifesavers when you're on a deadline.

Digi-Key Corporation

701 Brooks Avenue South

PO Box 677

Thief River Falls, MN 56701-0677

Phone: (800) 344-4539 (or [800] DIGI-KEY)

Fax: (218) 681-3380

Web: www.digi-key.com/

Mondo-tronics' Robot Store

Mondo-tronics has proclaimed itself the "world's biggest collection of miniature robots and supplies," and I have to agree with them. This is a great source for servos, tracked vehicles, and robot arms.

Mondo-tronics, Inc.

Order Desk

524 San Anselmo Ave. #107-13

San Anselmo, CA 94960

Toll free in the United States and Canada: (800) 374-5764

Fax: (415) 455-9333

Web: www.robotstore.com/

Tower Hobbies

This is an excellent source for servos and R/C parts useful in home-built robots.

Tower Hobbies

PO Box 9078

Champaign, IL 61826-9078

Phone: (217) 398-3636

Toll-free ordering in the United States and Canada: (800) 637-4989

Toll-free support in the United States and Canada: (800) 637-6050

Fax: (217) 356-6608

Toll-free fax in the United States and Canada: (800) 637-7303

Web: www.towerhobbies.com/

E-mail: orders@towerhobbies.com

Jameco Electronics

Jameco offers components, PC parts/accessories, and hard-to-find connectors.

Jameco

1355 Shoreway Road

Belmont, CA 94002-4100

Toll free in the United States and Canada: (800) 831-4242

Web: www.jameco.com/

JDR Microdevices

JDR offers components, PC parts/accessories, and rare connectors.

JDR Microdevices

1850 South 10th Street

San Jose, CA 95112-4108

Phone: (408) 494-1400

Toll-free in the United States and Canada: (800) 538-5000

BBS: (408) 494-1430

Toll-free fax in the United States and Canada: (800) 538-5005

Web: www.jdr.com

E-mail: techsupport@jdr.com

Compuserve: 70007,1561

Newark Electronics

Newark sells components, including the Dallas Line of semiconductors. (The DS87C520 and DS275 are used for RS-232-level conversion in this book.)

Toll free in the United States and Canada: (800) 463-9275
([800] 4-NEWARK)

Web: www.newark.com/

Marshall Industries

Marshall is a full-service distributor of Philips microcontrollers as well as other parts.

Marshall Industries
9320 Telstar Avenue
El Monte, CA 91731
Phone: (800) 833-9910
Web: www.marshall.com

Mouser Electronics

Mouser is the distributor for the Seiko S7600A TCP/IP stack chips.

Mouser Electronics, Inc.
958 North Main Street
Mansfield, Texas 76063
Phone (sales): (817) 483-6888
Toll free (sales): (800) 346-6873
Fax: (817) 483-6899
Web: www.mouser.com
E-mail: sales@mouser.com

AP Circuits

AP Circuits will build prototype bare boards from your Gerber files. Boards are available within three days. I have been a customer for several years and they have always produced excellent quality and have been helpful in providing direction for developing my own bare boards. Their web site contains the EasyTrax and GCPrevue MS-DOS tools necessary to develop your own Gerber files.

Alberta Printed Circuits, Ltd.
#3, 1112-40th Avenue N.E.
Calgary, Alberta T2E 5T8
Phone: (403) 250-3406
BBS: (403) 291-9342
Web: www.apcircuits.com/
E-mail: staff@apcircuits.com

■■■ Good Reference Web Sites

Although none of these are robot specific, they are a good source of ideas, information, and products that will make your life a bit more interesting and maybe give you some ideas for how you would implement your own robot.

List of Stamp Applications (L.O.S.A.)

www.hth.com/losa/

The list of Parallax basic stamp applications will give you an idea of what can be done with the basic stamp (and other microcontrollers, such as the PICmicro). The list contains projects that are of interest to the robot designer.

gpsim

www.dattalo.com/gnupic/gpsim.html

This is a software simulator for PICmicro GNU *General Public License* (GPL) development tools for Linux, which can be downloaded from the web site.

Dontronics Home Page

www.dontronics.com/

Don McKenzie has a wealth of information on the PICmicro microcontroller as well as other electronic products. His page has lots of useful links to other sites and it is the home of the SimmStick.

MicroTronics

www.eedevl.com/index.html

This site lists programmers and application reviews.

Alexy Vladimirov's outstanding list of PICmicro MCU resource pages

www.geocities.com/SiliconValley/Way/5807/

Over 900 pages have been listed as of February 2002.

Adobe PDF viewers

www.adobe.com

The Adobe .pdf file format is used for virtually all vendor datasheets, including the devices presented in this book (and their datasheets on the CD-ROM).

PKZIP and PKUNZIP

www.pkware.com

PKWARE's .zip file compression format is a standard for combining and compressing files for transfer.

INDEX

Symbols

A

B